AF789506

Multifrequency Oscillations of Nonlinear Systems

Mathematics and Its Applications

Volume 567

Multifrequency Oscillations of Nonlinear Systems

by

A. Samoilenko

Institute of Mathematics,
National Academy of Sciences of Ukraine, Kiev, Ukraine

and

R. Petryshyn

Chernivtsi National University,
Chernivtsi, Ukraine

SPRINGER-SCIENCE+BUSINESS MEDIA, B.V.

A C.I.P. Catalogue record for this book is available from the Library of Congress.

ISBN 978-90-481-6574-2 ISBN 978-1-4020-2031-5 (eBook)
DOI 10.1007/978-1-4020-2031-5

Printed on acid-free paper

Contents

INTRODUCTION

Among processes studied by various natural sciences (mechanics, physics, ecology, etc.), an important place is occupied by oscillation processes. By now, numerous efficient methods for the investigation of oscillation phenomena described by linear and nonlinear differential equations have been developed and mathematically justified [AVK, Bib, BNF, KBK, LiR, MiR, Nei, Pero, Pli1, Pros, FiS, Hal2]. It turned out that, among these methods, the most efficient are asymptotic methods, in particular, the averaging method, the method of integral manifolds, and iterative methods developed by mathematicians of the Kiev Mathematical School (Krylov, Bogolyubov, Mitropol'skii, Samoilenko, and their disciples) [Bog, BoZ, BoM1, BoM2, BMS, GGP1, GGP2, GoP, Gre, GrR1–GrR3, KrB, Kul, Luc, Lyk, LyB, MaS, MaT, Mit1–Mit4, MiLo, MiLy, MiS1–MiS5, MSK, MSM1, MSM2, MiK, Par, Pere, PeA, Pet1–Pet10, PeL, PeP, Sam1–Sam10, SaP1, SaP2, SPe1–SPe7, SPet1, SPet2, SaR, SaS, SaT, SSh, TeA, Tro, FBKCY, SVL, SSY].

The foundations of the averaging method were laid in works of the founders of celestial mechanics in the times of Lagrange and Laplace. The idea of this method is as follows: Using a special operator, one replaces the system of differential equations under investigation by another system (the so-called averaged system). The averaged system, on the one hand, should be simpler in a certain sense than the original one, and, on the other hand, it must describe the main features of the phenomenon under investigation. In this case, there naturally arises the problem of justification of the averaging method, i.e., the problem of finding efficient estimates for the norm of the difference of solutions of the original and the averaged equations on a finite or an infinite time interval.

Although the averaging method has been used for the solution of numerous problems for almost two centuries, the problem of its justification remained unsolved for a long time. Only in the 1930–1940s, beginning with Fatou's work [Fat], the first fundamental results were obtained in this direction. Thus, Bogolyubov showed [Bog, Mit2] that, for systems of the standard form, the averaging method is closely related to the problem of the existence of a change of variables that enables one to exclude the time variable on the right-hand side of the system. Furthermore, Bogolyubov investigated systems of equations of higher approximations whose solutions approximate the solutions of the original system of equations to within values proportional to integer powers of a small parameter ε.

The averaging method was further developed by Mitropol'skii and other authors for various classes of differential equations with large and small parameters. In particular, it was extended to equations with nondifferentiable right-hand sides, integro-differential and stochastic differential equations, partial differential equations, and linear differential equations with slowly varying parameters that describe nonstationary oscillation processes. These results are presented in [Mit2]. Mitropol'skii and Samoilenko developed the axiomatic theory of the averaging method [MiS4], which includes the classical version of the averaging method and, in particular, leads to the method of normal forms [Bry1]. Important results on the problem of justification of the averaging method were also obtained in [Aku, Vol, VoM, ZaL, MiK, Plo, PlB, PlZ, PlL, Sam1, Sam5, Fil, Kha1, Kha2, KhF].

The last decades were marked by the extensive investigation of multifrequency nonlinear systems of differential equations appearing in various problems of classical and celestial mechanics, radio engineering, and physics. In this connection, the development of algorithms for the asymptotic integration of oscillation systems with many degrees of freedom and their mathematical justification has become an urgent problem.

In the case of systems with constant frequency vector, this important problem was solved by Mitropol'skii and Samoilenko in [MiS1–MiS5, Sam2–Sam4]. In particular, they thoroughly investigated an important phenomenon appearing in multifrequency systems, namely, quasiperiodic oscillations.

For systems with variable frequency vector, the best-studied cases are the one- and two-dimensional cases, which were investigated by Arnol'd [Arn2, Arn4], Bakhtin [Bak2], Neishtadt [Neis1], and Pronchatov [Pron]. In the works indicated, efficient estimates were obtained for the error of the averaging method on a finite time segment and for the measure of the set of initial data for the equations under investigation for which the resonance phenomenon takes place. If the number of frequencies is greater than two, then the investigation of oscillation systems leads to considerable difficulties because, in this case, the structure of resonance surfaces is very complicated [Arn4]. Certain problems of justification of averaging schemes in multifrequency systems and their applications to the solution of practical problems were studied by Anosov [3], Bakhtin [Bak1], Grebenikov [Gre], Neishtadt [Neis2], Plotnikov [Plo, PlL], and Khapaev [Kha1, Kha2].

The problem of averaging in Hamiltonian systems has been fairly thoroughly studied. Note that, for such systems, a solution of the averaged equations for slow variables is always stationary. Kolmogorov [Kol] and Arnol'd [Arn1] solved the problem of the stability of Hamiltonian systems on an infinite time segment, and Nekhoroshev [Nek1, Nek2] obtained an exponential estimate for stability time for almost all Hamiltonian functions.

The averaging method is often used for the solution of boundary-value problems appearing in the simulation of the behavior of real processes and in problems of optimal control. In the course of investigation of such processes, in many cases it is rather difficult to specify initial data that uniquely describe the process (the Cauchy problem), but it is possible to determine the values of some parameters of this process at certain times by using various devices. There is a fairly rich theory of multipoint boundary-value problems for ordinary differential equations developed in works of many authors [Boi, VaK, VaB, VaD, Gab, DmK, ZhK, Kig, LeL, Luc, Pta, SaR, ChH]. This theory is based on both analytic methods, by using which one studies the problem of the existence and uniqueness of solutions and their continuous dependence on parameters, and numerical methods, which enable one to calculate approximate values of solutions. Analytic and numerical methods are often combined, which allows one to efficiently solve the problem of the existence of solutions and their construction. An extensive bibliography on this problem can be found in [SaR]. Applications of the averaging method to the solution of boundary-value problems were studied by Akulenko [Aku], Chernous'ko [Che], Bainov and Milusheva [BaM], Plotnikov, Zverkova, and Bardai [Plo, PlB, PlZ], and others.

Another powerful and convenient method for the investigation of nonlinear systems of differential equations is the method of integral manifolds. The first deep results on integral manifolds of toroidal type were obtained by Krylov and Bogolyubov [Bog, BoM1, KrB] in the process of justification of asymptotic methods in nonlinear mechanics. Later, the ideas of these works were generalized in [BoM2] and extensively developed in the study of differential equations in various functional spaces. They also affected the character of new developments in perturbation theory for toroidal manifolds and led to deep results of Diliberto [Dil1, Dil2], Hale [Hal1], Moser [Mos1–Mos5], Sacker [Sac1, Sac2], and Sell [Sel1, Sel2].

A new impulse toward the development of the theory of perturbations and stability of invariant manifolds was given by the concept of Green function in the problem of invariant tori of a linear extension of a dynamical system on a torus, which led to new results in this theory [Sam8].

The method of integral manifolds was extended to systems of differential equations with slow and fast variables (in particular, singularly perturbed ones), impulsive systems, systems close to integrable ones, etc. Important results concerning the existence and properties of integral manifolds can be found in the monographs of Bakai and Stepanovskii [BaS], Bibikov [Bib], Mitropol'skii and Lykova [MiLy], Pliss [Pli2], Samoilenko and Perestyuk [SaP2], and Strygin and Sobolev [StS].

In the present work, we investigate multifrequency nonlinear systems of ordinary differential equations of the form

$$\frac{dx}{d\tau} = a(x, \varphi, \tau, \varepsilon),$$
$$\frac{d\varphi}{d\tau} = \frac{\omega(x, \tau)}{\varepsilon} + b\,(x, \varphi, \tau, \varepsilon), \tag{1}$$

where x and φ are n- and m-dimensional vectors, respectively, τ is "slow" time, ε is a small positive parameter, and the real vector functions a, b, and ω belong to certain classes of smooth functions 2π-periodic in φ. Systems of this type appear in the course of investigation of oscillation processes in numerous problems of mechanics, electrical engineering, biology, etc. [GrR2, GrR3, Mit3, Mit4, Hal2]. The main problem arising in the study of properties of solutions of system (1) is the problem of resonance relations between the components of the variable frequency vector $\omega(x, \tau)$. Here, the resonance case is understood as the case where the scalar product of the vector $\omega(x, \tau)$ and a nonzero vector with integer-valued coordinates turns into zero or becomes close to zero for certain values of x and τ. This leads to the appearance of slowly varying harmonics in the Fourier series on the right-hand sides of Eqs. (1) and generates the problem of small denominators [Arn1, BMS, GrR2]. At present, there is a fairly complete and rich theory of one- and two-frequency systems. Note that, in the case of two-frequency systems, the resonance surfaces form, generally speaking, a collection of level surfaces; therefore, the main effect in these systems is the passage through resonances in the course of evolution. If an oscillation system has a greater number of frequencies, then its solutions can stay in a neighborhood of resonance surfaces for a fairly long time or intersect these surfaces at arbitrarily small angles, which substantially complicates the investigation of oscillations. The case $m \geq 3$ has been studied to a significantly lesser extent than the one- and two-frequency cases, and, therefore, the investigation of various aspects of the theory of multifrequency nonlinear oscillation systems is an urgent problem. The present monograph is devoted to the solution of certain problems in this theory.

In Chapter 1, we establish uniform estimates for certain oscillation integrals depending on parameters, which are used in the proof of new theorems on the justification of the averaging method. The averaged (with respect to all angular variables φ) system corresponding to system (1) has the form

$$\frac{d\overline{x}}{d\tau} = \overline{a}\,(\overline{x}, \tau, \varepsilon),$$
$$\frac{d\overline{\varphi}}{d\tau} = \frac{\omega(\overline{x}, \tau)}{\varepsilon} + \overline{b}\,(\overline{x}, \tau, \varepsilon), \tag{2}$$

where

$$[\overline{a}; \overline{b}] = (2\pi)^{-m} \int_0^{2\pi} \dots \int_0^{2\pi} \Big[a(\overline{x}, \varphi, \tau, \varepsilon); b(\overline{x}, \varphi, \tau, \varepsilon)\Big] d\varphi_1 \dots d\varphi_m.$$

The averaged system (2) is simpler than (1) because it does not contain oscillation terms on its right-hand side, and, therefore, for the construction of its solution, one can use numerical methods with step greater than in the case of (1). The problem of the justification of the averaging method reduces to the proof of the estimate

$$\|x(\tau, \varepsilon) - \overline{x}(\tau, \varepsilon)\| + \|\varphi(\tau, \varepsilon) - \overline{\varphi}(\tau, \varepsilon)\| \le C\varepsilon^{\alpha}$$

or

$$\|x(\tau, \varepsilon) - \overline{x}(\tau, \varepsilon)\| \le C\varepsilon^{\alpha}$$

for all $\tau \in I$. Here, C and α are certain positive constants, $x(0, \varepsilon) = \overline{x}(0, \varepsilon)$, $\varphi(0, \varepsilon) = \overline{\varphi}(0, \varepsilon)$, and either $I = [0, L]$, or $I = [0, T(\varepsilon)]$ $(T(\varepsilon) \to \infty$ as $\varepsilon \to 0)$, or $I = [0, \infty)$. We obtain efficient estimates for partial derivatives of the difference of solutions of systems (1) and (2) with respect to the initial data, prove an analog of Banfi–Filatov theorem [Fil, Ban], and investigate multifrequency systems of higher approximations.

Chapter 2 is devoted to the application of the averaging method to the solution of boundary-value problems. In the case of oscillation systems with $\omega = \omega(\tau)$, we prove the solvability of two-point boundary-value problems and establish the quantitative dependence of the norm of the difference of solutions of original and averaged problems on the value of the small parameter ε. Combining the averaging method with the solution of boundary-value problems, we prove the existence of solutions of system (1) defined on the entire axis the slow variables $x(\tau, \varepsilon)$ of which are uniformly bounded. It is important to note that this result is established without using the method of integral manifolds, which requires additional restrictions on a multifrequency system. The solvability of multipoint boundary-value problems in the case $\omega = \omega(x,\tau)$ is studied, and the existence of solutions of boundary-value problems with parameters is proved.

In Chapter 3, we establish conditions for the existence of an integral manifold of system (1) in the case where the frequencies of the system depend on τ. The smoothness properties are studied, estimates for partial derivatives of the function defining the integral manifold are obtained, and a theorem on the conditional asymptotic stability of an integral manifold is proved. In a small neighborhood of an asymptotically stable integral manifold $x = X(\psi,\tau,\varepsilon)$, we decompose the equations for slow and fast variables, i.e., we construct a change of variables

$$x = y + X(\varphi,\tau,\varepsilon), \qquad \varphi = \psi + \Phi(y,\psi,\tau,\varepsilon)$$

that reduces system (1) to the form

$$\frac{dy}{d\tau} = Y(y,\psi,\tau,\varepsilon),$$

$$\frac{d\psi}{d\tau} = \frac{\omega(\tau)}{\varepsilon} + b\big(X(\psi,\tau,\varepsilon),\psi,\tau,\varepsilon\big).$$

The results obtained are used for the investigation of a system of weakly connected oscillators with slowly varying frequencies.

In Chapter 4, we study the behavior of a dynamical system

$$\frac{dx}{dt} = X(x), \quad x \in R^n, \tag{3}$$

in a neighborhood of a toroidal manifold M filled by a quasiperiodic trajectory of the system. By passing to local coordinates, we establish conditions for the reducibility of system (3) in such a neighborhood to a system with quasiperiodic coefficients and investigate the smoothness of the corresponding change of variables. We prove a statement on the exponential attraction as $t \to \infty$ of a solution of system (3) originating in a small neighborhood of the invariant manifold to the corresponding solution of this system that lies on M. We also establish the invariance of the behavior of trajectories of a dynamical system in a neighborhood of the manifold M under small perturbations of system (3). The results obtained are extended to the case of discrete dynamical systems.

The present monograph is based on the investigations carried out by the authors themselves [Pet1–Pet10, Sam5–Sam7, Sam9, SPe1–SPe7] and in collaboration with their disciples [PeL, PeP, SPet1, SPet2].

The authors hope that the ideas and methods proposed in this monograph will be further developed and applied to new classes of problems in the theory of nonlinear oscillations.

1. AVERAGING METHOD IN OSCILLATION SYSTEMS WITH VARIABLE FREQUENCIES

1. Uniform Estimates for One-Dimensional Oscillation Integrals

In the present section, we study properties of oscillation integrals of the form

$$\mathcal{I}(\tau, \varepsilon) = \int_0^\tau f(t) \exp\left\{\frac{i}{\varepsilon}\int_0^t a(z)dz\right\} dt, \qquad \tau \in [0, L], \tag{1.1}$$

where $f(\tau) = (f_1(\tau), \ldots, f_n(\tau))$, $f_j(\tau)$ and $a(\tau)$ are real functions, $j = \overline{1, n}$, L is a positive constant, $\varepsilon_0 \geq \varepsilon$ is a positive small parameter, and i is the imaginary unit. Integrals of this type appear in the study of oscillation phenomena in various problems of classical and celestial mechanics, physics, and engineering [Arn2, Arn3, Gre, Mit4]. In [PlL, Sam5, SPe3, SPe4, Kha1, Kha2], estimates of integrals of the type (1.1) were considered for the justification of the averaging method in multifrequency systems with slow and fast variables.

In what follows, we investigate the dependence of oscillation integrals on the value of the small parameter ε and on properties of the functions $f(\tau)$ and $a(\tau)$. The analysis of integral (1.1) shows that, for $f(\tau) \not\equiv 0$, an estimate of $\mathcal{I}(\tau, \varepsilon)$ substantially depends on the character of zeros of the function $a(\tau)$. In particular, if $a(\tau) \equiv 0$ and $f_j(\tau) \equiv 1$ for any $\tau \in [0, L]$ and certain j, then $\|\mathcal{I}(L, \varepsilon)\| \geq L$. In what follows, unless otherwise stated, the norm of a matrix is understood as the sum of the absolute values of its elements.

Theorem 1.1. *Let* $a(\tau) \in C^p_{[0,L]}$, $p \geq 1$, *let* $f(\tau) \in C^1_{[0,L]}$, *and let* $a(\tau)$ *have zeros of multiplicity not higher than* p *on* $[0, L]$. *Then there exist a constant*

$\varepsilon_1 > 0$ *and a constant* $c_1 > 0$ *independent of* ε *such that*

$$\|\mathcal{I}(\tau, \varepsilon)\| \leq c_1 \varepsilon^{\frac{1}{p+1}} \tag{1.2}$$

for all $\tau \in [0, L]$ *and* $\varepsilon \in (0, \varepsilon_1]$.

Proof. It is known [Sam5] that, under the assumptions made above, $a(t)$ has finitely many zeros $t_1 < t_2 < \ldots < t_s$ of multiplicities $r_1, r_2, \ldots, r_s$, respectively, on $[0, L]$; here, $r_j \leq p$ for all $j = \overline{1, s}$. Since

$$|a^{(r_j)}(t_j)| \equiv c^{(j)} > 0 \quad \forall j = \overline{1, s}, \qquad a^{(r_j)}(t) = \frac{d^{r_j} a(t)}{dt^{r_j}},$$

taking into account the continuity of the functions $a^{(r_j)}(t)$ we establish that there exists a number $\delta > 0$ independent of j and such that

$$|a^{(r_j)}(t)| \geq \frac{1}{2} c^{(j)} \geq c_2 = \frac{1}{2} \min_j \, c^{(j)}$$

for $|t - t_j| \leq \delta$, $t \in [0, L]$. We choose $\delta < \frac{1}{2} \min\limits_{1 \leq j \leq s-1} (t_{j+1} - t_j)$ and denote by $B(\tau)$ the set $\bigcup\limits_{j=1}^{s} [t_j - \delta, t_j + \delta] \cap [0, \tau]$ and by $A(\tau)$ the closure of the set $[0, \tau] \backslash B(\tau)$. Then $[0, \tau] = A(\tau) \cup B(\tau)$ and, furthermore, the function $a(t)$ is nonzero at every point of the set $A(\tau)$. Therefore,

$$\min_{t \in A(\tau)} |a(t)| \geq \min_{t \in A(L)} |a(t)| = c_3 > 0 \tag{1.3}$$

and the inequality

$$|a^{(r_j)}(t)| \geq c_2 \tag{1.4}$$

holds on each segment $T_j = [t_j - \delta, t_j + \delta] \cap [0, \tau]$, $j = \overline{1, s}$, of the set $B(\tau)$. It follows from (1.4) that the function $a^{(r_j - 1)}(t)$ vanishes on T_j at at most one point $t_{1,j}$; moreover, for $t \in T_j \backslash [t_{1,j} - \mu, t_{1,j} + \mu]$ and $0 < \mu < \min\{1; \delta\}$, the inequality $|a^{(r_j - 1)}(t)| \geq c_2 \mu$ is satisfied. If $a^{(r_j - 1)}(t)$ does not change its sign on T_j, then, as $t_{1,j}$, we choose, respectively, the left or the right endpoint of this segment, depending on whether the function $|a^{(r_j - 1)}(t)|$ is increasing or decreasing. We assume that $T_j \cap [t_{1,j} - \mu, t_{1,j} + \mu]$ belongs to the set $A(t_j, \mu)$ and use analogous arguments for the functions $a^{(l)}(t)$, $l = \overline{0, r_j - 1}$. As a result, we establish that the set $A(t_j, \mu)$ consists of $d_1(t_j, \mu) \leq 2^p - 1$ segments of length not greater than 2μ, and the set $B(t_j, \mu)$, which is the closure of the set

$T_j \backslash A(t_j, \mu)$, consists of $d_2(t_j, \mu) \leq 2^p$ segments on each of which the following inequality is true:

$$|a(t)| \geq c_2\, \mu^p. \tag{1.5}$$

Note that the function $a^(1)(t)$ does not change its sign on each segment of the set $B(t_j, \mu)$.

We represent $I(\tau, \varepsilon)$ in the form of a sum, namely,

$$I(\tau, \varepsilon) = \int\limits_{A(\tau)} F(t, \varepsilon)\, dt + \sum_j \Bigg(\int\limits_{A(\tau_j, \mu)} F(t, \varepsilon) dt + \int\limits_{B(\tau_j, \mu)} F(t, \varepsilon)\, dt \Bigg), \tag{1.6}$$

where $F(t, \varepsilon)$ is the integrand of $I(\tau, \varepsilon)$. According to the definition of the set $A(\tau_j, \mu)$, we have

$$\Bigg\| \int\limits_{A(\tau_j, \mu)} F(t, \varepsilon) dt \Bigg\| \leq 2\mu(2^p - 1) \max_{[0,L]} \|f(t)\| \equiv c_4^{(1)} \mu. \tag{1.7}$$

Let $[\alpha, \beta]$ be a segment from the set $B(t_j, \mu)$. Then, integrating by parts and taking (1.5) into account, we obtain

$$\begin{aligned} \Bigg\| \int\limits_\alpha^\beta F(t, \varepsilon) dt \Bigg\| &= \varepsilon \Bigg\| \int\limits_\alpha^\beta \frac{f(t)}{a(t)}\, d \Bigg(\exp \Bigg\{ \frac{i}{\varepsilon} \int\limits_0^t a(z) dz \Bigg\} \Bigg) \Bigg\| \\ &\leq \varepsilon \Bigg(\frac{2}{c_2 \mu^p} + \int\limits_\alpha^\beta \frac{|a^{(1)}(t)|}{a^2(t)} dt \Bigg) \max_{[0,L]} \|f(t)\| \\ &\quad + \varepsilon \int\limits_\alpha^\beta \frac{1}{c_2 \mu^p} \|f^{(1)}(t)\| dt. \end{aligned}$$

Since $a^{(1)}(t)$ does not change its sign on $[\alpha, \beta]$, the relations

$$\int\limits_\alpha^\beta \frac{|a^{(1)}(t)|}{a^2(t)} dt = \Bigg| \int\limits_\alpha^\beta \frac{d}{dt} \Big(\frac{1}{a(t)} \Big) dt \Bigg| \leq \Bigg| \frac{1}{a(\beta)} - \frac{1}{a(\alpha)} \Bigg| \leq \frac{2}{c_2 \mu^p}$$

yield the estimate

$$\Bigg\| \int\limits_\alpha^\beta F(t, \varepsilon) dt \Bigg\| \leq \frac{\varepsilon}{c_2 \mu^p} \Big[4 \max_{[0,L]} \|f(t)\| + (\beta - \alpha) \max_{[0,L]} \|f^{(1)}(t)\| \Big].$$

Thus,

$$\left\| \int\limits_{B(t_j,\mu)} F(t,\varepsilon)dt \right\| \le \frac{2^p}{c_2}\Big[4\max_{[0,L]}\|f(t)\| + L\max_{[0,L]}\|f^{(1)}(t)\|\Big]\varepsilon\mu^{-p}$$

$$\equiv c_4^{(2)}\varepsilon\mu^{-p}. \tag{1.8}$$

It remains to estimate the first term on the right-hand side of equality (1.6). Since the function $a(t)$ satisfies inequality (1.3) on every segment of the set $A(\tau)$, we establish the following estimate by integrating by parts:

$$\left\| \int\limits_{A(\tau)} F(t,\varepsilon)dt \right\|$$

$$\le \frac{s+1}{c_3^2}\Big[\Big(2c_3 + \max_{[0,L]}|a^{(1)}(t)|\Big)\max_{[0,L]}\|f(t)\| + Lc_3\max_{[0,L]}\|f^{(1)}(t)\|\Big]\varepsilon$$

$$\equiv c_4^{(3)}\varepsilon\,. \tag{1.9}$$

Combining (1.7)–(1.9) and using (1.6), we get

$$\|I(\tau,\varepsilon)\| \le sc_4^{(1)}\mu + sc_4^{(2)}\varepsilon\mu^{-p} + c_4^{(3)}\varepsilon \tag{1.10}$$

for all $\tau \in [0,L]$, $\varepsilon \in (0,\varepsilon_0]$, and $0 < \mu < \min\{1;\delta\}$. It is clear that the last estimate is the best order estimate with respect to ε in the case where $\varepsilon = \mu^{p+1}$. Setting

$$c_1 = (c_4^{(1)} + c_4^{(2)})s + c_4^{(3)} \qquad \text{and} \qquad \varepsilon_1 = \min\Big\{\varepsilon_0;\Big(\frac{\delta}{2}\Big)^{p+1}\Big\},$$

we deduce (1.2) from (1.10). Theorem 1.1 is proved.

We now consider an oscillation integral of the form

$$I_\lambda(t,\bar{t},\tau,\varepsilon) = \int\limits_t^{t+\tau} f(y)\exp\Big\{\frac{i}{\varepsilon}\int\limits_{\bar{t}}^{y}(\lambda,\omega(z))\,dz\Big\}\,dy, \tag{1.11}$$

where $\tau \in [0,L]$, $t \in R = (-\infty;\infty)$, $\bar{t} \in R$, $\lambda = (\lambda_1,\ldots,\lambda_m)$ is a positive nonzero m-dimensional vector, $m \ge 2$, $\omega(t) = (\omega_1(t),\ldots,\omega_m(t)) \in C_R^{p-1}$,

$p \geq m$, $f(t) = (f_1(t), \ldots, f_n(t)) \in C_R^1$, (λ, ω) is the scalar product of vectors, and L is a positive constant.

In what follows, we establish sufficient conditions that guarantee the uniform estimate $\|I_\lambda(t, \overline{t}, \tau, \varepsilon)\| \leq c\varepsilon^{\frac{1}{p}}$ with a constant c independent of λ. In the case $m = 2$, the behavior of integral (1.11) is determined by the inequality $\|I_\lambda(t, \overline{t}, \tau, \varepsilon)\| \leq c\sqrt{\varepsilon}$ [Arn4, Neis1]. In the case $m \geq 3$, the investigation of the behavior of integral (1.11) as $\varepsilon \to 0$ becomes more complicated because there appear resonance relations between the components $\omega_\nu(t)$ of the vector $\omega(t)$ [Bak1, GrR2, Kha2].

By $W_p(t)$ and $W_p^T(t)$ we denote the matrix

$$(\omega_\nu^{(j-1)}(t))_{\nu,j=1}^{m,p}$$

and its transpose, respectively.

Theorem 1.2. *Let* $\|(W_p^T(t)W_p(t))^{-1}W_p^T(t)\|$ *be uniformly bounded by a constant* σ_1 *and let the functions* $\omega_\nu^{(j-1)}(t)$, $\nu = \overline{1, m}$, $j = \overline{1, p}$, *be uniformly continuous for* $t \in R$. *Then one can indicate constants* $\overline{\varepsilon}_1 > 0$ *and* $\sigma_2 > 0$ *independent of* λ, t, $\overline{t}$, τ, *and* ε *and such that the following estimate holds for all* $\lambda \neq 0$, $t \in R$, $\overline{t} \in R$, $\tau \in [0, L]$, *and* $\varepsilon \in (0, \overline{\varepsilon}_1]$:

$$\|I_\lambda(t, \overline{t}, \tau, \varepsilon)\|$$
$$\leq \sigma_2 \varepsilon^{\frac{1}{p}} \left[\left(1 + \frac{1}{\|\lambda\|}\right) \max_{[t, t+L]} \|f(y)\| + \frac{1}{\|\lambda\|} \max_{[t, t+L]} \|f^{(1)}(y)\| \right]. \qquad (1.12)$$

Proof. For an arbitrary vector $\lambda = (\lambda_1, \ldots, \lambda_m) \neq 0$, we consider the obvious equality $W_p(t)\lambda = \Omega$, where

$$\Omega = (\Omega_0, \ldots, \Omega_{p-1}), \Omega_j = \sum_{\nu=1}^{m} \lambda_\nu \omega_\nu^{(j)}(t) = (\lambda, \omega^{(j)}(t)), \quad j = \overline{0, p-1}.$$

Hence, we get

$$\|\Omega\| \geq \|\lambda\| \ \|(W_p^T(t)W_p(t))^{-1}W_p^T(t)\|^{-1} \geq \frac{\|\lambda\|}{\sigma_1},$$

which implies that, for every $\overline{y} \in R$ and $\lambda \neq 0$, there is an integer $r = r(\overline{y}, \lambda)$, $0 \leq r \leq p - 1$, for which

$$|(\lambda, \omega^{(r)}(\overline{y}))| = \max_{0 \leq j \leq p-1} |(\lambda, \omega^{(j)}(\overline{y}))| \geq \frac{\|\lambda\|}{p\sigma_1}. \qquad (1.13)$$

Since the functions $\omega_\nu^{(j)}(t),\ \nu = \overline{1,m},\ j = \overline{0,p-1}$, are uniformly continuous on the entire axis, it is obvious that we can choose a constant $\delta > 0$ independent of $\overline{y},\ \lambda$, and j and such that the following inequalities hold for any $y \in [\overline{y} - \delta, \overline{y} + \delta]$ and $j = \overline{0,p-1}$:

$$|(\lambda, \omega^{(r)}(y))| \geq \frac{\|\lambda\|}{2p\sigma_1}, \quad |(\lambda, \omega^{(j)}(y))| \leq 4|(\lambda, \omega^{(r)}(y))|. \tag{1.14}$$

Indeed, according to the definition of uniform continuity, for every $\nu = \overline{1,m}$ and $j = \overline{0,p-1}$ there exists $\delta_\nu^{(j)} > 0$ such that, for any $y', y'' \in R$ satisfying the inequality $|y' - y''| < \delta_\nu^{(j)}$, the following estimate is true:

$$|\omega_\nu^{(j)}(y') - \omega_\nu^{(j)}(y'')| < \overline{\sigma}_1 \equiv \frac{1}{2p\sigma_1}. \tag{1.15}$$

Denote $\delta = \min\limits_{\nu,j} \delta_\nu^{(j)}$. Then estimate (1.15) is valid for $\|y' - y''\| < \delta,\ 0 \leq j \leq p-1$, and $1 \leq \nu \leq m$, and the relations

$$|(\lambda, \omega^{(j)}(y) - \omega^{(j)}(\overline{y}))| \leq \sum_{\nu=1}^{m} |\lambda_\nu||\omega_\nu^{(j)}(y) - \omega_\nu^{(j)}(\overline{y})| < \overline{\sigma}_1\|\lambda\|,$$

which are true for $|y - \overline{y}| < \delta$ and $j = \overline{0,p-1}$, lead to inequalities (1.14).

We denote by s the integer part of the number $\dfrac{\tau}{2\delta}$ $\left(s \leq \dfrac{L}{2\delta}\right)$ and represent integral (1.11) in the form of a sum, namely,

$$\mathcal{I}(t, \overline{t}, \tau, \varepsilon) = \sum_{k=0}^{s-1} \int\limits_{t+2\delta k}^{t+2\delta(k+1)} F dy + \int\limits_{t+2\delta s}^{t+\tau} F dy, \tag{1.16}$$

where

$$F = f(y) \exp\left\{\frac{i}{\varepsilon} \int\limits_{\overline{t}}^{y} (\lambda, \omega(z))\, dz\right\}.$$

To estimate the integral

$$P_k = \int\limits_{t+2\delta k}^{t+2\delta(k+1)} F dy, \tag{1.17}$$

we use inequalities (1.14) and the methods used in the proof of Theorem 1.1. As a result, we obtain

$$\|P_k\| \leq \left[(2^p - 2)\,\mu + \frac{2^{p+1}}{\overline{\sigma}_1 \|\lambda\|}\,\varepsilon\mu^{1-p}\right] \max_{[t,t+L]} \|f(y)\| + \frac{2\delta}{\overline{\sigma}_1\|\lambda\|}\,\varepsilon\mu^{1-p} \max_{[t,t+L]} \|f^{(1)}(y)\| \tag{1.18}$$

for $r(t + 2\delta k + \delta, \lambda) \geq 1$ and $0 < \mu < \min\{1, \delta\}$. If $r(t + 2\delta k + \delta, \lambda) = 0$, then, integrating by parts, we get

$$\|P_k\| \leq \frac{2}{\overline{\sigma}_1\|\lambda\|}(1 + 4\delta)\,\varepsilon \max_{[t,t+L]} \|f(y)\| + \frac{2\delta}{\overline{\sigma}_1\|\lambda\|}\,\varepsilon \max_{[t,t+L]} \|f^{(1)}(y)\|. \tag{1.19}$$

Analyzing relations (1.17) and (1.18), we conclude that, in the case where $\mu^{p-1} \leq 2^p(1 + 4\delta)^{-1}$, integral (1.17) satisfies inequality (1.18) for all $r(t + 2\delta k + \delta, \lambda) \geq 0$. The same inequality is also satisfied by the last integral on the right-hand side of (1.16). Therefore, for

$$\varepsilon = \mu^p, \quad 0 < \varepsilon \leq \overline{\varepsilon}_1 = \min\left\{\varepsilon_0;\ \left(\frac{1}{2}\delta\right)^p;\ \left(\frac{2^p}{1 + 4\delta}\right)^{\frac{p}{p-1}}\right\},$$

relation (1.16) yields inequality (1.12) with the constant

$$\sigma_2 = \left(2^p - 2 + \frac{2^{p+1}}{\overline{\sigma}_1} + \frac{2\delta}{\overline{\sigma}_1}\right)\left(1 + \frac{L}{2\delta}\right).$$

Theorem 1.2 is proved.

Corollary 1. *If* $\|\lambda\| = 1$, *i.e.,* λ *is an arbitrary point of the unit sphere, then inequality (1.12) yields a uniform estimate of the integral* $\mathcal{I}_\lambda$ *of the form*

$$\|\mathcal{I}_\lambda(t, \overline{t}, \tau, \varepsilon)\| \leq 2\sigma_2\varepsilon^{\frac{1}{p}}\left[\max_{[t,t+L]} \|f(y)\| + \max_{[t,t+L]} \|f^{(1)}(y)\|\right].$$

Corollary 2. *If* $\lambda = k = (k_1, \ldots, k_m)$ *is an arbitrary nonzero vector with integer coordinates, then, for* $\sigma_3 = 2\sigma_2$, *estimate (1.12) takes the following form:*

$$\|\mathcal{I}_\lambda(t, \overline{t}, \tau, \varepsilon)\| \leq \sigma_3\varepsilon^{\frac{1}{p}}\left[\max_{[t,t+L]} \|f(y)\| + \frac{1}{\|k\|} \max_{[t,t+L]} \|f^{(1)}(y)\|\right]. \tag{1.20}$$

Note that estimate (1.20) is often used in what follows for the investigation of properties of solutions of oscillation systems on finite and infinite time intervals.

Let us analyze in more detail the conditions imposed on the function $\omega(\tau) = (\omega_1(\tau), \ldots, \omega_m(\tau))$ in Theorem 1.2. Assume that $t = \bar{t} = 0,\ \tau \in [0, L],\ \varepsilon \in (0, \varepsilon_0],\ \lambda \neq 0$, and $\mathcal{I}_\lambda(0, 0, \tau, \varepsilon) \equiv \mathcal{I}_\lambda(\tau, \varepsilon)$ in integral (1.11). Since $\omega(\tau) \in C^{p-1}_{[0,L]}$, we conclude that the functions $\omega_\nu(\tau),\ \nu = \overline{1, m}$, and their derivatives up to the order $p - 1$ inclusive are uniformly continuous and bounded on $[0, L]$. Thus, the condition of the boundedness of the matrix $(W_p^T(\tau)W_p(\tau))^{-1}W_p^T(\tau)$ is equivalent in this case to the condition $\det(W_p^T(\tau)W_p(\tau)) \neq 0 \quad \forall \tau \in [0, L]$. To calculate the determinant of the product of the matrices $W_p^T(\tau)$ and $W_p(\tau)$, we use the Binet–Cauchy formula [Gan, Lan]

$$\begin{aligned}
&\det(W_p^T(\tau)W_p(\tau)) \\
&\quad = \det \begin{pmatrix} \omega_1(\tau) & \ldots & \omega_1^{(p-1)}(\tau) \\ \ldots & \ldots & \ldots \\ \omega_m(\tau) & \ldots & \omega_m^{(p-1)}(\tau) \end{pmatrix} \begin{pmatrix} \omega_1(\tau) & \ldots & \omega_m(\tau) \\ \ldots & \ldots & \ldots \\ \omega_1^{(p-1)}(\tau) & \ldots & \omega_m^{(p-1)}(\tau) \end{pmatrix} \\
&\quad = \sum_{0 \le k_1 < k_2 < \ldots < k_m \le p-1} \Delta^2_{k_1 \ldots k_m}(\tau),
\end{aligned}$$

where

$$\Delta_{k_1 \ldots k_m}(\tau) = \det(\omega_\nu^{(k_j)}(\tau))_{\nu, j = 1}^m.$$

It follows from the above relations that

$$\det(W_p^T(\tau)W_p(\tau)) \neq 0 \quad \forall \tau \in [0, L]$$

if and only if at least one mth-order minor of the matrix $W_p(\tau)$ is nonzero at every point $\tau \in [0, L]$. Thus, the following statement is true:

Theorem 1.3. *Suppose that $\omega(\tau) \in C^{p-1}_{[0,L]},\ p \ge m,\ f(\tau) \in C^1_{[0,L]}$, and, at every point $\tau \in [0, L]$, at least one mth order minor of the matrix $W_p(\tau)$ is nonzero. Then, for all $\lambda \neq 0,\ \tau \in [0, L]$, and $\varepsilon \in (0, \varepsilon_0]$ (ε_0 is sufficiently small), the following estimate is true:*

$$\|\mathcal{I}_\lambda(\tau, \varepsilon)\| \le \overline{\sigma}_2 \varepsilon^{\frac{1}{p}} \left[\left(1 + \frac{1}{\|\lambda\|}\right) \max_{[0,L]} \|f(\tau)\| + \frac{1}{\|\lambda\|} \max_{[0,L]} \|f^{(1)}(\tau)\| \right],$$

where the constant $\overline{\sigma}_2$ is independent of $\lambda,\ \tau$, and ε.

The above analysis shows that the conditions

$$|\,\omega_\nu^{(j)}(\tau)\,| \le c \quad \forall \tau \in R,\ \ \nu = \overline{1,m},\ \ j = \overline{0,p},$$

$$|\det\left(W_p^T(\tau)W_p(\tau)\right)| \ge \overline{c} > 0 \quad \forall \tau \in R,$$

(where c and $\overline{c}$ are certain constants) guarantee the uniform continuity of the functions $\omega_\nu^{(j)}(\tau),\ \ \nu = \overline{1,m},\ \ j = \overline{0,p-1},$ on the entire axis and the uniform boundedness

$$\|(W_p^T(\tau)W_p(\tau))^{-1}W_p^T(\tau)\| \le \underline{c} = \text{const} \quad \forall \tau \in R.$$

The following question arises: Is the assumption on the boundedness of the functions $\omega_\nu(\tau)$ and their derivatives necessary? The following example shows that, generally speaking, the answer to this question is negative: For all $\tau \in R,$ consider the functions

$$\omega_1(\tau) = \sin\tau,\ \ \omega_2(\tau) = \cos\tau,\ \ \omega_3(\tau) = \begin{cases} \tau, & \tau \in [\pi,\infty), \\ \pi - \sin\tau, & \tau \in (-\infty;\pi). \end{cases}$$

It is obvious that $\omega_3(\tau)$ is not bounded for any $\tau \in R,$ and the functions $\omega_\nu(\tau),$ $\nu = 1,2,3,$ and their derivatives up to the second order inclusive are uniformly continuous on the entire axis. By direct calculation, one can easily verify that

$$\det\, W_3(\tau) = \begin{cases} -\tau, & \tau \in [\pi,\infty), \\ -\pi, & \tau \in (-\infty,\pi), \end{cases}$$

$$\|(W_3^T(\tau)W_3(\tau))^{-1}W_3^T(\tau)| \le 6 + \frac{4}{\pi} \quad \forall \tau \in R.$$

Hence, the indicated collection of functions $\omega_1(\tau),\ \omega_2(\tau),$ and $\omega_3(\tau)$ satisfies all conditions of Theorem 1.2.

Remark 1. If $p = m,$ then

$$\det\left(W_m^T(\tau)W_m(\tau)\right) = (\det\, W_m(\tau))^2.$$

Therefore, in this case, the condition that the Wronskian determinant of the functions $\omega_1(\tau),\ldots,\omega_m(\tau)$ is nonzero on $[0,L]$ is a sufficient condition for finding an efficient estimate for the oscillation integral $I_\lambda(\tau,\varepsilon).$

We now assume that this condition is not satisfied at finitely many points of the segment $[0,L]$ and investigate how this assumption affects the estimate of the oscillation integral.

Theorem 1.4. *Suppose that* $f(\tau) \in C^1_{[0,L]}$, $\omega(\tau) \in C^{m-1+r}_{[0,L]}$, *and the function* $\Delta(\tau) = \det W_m(\tau)$ *has zeros of multiplicity not higher than* r, $r \geq 1$, *on* $[0, L]$. *Then one can choose constants* $\sigma_4 > 0$ *and* $\underline{\varepsilon}_1 \in (0, \varepsilon_0]$ *independent of* λ, τ, *and* ε *and such that the following inequality holds for all* $\lambda \neq 0$, $\tau \in [0, L]$, *and* $\varepsilon \in (0, \underline{\varepsilon}_1]$:

$$\|I_\lambda(\tau, \varepsilon)\| \leq \sigma_4 \varepsilon^{\frac{1}{m+r}} \left[\left(1 + \frac{1}{\|\lambda\|}\right) \max_{[0,L]} \|f(\tau)\| + \frac{1}{\|\lambda\|} \max_{[0,L]} \|f^{(1)}(\tau)\| \right]. \quad (1.21)$$

Proof. Under the assumptions made above, the function $\Delta(\tau)$ has finitely many zeros $0 \leq \tau_1 < \tau_2 < \ldots < \tau_s \leq L$ of multiplicities $r_1, r_2, \ldots, r_s$, $r_j \leq r$ $\forall j = \overline{1, s}$, on $[0, L]$. We fix an arbitrary positive $\overline{\delta} < \dfrac{1}{2} \min\limits_{1 \leq j \leq s-1} (\tau_{j+1} - \tau_j)$ and divide the segment $[0, \tau]$ into two sets of points $A_{\overline{\delta}}(\tau)$ and $B_{\overline{\delta}}(\tau)$ such that

$$B_{\overline{\delta}}(\tau) = \bigcup_{j=1}^{s} [\tau_j - \overline{\delta}, \tau_j + \overline{\delta}] \cap [0, \tau]$$

and $A_{\overline{\delta}}(\tau)$ is the closure of the set $[0, \tau] \backslash B_{\overline{\delta}}(\tau)$. It is obvious that the set $A_{\overline{\delta}}(\tau)$ consists of $d_1 \leq s + 1$ segments on each of which we have $\Delta(y) \neq 0$. Then, by virtue of the continuity of $\Delta(y)$, the following inequality holds for all $y \in A_{\overline{\delta}}(\tau)$:

$$|\Delta(y)| \geq \min_{y \in A_{\overline{\delta}}(\tau)} |\Delta(y)| \geq \min_{y \in A_{\overline{\delta}}(L)} |\Delta(y)| = c_{\overline{\delta}} > 0.$$

We now consider an arbitrary nonzero vector $\lambda = (\lambda_1, \ldots, \lambda_m)$ and write the following identity for it:

$$(\lambda, \omega(y)) \Delta_{i_0,1}(y) + (\lambda, \omega^{(1)}(y)) \Delta_{i_0,2}(y)$$
$$+ \ldots + (\lambda, \omega^{(m-1)}(y)) \Delta_{i_0,m}(y) = \lambda_{i_0} \Delta(y), \quad y \in [0, L], \quad (1.22)$$

where $|\lambda_{i_0}| = \max\limits_{\nu} |\lambda_\nu|$, $\Delta_{i_0,\nu}(y)$ is the cofactor of the element $\omega_{i_0}^{(\nu-1)}(y)$ in the determinant $\Delta(y)$, and $|\Delta_{i_0,\nu}(y)| \leq M = \text{const}$ $\forall y \in [0, L]$. Differentiating equality (1.22) $r_j \leq r$ times with respect to y, we get

$$(\lambda, \omega(y)) \Delta_{i_0,1,j}(y) + (\lambda, \omega^{(1)}(y)) \Delta_{i_0,2,j}(y)$$
$$+ \ldots + (\lambda, \omega^{(m-1+r_j)}(y)) \Delta_{i_0,m+r_j,j}(y) = \lambda_{i_0} \Delta^{(r_j)}(y). \quad (1.23)$$

Here, $\Delta_{i_0,l,j}(y),\ l=\overline{1,m+r_j}$, can be linearly expressed in terms of $\Delta_{i_0,\nu}(y)$, $\nu=\overline{1,m}$, and their derivatives with respect to y up to the order r_j, and, therefore,

$$|\Delta_{i_0,l,j}(y)| \le M$$

$$\forall i_0=\overline{1,m}, \quad l=\overline{1,m+r_j}, \quad j=\overline{1,s}, \quad y\in[0,L].$$

Since $|\Delta^{(r_j)}(\tau_j)| \equiv c^{(j)} > 0$, it follows from (1.23) that there exists an integer $q_j=q_j(\tau_j,\lambda),\ 0\le q_j\le m-1+r_j$, such that

$$|(\lambda,\omega^{(q_j)}(\tau_j))| = \max_{0\le l\le m-1+r_j} |(\lambda,\omega^{(l)}(\tau_j))| \ge \frac{|\lambda_{i_0}|c^{(j)}}{(m+p)M}$$

$$\ge \frac{\|\lambda\|}{m(m+p)M}\min_{1\le j\le s} c^{(j)} \equiv \sigma_5\|\lambda\|.$$

It follows from the last inequality and the condition of the continuity of the functions $\omega_\nu^{(l)}(y),\ \nu=\overline{1,m},\ l=\overline{0,m-1+r},\ y\in[0,L]$, that the following estimates hold for all $y\in[\tau_j-\underline{\delta},\tau_j+\underline{\delta}]\cap[0,\tau],\ 0\le l\le m-1+r_j$:

$$|(\lambda,\omega^{(q_j)}(y))| \ge \frac{1}{2}\sigma_5\|\lambda\|,\ \ |(\lambda,\omega^{(l)}(y))| \le 4|(\lambda,\omega^{(q_j)}(y))|,\ \ j=\overline{1,s}, \tag{1.24}$$

where $\underline{\delta}>0$ is a certain constant independent of λ and j.

We set $\delta=\min\{\overline{\delta},\underline{\delta}\}$ and represent the integral $I_\lambda(\tau,\varepsilon)$ in the form

$$I_\lambda(\tau,\varepsilon) = \int\limits_{A_\delta(\tau)} F(y,0,\lambda,\varepsilon)\,dy + \int\limits_{B_\delta(\tau)} F(y,0,\lambda,\varepsilon)\,dy\,. \tag{1.25}$$

For all $y\in A_\delta(\tau)$, we have $|\Delta(y)|\ge c_\delta>0$, and the set $A_\delta(\tau)$ consists of $d_1\le s+1$ segments. Consequently, according to Theorem 1.2, we get

$$\Bigg\|\int\limits_{A_\delta(\tau)} F(y,0,\lambda,\varepsilon)\,dy\Bigg\|$$

$$\le \sigma_6\varepsilon^{\frac{1}{m}}\left[\left(1+\frac{1}{\|\lambda\|}\right)\max_{[0,L]}\|f(\tau)\| + \frac{1}{\|\lambda\|}\max_{[0,L]}\|f^{(1)}(\tau)\|\right] \tag{1.26}$$

where the constant σ_6 is independent of λ and ε if $\varepsilon\in(0,\underline{\varepsilon}_1]$ ($\underline{\varepsilon}_1$ is sufficiently small).

The set $B_\delta(\tau)$ consists of $d_2 \leq s$ segments on each of which inequalities (1.24) are satisfied. Using the scheme of the proof of Theorem 1.2, we obtain

$$\left\| \int\limits_{B_\delta(\tau)} F(y,0,\lambda,\varepsilon)dy \right\|$$

$$\leq s\left[(2^{m+r}-2)\mu + \frac{2^{m+r+1}}{\sigma_5\|\lambda\|}\,\varepsilon\mu^{1-(m+r)}\right]$$

$$\times \max_{[0,L]}\|f(\tau)\| + \frac{2\delta s}{\sigma_5\|\lambda\|}\varepsilon\mu^{1-(m+r)}\max_{[0,L]}\|f^{(1)}(\tau)\|. \quad (1.27)$$

We set

$$\varepsilon = \mu^{m+r}, \quad \sigma_4 = \sigma_6 + s\left[2^{m+r} - 2 + \frac{2^{m+r-1}}{\sigma_5} + \frac{2\delta}{\sigma_5}\right],$$

$$\underline{\varepsilon}_1 \leq \min\left\{\left(\frac{1}{2}\delta\right)^{m+r};\ \left(\frac{2^{m+r}}{1+4\delta}\right)^{\frac{m+r}{m+r-1}}\right\}.$$

Then, combining inequalities (1.26) and (1.27), we deduce estimate (1.21) from (1.25). Theorem 1.4 is proved.

The example of the integral $I_\lambda(\tau,\varepsilon)$ for $\tau \in [0,1]$, $f(\tau) \equiv 1$, and

$$\omega(\tau) = \left(1, \tau, \frac{\tau^2}{2!}, \ldots, \frac{\tau^{m-2}}{(m-2)!}, \tau^{m-1+r}\right), \quad \lambda = (0,\ldots,0,m+r),$$

shows that estimate (1.21) cannot be improved in order with respect to ε under the assumptions made in Theorem 1.4. Indeed, in this case, $\tau = 0$ is a zero of multiplicity r of the function $\Delta(\tau) = \tau^r \dfrac{(m+r-1)!}{r!}$ and

$$|I_\lambda(\tau_1,\varepsilon)| = \left|\int\limits_0^{\tau_1} e^{\frac{i}{\varepsilon}y^{m+r}}dy\right| \geq \int\limits_0^{\tau_1}\left|\cos\frac{y^{m+r}}{\varepsilon}\right|dy \geq \frac{1}{2}\tau_1 = \frac{1}{2}\left(\frac{\pi}{3}\right)^{\frac{1}{m+r}}\varepsilon^{\frac{1}{m+r}},$$

$$\tau_1 = \left(\frac{\pi\varepsilon}{3}\right)^{\frac{1}{m+r}}.$$

2. Justification of Averaging Method for Oscillation Systems with $\omega = \omega(\tau)$

Consider the nonlinear system of ordinary differential equations

$$\frac{dx}{d\tau} = a(x, \varphi, \tau, \varepsilon),$$

$$\frac{d\varphi}{d\tau} = \frac{\omega(\tau)}{\varepsilon} + b(x, \varphi, \tau, \varepsilon), \tag{2.1}$$

where $x = (x_1, \ldots, x_n) \in \mathcal{D}$, $\varphi = (\varphi_1, \ldots, \varphi_m) \in R^m$, $n \geq 1$, $m \geq 2$, $\tau \in [0, L]$, L is a positive constant, $(0, \varepsilon_0] \ni \varepsilon$ is a small parameter, $\mathcal{D}$ is a bounded domain, and R^m is the m-dimensional real Euclidean space.

Let $\omega(\tau) \in C^l_{[0,L]}$, $l \geq m-1$. Furthermore, assume that the function $c(x, \varphi, \tau, \varepsilon) = [a(x, \varphi, \tau, \varepsilon); b(x, \varphi, \tau, \varepsilon)]$ is continuously differentiable with respect to $(x, \varphi, \tau) \in \mathcal{D} \times R^m \times [0, L]$ for every fixed $\varepsilon \in (0, \varepsilon_0]$, 2π-periodic in each variable φ_ν, $\nu = \overline{1, m}$, and such that

$$\sup_G \|c_0\| + \sup_G \left\| \frac{\partial c_0}{\partial x} \right\| + \sup_G \left\| \frac{\partial c_0}{\partial \tau} \right\|$$

$$+ \sum_{\|k\|>0} \left[\sup_G \|c_k\| + \frac{1}{\|k\|} \left(\sup_G \left\| \frac{\partial c_k}{\partial x} \right\| + \sup_G \left\| \frac{\partial c_k}{\partial \tau} \right\| \right) \right] \|k\|^q$$

$$\leq \sigma_1 = \text{const}, \qquad q \geq 0. \tag{2.2}$$

Here, $G = \mathcal{D} \times [0, L] \times [0, \varepsilon_0]$, $c_k = c(x, \tau, \varepsilon)$ are the Fourier coefficients of the harmonics $\exp\{i(k, \varphi)\}$ of the Fourier expansion of the function $c(x, \varphi, \tau, \varepsilon)$, and $k = (k_1, \ldots, k_m)$ is a vector with integer coordinates.

The following conditions are sufficient for the validity of (2.2):

$$c(x, \varphi, \tau, \varepsilon) \in C^{l_1}_\varphi(\overline{G}, \sigma), \quad \frac{\partial}{\partial \tau} c(x, \varphi, \tau, \varepsilon) \in C^{l_2}_\varphi(\overline{G}, \sigma),$$

$$\frac{\partial}{\partial x} c(x, \varphi, \tau, \varepsilon) \in C^{l_3}_\varphi(\overline{G}, \sigma), \quad \min\{l_1 - 1; l_2; l_3\} \geq m + q,$$

where $C^l_\varphi(\overline{G}, \sigma)$ denotes the set of functions $f(x, \varphi, \tau, \varepsilon)$ that, for every fixed ε, have partial derivatives up to the order l inclusive continuous with respect to x, φ, and τ, and uniformly bounded by a constant σ on the set $(x, \varphi, \tau, \varepsilon) \in \mathcal{D} \times$

$R^m[0, L] \times [0, \varepsilon_0] \equiv \overline{G}$. Indeed, under this assumption, the following estimates hold for all $k \neq 0$ [BMS]:

$$\sup_G \|c_k\| \leq \frac{\sigma m^{l_1}}{\|k\|^{l_1}}, \quad \sup_G \left\| \frac{\partial c_k}{\partial \tau} \right\| \leq \frac{\sigma m^{l_2}}{\|k\|^{l_2}}, \quad \sup_G \left\| \frac{\partial c_k}{\partial x} \right\| \leq \frac{\sigma m^{l_3}}{\|k\|^{l_3}}.$$

Consequently,

$$\sum_{k\neq 0} \|k\|^q \sup_G \|c_k\| \leq \sigma m^{l_1} \sum_{k\neq 0} \|k\|^{q-l_1} \leq \sigma m^{l_1} \sum_{s=1}^{\infty} s^{q-l_1} \Big(\sum_{\|k\|=s} 1 \Big)$$

$$\leq \sigma m^{l_1} 2^m \sum_{s=1}^{\infty} s^{q-l_1+m-1} \leq \sigma m^{l_1} 2^m \Big(1 + \int_1^{\infty} t^{q-l_1+m-1} dt \Big)$$

$$= \sigma m^{l_1} 2^m \left(1 + \frac{1}{l_1 - q - m} \right),$$

$$\sum_{k\neq 0} \|k\|^{q-1} \Big(\sup_G \left\| \frac{\partial c_k}{\partial \tau} \right\| + \sup_G \left\| \frac{\partial c_k}{\partial x} \right\| \Big)$$

$$\leq \sigma 2^m \left[m^{l_2} \left(1 + \frac{1}{l_2 - q - m + 1} \right) + m^{l_3} \left(1 + \frac{1}{l_3 - q - m + 1} \right) \right].$$

In the proof of the last inequalities, we have used the fact that the number of m-dimensional vectors with integer coordinates whose norm is equal to s does not exceed $2^m s^{m-1}$ [GrR3].

Along with (2.1), we consider the following system averaged over all angular variables φ:

$$\frac{d\overline{x}}{d\tau} = \overline{a}\,(\overline{x}, \tau, \varepsilon), \quad \frac{d\overline{\varphi}}{d\tau} = \frac{\omega(\tau)}{\varepsilon} + \overline{b}\,(\overline{x}, \tau, \varepsilon)\,, \tag{2.3}$$

where

$$[\overline{a}(x, \tau, \varepsilon); \overline{b}(x, \tau, \varepsilon)] = (2\pi)^{-m} \int_0^{2\pi} \ldots \int_0^{2\pi} [a(x, \varphi, \tau, \varepsilon),$$

$$b(x, \varphi, \tau, \varepsilon)]\, d\varphi_1 \ldots d\varphi_m = [\overline{a}_0(x, \tau, \varepsilon); b_0(x, \tau, \varepsilon)] = c_0(x, \tau, \varepsilon).$$

For Eqs. (2.1) and (2.2), we specify the initial conditions

$$x|_{\tau=0} = y \in \mathcal{D}_1 \subset \mathcal{D}, \quad \varphi|_{\tau=0} = \psi \in R^m, \tag{2.4}$$

where $\mathcal{D}_1$ is a certain domain, and denote by $(x(\tau, y, \psi, \varepsilon); \varphi(\tau, y, \psi, \varepsilon))$ and $(\overline{x}(\tau, y, \varepsilon); \overline{\varphi}(\tau, y, \psi, \varepsilon))$ solutions of problems (2.1), (2.4) and (2.3), (2.4), respectively.

Theorem 2.1. *Suppose that the following conditions are satisfied:*

(i) $\det\left(W_p^T(\tau) W_p(\tau)\right) \neq 0 \;\; \forall \tau \in [0, L]$ *for certain minimal* $p \geq m, \;\; p \leq l + 1$;

(ii) condition (2.2) is satisfied for $q = 0$;

(iii) for all $\tau \in [0, L], \;\; y \in D_1,$ *and* $\varepsilon \in (0, \varepsilon_0],$ *the curve* $x = \overline{x}(\tau, y, \varepsilon)$ *lies in* $\mathcal{D}$ *together with its* ρ*-neighborhood.*

Then one can find a constant σ_2 *independent of* ε *and such that, for sufficiently small* $\varepsilon_0 > 0$ *and every* $\tau \in [0, L], \;\; y \in \mathcal{D}_1, \;\; \psi \in R^m,$ *and* $\varepsilon \in (0, \varepsilon_0]$, *the following estimate holds:*

$$\|U(\tau, y, \psi, \varepsilon)\| \leq \sigma_2 \, \varepsilon^{1/p}, \tag{2.5}$$

where $U = (x(\tau, y, \psi, \varepsilon) - \overline{x}(\tau, y, \varepsilon); \varphi(\tau, y, \psi, \varepsilon) - \overline{\varphi}(\tau, y, \psi, \varepsilon)).$

Proof. Since the right-hand side of system (2.1) is smooth, a solution of the Cauchy problem (2.1), (2.4) exists. Denote by $[0, T), \;\; T = T(y, \psi, \varepsilon),$ the maximum half-interval of the segment $[0, L]$ for which the curve $x = x(\tau, y, \psi, \varepsilon)$ lies in the ρ-neighborhood of the curve $x = \overline{x}(\tau, y, \varepsilon)$. Then Eqs. (2.1) and (2.3) and the Gronwall–Bellman lemma [FiS] imply that, for any $\tau \in [0, T)$,

$$\|U(\tau, y, \psi, \varepsilon)\| \leq e^{\sigma_1 L} \sup_{\tau \in [0, L]} \sum_{k \neq 0} \Bigg\| \int_0^{\tau} c_k(\overline{x}(t, y, \varepsilon), t, \varepsilon)$$

$$\times \exp\{i(k, \overline{\theta})\} \exp\left\{ \frac{i}{\varepsilon} \int_0^t (k, \omega(z)) dz \right\} dt \Bigg\|,$$

where

$$\overline{\theta} = \overline{\varphi}(t, y, \psi, \varepsilon) - \frac{1}{\varepsilon}\int_0^t \omega(z)\,dz.$$

Note that, under the conditions imposed on the functions $\omega(t)$ and $f(t) = c_k(\overline{x}(t, y, \varepsilon), t, \varepsilon)\exp\{i(k, \overline{\theta})\}$, all conditions of Theorem 1.3 are satisfied. Consequently, using this theorem for the estimation of each integral on the right-hand side of the last inequality, we get

$$\Big\|\int_0^\tau c_k(\overline{x}(t, y, \varepsilon), t, \varepsilon)\exp\{i(k, \overline{\theta})\}dt\Big\|$$

$$\leq \overline{\sigma}_2\varepsilon^{\frac{1}{p}}\Big[\Big(2 + \sup_G\|\overline{b}(x, \tau, \varepsilon)\|\Big)\Big(1 + \sup_G\|\overline{a}(x, \tau, \varepsilon)\|\Big)$$

$$\times\Big(\sup_G\|c_k\| + \frac{1}{\|k\|}\sup_G\Big\|\frac{\partial c_k}{\partial\tau}\Big\| + \frac{1}{\|k\|}\sup_G\Big\|\frac{\partial c_k}{\partial x}\Big\|\Big)\Big],$$

$$\|U(\tau, y, \psi, \varepsilon)\| \leq e^{\sigma_1 L}(2 + \sigma_1)(1 + \sigma_1)\sigma_1\overline{\sigma}_2\varepsilon^{\frac{1}{p}} \equiv \sigma_2\varepsilon^{\frac{1}{p}} \quad \forall\tau\in[0, T).$$

Let $\sigma_2\varepsilon_0^{\frac{1}{p}} \leq \frac{1}{2}\rho$. Then inequality (2.5) implies that $x(T, y, \psi, \varepsilon) \in \mathcal{D}$ together with its $\frac{1}{2}\rho$-neighborhood. Therefore, $T = L$ and estimate (2.5) holds for all $\tau\in[0, L]$. Theorem 2.1 is proved.

We now study in more detail the dependence of the function $U(\tau, y, \psi, \varepsilon)$ on the initial data y and ψ. Below, using properties of oscillation integrals, we establish estimates for the partial derivatives $\frac{\partial}{\partial y}U$ and $\frac{\partial}{\partial\psi}U$, which are substantially used in the Chapter 2 for the solution of boundary-value problems. For this purpose, we impose on $c(x, \varphi, \tau, \varepsilon)$ a stronger restriction than (2.2). Assume that the function $c(x, \varphi, \tau, \varepsilon)$ is twice continuously differentiable with respect to x, φ, and τ for every fixed ε, and its Fourier coefficients satisfy the inequality

$$\sup\|c_0\| + \sup\Big\|\frac{\partial c_0}{\partial\tau}\Big\| + \sup\Big\|\frac{\partial c_0}{\partial x}\Big\| + \sum_{j=1}^n\sup\Big\|\frac{\partial^2 c_0}{\partial x\partial x_j}\Big\|$$

$$+ \sum_{k \neq 0} \Big[\|k\| \sup \|c_k\| + \sup \Big\| \frac{\partial c_k}{\partial \tau} \Big\| + \sup \Big\| \frac{\partial c_k}{\partial x} \Big\|$$

$$+ \frac{1}{\|k\|} \Big(\sup \Big\| \frac{\partial^2 c_k}{\partial x \partial \tau} \Big\| + \sum_{j=1}^{n} \sup \Big\| \frac{\partial^2 c_k}{\partial x \partial x_j} \Big\| \Big) \Big] \leq \sigma_1. \qquad (2.6)$$

Here, the supremum is taken over all $(x, \tau, \varepsilon) \in G$.

Theorem 2.2. *If conditions (i) and (iii) of Theorem 2.1 and inequality (2.6) are satisfied, then, for all* $\tau \in [0, L]$, $y \in \mathcal{D}_1$, $\psi \in R^m$, *and* $\varepsilon \in (0, \varepsilon_0]$ *(where* ε_0 *is positive and sufficiently small), the following estimate holds:*

$$\Big\| \frac{\partial}{\partial y} U(\tau, y, \psi, \varepsilon) \Big\| + \Big\| \frac{\partial}{\partial \psi} U(\tau, y, \psi, \varepsilon) \Big\| \leq \sigma_3 \varepsilon^{\frac{1}{p}}, \qquad (2.7)$$

where the constant σ_3 *is independent of* ε.

Proof. First, we establish estimates for the first-order partial derivatives with respect to y and ψ for a solution of the Cauchy problem (2.3), (2.4). The smoothness conditions for the right-hand side of system (2.3) yield

$$\overline{x}(\tau, y, \varepsilon) = y + \int_0^\tau \overline{a}(\overline{x}(t, y, \varepsilon), t, \varepsilon)\, dt,$$

$$\frac{\partial \overline{x}(\tau, y, \varepsilon)}{\partial y} = E_n + \int_0^\tau \frac{\partial}{\partial \overline{x}} \overline{a}(\overline{x}(t, y, \varepsilon), t, \varepsilon) \frac{\partial \overline{x}(t, y, \varepsilon)}{\partial y}\, dt,$$

whence

$$\Big\| \frac{\partial \overline{x}(\tau, y, \varepsilon)}{\partial y} \Big\| \leq n + \sigma_1 \int_0^\tau \Big\| \frac{\partial \overline{x}(\tau, y, \varepsilon)}{\partial y} \Big\|\, dt.$$

Solving this inequality, for all $\tau \in [0, L]$, $y \in \mathcal{D}_1$ and $\varepsilon \in (0, \varepsilon_0]$ we get the estimate

$$\Big\| \frac{\partial \overline{x}(\tau, y, \varepsilon)}{\partial y} \Big\| \leq n e^{\sigma_1 L}, \qquad (2.8)$$

which, together with the first equation in (2.3), yields

$$\Big\| \frac{d}{d\tau} \frac{\partial \overline{x}(\tau, y, \varepsilon)}{\partial y} \Big\| \leq n \sigma_1 e^{\sigma_1 L}. \qquad (2.9)$$

Since

$$\overline{\varphi}(\tau, y, \psi, \varepsilon) = \psi + \int_0^\tau \overline{b}(\overline{x}(t, y, \varepsilon), t, \varepsilon)dt + \frac{1}{\varepsilon}\int_0^\tau \omega(t)dt,$$

we have

$$\left\|\frac{\partial\overline{\varphi}(\tau, y, \psi, \varepsilon)}{\partial y}\right\| \le L\sigma_1 n e^{\sigma_1 L}, \qquad \left\|\frac{\partial\overline{\varphi}(\tau, y, \psi, \varepsilon)}{\partial \psi}\right\| = m,$$

$$\frac{d}{d\tau}\,\frac{\partial\overline{\varphi}(\tau, y, \psi, \varepsilon)}{\partial \psi} \equiv 0, \qquad \left\|\frac{d}{d\tau}\,\frac{\partial\overline{\varphi}(\tau, y, \psi, \varepsilon)}{\partial y}\right\| \le \sigma_1 n e^{\sigma_1 L} \tag{2.10}$$

for all $\tau \in [0, L]$, $y \in \mathcal{D}_1$, $\psi \in R^m$, and $\varepsilon \in (0, \varepsilon_0]$.

We now differentiate Eqs. (2.1) and (2.3) with respect to y. Then we obtain the following integral equation for $\dfrac{\partial}{\partial y}U(\tau, y, \psi, \varepsilon)$:

$$\begin{aligned}\frac{\partial}{\partial y}U(\tau, y, \psi, \varepsilon) =& \int_0^\tau A(x(t, y, \psi, \varepsilon), \varphi(t, y, \psi, \varepsilon), t, \varepsilon)\frac{\partial}{\partial y}U(t, y, \psi, \varepsilon)dt \\ &+ \int_0^\tau \left[\frac{\partial}{\partial x}c_0(x(t, y, \psi, \varepsilon), t, \varepsilon) - \frac{\partial}{\partial x}c_0(\overline{x}(t, y, \varepsilon), t, \varepsilon)\right] \frac{\partial}{\partial y}\overline{x}(t, y, \varepsilon)\, dt \\ &+ \int_0^\tau \frac{\partial}{\partial x}\tilde{c}(x(t, y, \psi, \varepsilon), \varphi(t, y, \psi, \varepsilon), t, \varepsilon)\, \frac{\partial}{\partial y}\,\overline{x}\,(t, y, \varepsilon)\, dt \\ &+ \int_0^\tau \frac{\partial}{\partial \varphi}\,\tilde{c}\,(x(t, y, \psi, \varepsilon), \varphi(t, y, \psi, \varepsilon), t, \varepsilon)\, \frac{\partial}{\partial y}\overline{\varphi}(t, y, \psi, \varepsilon)\, dt,\end{aligned} \tag{2.11}$$

where

$$\tilde{c}\,(x, \varphi, t, \varepsilon) = c\,(x, \varphi, t, \varepsilon) - c_0\,(x, t, \varepsilon),$$

$$A(x, \varphi, t, \varepsilon) = \left(\frac{\partial}{\partial x}\,c\,(x, \varphi, t, \varepsilon);\, \frac{\partial}{\partial \varphi}\,c\,(x, \varphi, t, \varepsilon)\right).$$

Taking into account the estimate for the error of the averaging method (2.5) and condition (2.6), we obtain the inequalities

$$\Big\| \int_0^\tau \Big[\frac{\partial}{\partial x} c_0(x(t,y,\psi,\varepsilon),t,\varepsilon) - \frac{\partial}{\partial x} c_0(\overline{x}(t,y,\varepsilon),t,\varepsilon) \Big] \frac{\partial}{\partial y} \overline{x}(t,y,\varepsilon) dt \Big\|$$

$$\leq \sigma_1 \sigma_2 n L e^{\sigma_1 L} \varepsilon^{\frac{1}{p}},$$

$$\|A(x,\varphi,t,\varepsilon)\| \leq \sigma_1 \tag{2.12}$$

for all $\tau \in [0,L]$, $y \in \mathcal{D}_1$, $\psi \in R^m$, and $\varepsilon \in (0,\varepsilon_0]$.

We estimate the last two integrals on the right-hand side of (2.11) with the use of Theorem 1.3 for $\lambda = k$ and conditions (2.6) and (2.8)–(2.10). As a result, we obtain

$$\Big\| \int_0^\tau \frac{\partial \tilde{c}}{\partial x} \frac{\partial \overline{x}}{\partial y} dt \Big\| \leq \sum_{k \neq 0} \Big\| \int_0^\tau \frac{\partial}{\partial x} c_k(x(t,y,\psi,\varepsilon),t,\varepsilon) \frac{\partial}{\partial y} \overline{x}(t,y,\varepsilon)$$

$$\times \exp\{i(k,\theta)\} \exp\Big\{ \frac{i}{\varepsilon} \int_0^t (k,\omega(z))\, dz \Big\}\, dt \Big\|$$

$$\leq \overline{\sigma}_2 \varepsilon^{\frac{1}{p}} \sum_{k \neq 0} \Big[\sup_G \Big\| \frac{\partial c_k}{\partial x} \Big\| 2n e^{\sigma_1 L}(1+\sigma_1)$$

$$+ \frac{1}{\|k\|} \Big(\sup_G \Big\| \frac{\partial^2 c_k}{\partial x \partial \tau} \Big\| \sigma_1 \sum_{j=1}^n \sup_G \Big\| \frac{\partial^2 c_k}{\partial x \partial x_j} \Big\| \Big) n e^{\sigma_1 L} \Big]$$

$$\leq 3(1+\sigma_1)\sigma_1 \overline{\sigma}_2 n e^{\sigma_1 L} \varepsilon^{\frac{1}{p}}. \tag{2.13}$$

Similarly,

$$\Big\| \int_0^\tau \frac{\partial \tilde{c}}{\partial \varphi} \frac{\partial \overline{\varphi}}{\partial y} dt \Big\| \leq n \sigma_1^2 \overline{\sigma}_2 (1 + 2L + L\sigma_1)\, e^{L\sigma_1} \varepsilon^{\frac{1}{p}}. \tag{2.14}$$

In view of (2.12)–(2.14), Eq. (2.11) yields the integral inequality

$$\Big\| \frac{\partial}{\partial y} U(\tau,y,\psi,\varepsilon) \Big\| \leq \sigma_1 \int_0^\tau \Big\| \frac{\partial}{\partial y} U(t,y,\psi,\varepsilon) \Big\| dt + \varepsilon^{\frac{1}{p}} \underline{\sigma}_2,$$

whose solution satisfies the estimate

$$\Big\| \frac{\partial}{\partial y} U(\tau,y,\psi,\varepsilon) \Big\| \leq \underline{\sigma}_2 e^{\sigma_1 L} \varepsilon^{\frac{1}{p}} \equiv \overline{\sigma}_3 \varepsilon^{\frac{1}{p}} \tag{2.15}$$

$\forall(\tau, y, \psi, \varepsilon) \in [0, L] \times \mathcal{D}_1 \times R^m \times (0, \varepsilon_0]$. Here,

$$\underline{\sigma}_2 = n\sigma_1[\sigma_2 + \overline{\sigma}_2(3 + 4\sigma_1 + 2\sigma_1 L + \sigma_1^2 L)]e^{\sigma_1 L}.$$

Let us estimate the norm of the matrix $\dfrac{\partial}{\partial\psi}U(\tau, y, \psi, \varepsilon)$. Since

$$\frac{\partial}{\partial\psi}\,\overline{x}\,(\tau, y, \varepsilon) \equiv 0, \qquad \frac{\partial}{\partial\psi}\,\overline{\varphi}\,(\tau, y, \psi, \varepsilon) = E_m,$$

it follows from Eqs. (2.1) and (2.3) that

$$\frac{\partial}{\partial\psi}U(\tau, y, \psi, \varepsilon) = \int_0^\tau A(x(t, y, \psi, \varepsilon), \varphi(t, y, \psi, \varepsilon), t, \varepsilon)\frac{\partial}{\partial\psi}U(t, y, \psi, \varepsilon)dt$$
$$+ \int_0^\tau \frac{\partial}{\partial\varphi}\,c\,(x(t, y, \psi, \varepsilon), \varphi\,(t, y, \psi, \varepsilon), t, \varepsilon)\,dt.$$

This yields

$$\left\|\frac{\partial}{\partial\psi}U(\tau, y, \psi, \varepsilon)\right\| \le 2\sigma_1(1 + \sigma_1)\overline{\sigma}_2 e^{L\sigma_1}\varepsilon^{\frac{1}{p}} \equiv \underline{\sigma}_3\varepsilon^{\frac{1}{p}}. \tag{2.16}$$

Combining (2.15) and (2.16), we get estimate (2.7) with the constant $\sigma_3 = \overline{\sigma}_3 + \underline{\sigma}_3$. The smallness of $\varepsilon_0 > 0$ is determined by the possibility of the application of Theorems 1.3 and 2.1.

Remark 2. If $p = m$, then the condition $\det\left(W_p^T(\tau)W_p(\tau)\right) \neq 0 \ \ \forall\tau \in [0, L]$ can be reduced to the condition $\Delta(\tau) = \det\, W_m(\tau) \neq 0$. If the function $\Delta(\tau)$ has zeros of multiplicity not higher than r on the segment $[0, L]$, then, using Theorem 1.4, instead of estimates (2.5) and (2.7) we obtain estimates of the form

$$\|U(\tau, y, \psi, \varepsilon)\| + \left\|\frac{\partial}{\partial y}U(\tau, y, \psi, \varepsilon)\right\| + \left\|\frac{\partial}{\partial\psi}U(\tau, y, \psi, \varepsilon)\right\| \le \sigma_4\varepsilon^{\frac{1}{m+r}}.$$

The investigation of oscillation systems becomes more complicated if $\Delta(\tau)$ is identically equal to zero on some segment $[\alpha, \beta] \subset [0, L]$. In this case, the solution of system (2.1) may deviate from the solution of the averaged system (2.3) at time $\Delta\tau = L$ by a distance proportional to unit, i.e., the scheme of averaging over all angular variables is, generally speaking, inapplicable. As an example, we consider the Cauchy problem

$$\frac{dx}{d\tau} = 1 - \cos(\varphi_1 + \varphi_2 - \varphi_3), \quad \frac{d\varphi_1}{d\tau} = \frac{\omega_1(\tau)}{\varepsilon},$$
$$\frac{d\varphi_2}{d\tau} = \frac{\omega_2(\tau)}{\varepsilon}, \quad \frac{d\varphi_3}{d\tau} = \frac{\omega_1(\tau) + \omega_2(\tau)}{\varepsilon},$$
$$x(0) = \varphi_1(0) = \varphi_2(0) = \varphi_3(0) = 0$$

(where $\omega_1(\tau)$ and $\omega_2(\tau)$ are twice continuously differentiable functions on the segment $[0,1]$) and the corresponding problem averaged over all angular variables φ_1, φ_2, and φ_3, namely,

$$\frac{d\overline{x}}{d\tau} = 1, \quad \frac{d\overline{\varphi}_1}{d\tau} = \frac{\omega_1(\tau)}{\varepsilon}, \quad \frac{d\overline{\varphi}_2}{d\tau} = \frac{\omega_2(\tau)}{\varepsilon}, \quad \frac{d\overline{\varphi}_3}{d\tau} = \frac{\omega_1(\tau) + \omega_2(\tau)}{\varepsilon},$$
$$\overline{x}(0) = \overline{\varphi}_1(0) = \overline{\varphi}_2(0) = \overline{\varphi}_3(0) = 0.$$

It is obvious that $\overline{x}(\tau) = \tau$, $x(\tau) \equiv 0$, $\Delta(\tau) \equiv 0$ $\forall \tau \in [0,1]$, and $|x(1) - \overline{x}(1)| = 1$. Hence, the condition $\Delta(\tau) \equiv 0$ leads to the violation of the efficient estimate of the error of the method of averaging over all angular variables on the segment $[0,1]$.

In this case, it is convenient to perform averaging over a part of angular variables. Below, we describe this method in brief and give its justification.

Assume that

$$|\Delta(\tau)| \geq \sigma_5(\alpha - \tau)^{r_1} \quad \forall \tau \in [0, \alpha),$$
$$|\Delta(\tau)| \geq \sigma_5(\tau - \beta)^{r_2} \quad \forall \tau \in (\beta, L],$$
$$\Delta(\tau) \equiv 0 \quad \forall \tau \in [\alpha, \beta], \tag{2.17}$$

where r_1, r_2, and σ_5 are certain positive constants. The linear dependence of the functions $\omega_1(\tau), \ldots, \omega_m(\tau)$ on the segment $[\alpha, \beta]$ is a sufficient condition for the validity of the identity $\Delta(\tau) \equiv 0$ $\forall \tau \in [\alpha, \beta]$. Further, we assume that there exist $h < m$ linearly independent vectors $k^{(j)} = (k_1^{(j)}, \ldots, k_m^{(j)})$, $j = \overline{1, h}$, with integer coordinates for which

$$(k^{(j)}, \omega(\tau)) \equiv 0 \quad \forall \tau \in [\alpha, \beta]. \tag{2.18}$$

Without loss of generality, we can assume that the hth-order minor γ in the left upper corner of the matrix $K = \operatorname{colon}(k^{(1)}, \ldots, k^{(h)})$ is nonzero. Denote

$$\widetilde{K} = \operatorname{colon}(k^{(1)}, \ldots, k^{(h)}, e_{h+1}, \ldots, e_m),$$

where e_ν is a basis vector of the space R^m. Then, in the variables

$$\widetilde{K}\varphi = (\psi; \theta), \qquad \psi = (\psi_1, \ldots, \psi_h),$$
$$\theta = (\theta_1, \ldots, \theta_{m-h}) = (\varphi_{h+1}, \ldots, \varphi_m),$$
$$\varphi = \frac{1}{\gamma}(S_1\psi + S_2\theta),$$

where S_1 and S_2 are matrices whose elements are integer numbers, system (2.1) takes the form

$$\frac{dx}{d\tau} = A(x, \psi, \theta, \tau, \varepsilon), \quad \frac{d\psi}{d\tau} = \frac{1}{\varepsilon}K\omega(\tau) + B(x, \psi, \theta, \tau, \varepsilon),$$
$$\frac{d\theta}{d\tau} = \frac{1}{\varepsilon}\widetilde{\omega}(\tau) + C(x, \psi, \theta, \tau, \varepsilon). \tag{2.19}$$

Here,

$$\widetilde{\omega}(\tau) = (\omega_{h+1}(\tau), \ldots, \omega_m(\tau)),$$
$$A(x, \psi, \theta, \tau, \varepsilon) = a\left(x, \frac{1}{\gamma}S_1\psi + \frac{1}{\gamma}S_2\theta, \tau, \varepsilon\right),$$
$$B(x, \psi, \theta, \tau, \varepsilon) = Kb\left(x, \frac{1}{\gamma}S_1\psi + \frac{1}{\gamma}S_2\theta, \tau, \varepsilon\right),$$
$$C(x, \psi, \theta, \tau, \varepsilon) = \left(b_{h+1}\left(x, \frac{1}{\gamma}S_1\psi + \frac{1}{\gamma}S_2\theta, \tau, \varepsilon\right), \ldots, b_m\left(x, \frac{1}{\gamma}S_1\psi + \frac{1}{\gamma}S_2\theta, \tau, \varepsilon\right)\right).$$

The corresponding system averaged over all variables θ has the form

$$\frac{d\overline{x}}{d\tau} = \overline{A}(\overline{x}, \overline{\psi}, \theta, \tau, \varepsilon), \quad \frac{d\overline{\psi}}{d\tau} = \frac{1}{\varepsilon}K\omega(\tau) + \overline{B}(\overline{x}, \overline{\psi}, \tau, \varepsilon),$$
$$\frac{d\overline{\theta}}{d\tau} = \frac{1}{\varepsilon}\widetilde{\omega}(\tau) + \overline{C}(\overline{x}, \overline{\psi}, \tau, \varepsilon), \tag{2.20}$$

where

$$[\overline{A}; \overline{B}; \overline{C}] = (2\pi\gamma)^{h-m} \int_0^{2\pi\gamma} \dots \int_0^{2\pi\gamma} \Big[A(\overline{x}, \overline{\psi}, \theta, \tau, \varepsilon),$$

$$B(\overline{x}, \overline{\psi}, \theta, \tau, \varepsilon); C(\overline{x}, \overline{\psi}, \theta, \tau, \varepsilon)\Big] \, d\theta_1 \dots d\theta_{m-h}.$$

Theorem 2.3. *Suppose that the following conditions are satisfied:*

(i) there exists a solution $(\overline{x}(\tau, \varepsilon); \overline{\psi}(\tau, \varepsilon); \overline{\theta}(\tau, \varepsilon))$ *of the averaged system (2.20) that lies in* $\mathcal{D} \times R^m$ *together with its* ρ*-neighborhood* $\forall \tau \in [0, L]$, $\varepsilon \in (0, \varepsilon_0]$;

(ii) $\omega(\tau) \in C^{m-1}_{[0,L]}$;

(iii) conditions (2.2) for $q = 1$, *(2.17), and (2.18) are satisfied and, furthermore,*

$$\det\, (\omega_{h+\nu}^{(j-1)}(\tau))_{\nu,j=1}^{m-h} \neq 0 \quad \forall \tau \in [\alpha, \beta]. \tag{2.21}$$

Then there exist constants $\varepsilon^* > 0$ *and* $\sigma_6 > 0$ *such that* $\forall (\tau, \varepsilon) \in [0, L] \times (0, \varepsilon_0]$ $(\varepsilon_0 \leq \varepsilon^*)$, *the following inequality holds:*

$$\begin{aligned} v(\tau, \tau_1, \varepsilon) &\equiv \|x(\tau, \tau_1, \varepsilon) - \overline{x}(\tau, \varepsilon)\| + \|\psi(\tau, \tau_1, \varepsilon) - \overline{\psi}(\tau, \varepsilon)\| \\ &\qquad + \|\theta(\tau, \tau_1, \varepsilon) - \overline{\theta}(\tau, \varepsilon)\| \\ &\leq \sigma_6 g(\tau, \tau_1, \varepsilon), \end{aligned} \tag{2.22}$$

where $(x(\tau, \tau_1, \varepsilon); \psi(\tau, \tau_1, \varepsilon); \theta(\tau, \tau_1, \varepsilon))$ *is a solution of system (2.19) that coincides with the solution* $(\overline{x}(\tau, \varepsilon); \overline{\psi}(\tau, \varepsilon); \overline{\theta}(\tau, \varepsilon))$ *of the averaged system (2.20) for* $\tau = \tau_1 \in [0, L]$ *and, furthermore,* $g(\tau, \tau_1, \varepsilon) = \varepsilon^{\frac{1}{m+r}}$, $r = \max\{r_1; r_2\}$, *for all* $\tau \in [0, L]$ *and* $\tau_1 \notin [\alpha, \beta]$ *and*

$$g(\tau, \tau_1, \varepsilon) = \begin{cases} \varepsilon^{\frac{1}{m-h}}, & \tau \in [\alpha, \beta], \\ \varepsilon^{\frac{1}{m+r}}, & \tau \in [0, L] \setminus [\alpha, \beta], \end{cases}$$

for $\tau_1 \in [\alpha, \beta]$.

Proof. Denote

$$F(x,\psi,\theta,\tau,\varepsilon) = [A;B;C]$$

$$= \sum_k F_k(x,\tau,\varepsilon)\exp\Big\{\frac{i}{\gamma}(S_1^T k,\psi)\Big\}\exp\Big\{\frac{i}{\gamma}(S_2^T k,\theta)\Big\},$$

$$F_k(x,\tau,\varepsilon) = [a_k(x,\tau,\varepsilon); \widetilde{K} b_k(x,\tau,\varepsilon)],$$

$$\overline{F}(x,\psi,\tau,\varepsilon) = \sum_{S_2^T k=0} F_k(x,\tau,\varepsilon)\exp\Big\{\frac{i}{\gamma}(S_1^T k,\psi)\Big\} \equiv [\overline{A};\overline{B};\overline{C}].$$

Here, S_1^T and S_2^T are the transposed matrices. Using inequality (2.2) for $q=1$, one can easily obtain the estimate

$$\|F(x,\psi,\theta,\tau,\varepsilon) - F(\overline{x},\overline{\psi},\overline{\theta},\tau,\varepsilon)\| \le \sigma_7(\|x-\overline{x}\| + \|\psi-\overline{\psi}\| + \|\theta-\overline{\theta}\|),$$

$$\sigma_7 = \sigma_1\|\widetilde{K}\|\,\frac{1}{|\gamma|}\,(|\gamma| + \|S_1\| + \|S_2\|),$$

which, together with Eqs. (2.19) and (2.20), yields

$$v(\tau,\tau_1,\varepsilon) \le \sigma_7\Big|\int_{\tau_1}^{\tau} v(t,\tau_1,\varepsilon)\,dt\Big|$$

$$+ \sum_{S_2^T k\ne 0}\Big\|\int_{\tau_1}^{\tau} F_k(\overline{x}(t,\varepsilon),t,\varepsilon)$$

$$\times\exp\Big\{\frac{i}{\gamma}(S_1^T k,\overline{\psi}(t,\varepsilon))\Big\}\exp\Big\{\frac{i}{\gamma}(S_2^T k,\overline{\theta}(t,\varepsilon))\Big\}\,dt\Big\|. \qquad (2.23)$$

Let $\tau_1 \in [\alpha,\beta]$. Then inequality (2.21) and the identity $K\omega(\tau)\equiv 0$ are satisfied for $\tau\in[\alpha,\beta]$, i.e., $\overline{\psi}(\tau,\varepsilon)$ are slow variables. Therefore, each integral under the sum sign on the right-hand side of (2.23) can be interpreted as an oscillation integral. Setting

$$\widetilde{\theta}(t,\varepsilon) = \overline{\theta}(t,\varepsilon) - \frac{1}{\varepsilon}\int_{\tau_1}^{t}\widetilde{\omega}(z)\,dz,$$

$$f_k(t,\varepsilon) = F_k(\overline{x}(t,\varepsilon),t,\varepsilon)\exp\Big\{\frac{i}{\gamma}(S_1^T k,\overline{\psi}(t,\varepsilon))\Big\}\exp\Big\{\frac{i}{\gamma}(S_2^T k,\tilde{\theta}(t,\varepsilon))\Big\}$$

and using estimate (1.20) for $p = m - h$, we get

$$v(\tau, \tau_1, \varepsilon) \le \sigma_7 \Big| \int_{\tau_1}^{\tau} v(t, \tau_1, \varepsilon)\, dt \Big| + \sum_k \Big[(1 + \sigma_1) \|k\| \sup_G \|F_k\|$$

$$+ \sup_G \Big\| \frac{\partial}{\partial t} F_k \Big\| + \sigma_1 \sup_G \Big\| \frac{\partial}{\partial x} F_k \Big\| \Big] \sigma_8 \varepsilon^{\frac{1}{m-h}}$$

$$\le \sigma_7 \Big| \int_{\tau_1}^{\tau} v(t, \tau_1, \varepsilon)\, dt \Big| + (1 + \sigma_1)\sigma_1 \|\widetilde{K}\| \sigma_8 \varepsilon^{\frac{1}{m-h}},$$

where σ_8 is a constant corresponding to the constant σ_3 in inequality (1.20). The last inequality proves the following estimate for all $\tau \in [\alpha, \beta]$ and $\tau_1 \in [\alpha, \beta]$:

$$v(\tau, \tau_1, \varepsilon) \le (1 + \sigma_1) \|\widetilde{K}\| \sigma_8 e^{\sigma_7 L} \varepsilon^{\frac{1}{m-h}}. \tag{2.24}$$

Now let $\tau \in [0, \alpha)$ and $\tau_1 \in [\alpha, \beta]$. Then, taking into account condition (2.17) for $\tau \in [0, \alpha)$ and inequality (1.21) for $p = r_1$ and using (2.23), we get

$$v(\tau, \tau_1, \varepsilon) \le \sigma_7 \int_{\tau}^{\tau_1} v(t, \tau_1, \varepsilon) dt$$

$$+ \sum_{S_2^T k \ne 0} \Big[\Big\| \int_{\tau}^{\alpha} F_k(\overline{x}(t, \varepsilon), t, \varepsilon) \exp\{i(k, \tilde{\varphi}(t, \varepsilon))\}$$

$$\times \exp\Big\{ \frac{i}{\varepsilon} \int_{\tau_1}^{t} (k, \omega(z)) dz \Big\} dt \Big\|$$

$$+ \Big\| \int_{\alpha}^{\tau_1} f_k(t, \varepsilon) \exp\Big\{ \frac{i}{\gamma\varepsilon} \Big(S_2^T k, \int_{\tau_1}^{t} \omega(z) dz \Big) \Big\} dt \Big\| \Big]$$

$$\le \sigma_7 \int_{\tau}^{\tau_1} v(t, \tau_1, \varepsilon) dt + \varepsilon^{\frac{1}{m-h}} (1 + \sigma_1) \sigma_1 \|\widetilde{K}\| \sigma_8$$

$$+ \varepsilon^{\frac{1}{m+r_1}} (1 + \sigma_1) \sigma_1 \|\widetilde{K}\| \sigma_9,$$

$$\widetilde{\varphi}(t, \varepsilon) = \widetilde{K}^{-1} (\overline{\psi}(t, \varepsilon); \overline{\theta}(t, \varepsilon)) - \frac{1}{\varepsilon} \int_{\tau_1}^{t} \omega(z)\, dz,$$

or

$$v(\tau, \tau_1, \varepsilon) \leq (1+\sigma_1)\sigma_1 \|\widetilde{K}\| (\sigma_8 + \sigma_9) e^{\sigma_7 L} (\varepsilon^{\frac{1}{m-h}} + \varepsilon^{\frac{1}{m+r_1}}) \tag{2.25}$$

for all $\tau \in [0, \alpha)$. Here, σ_9 is a constant corresponding to the constant σ_4 in inequality (1.21) for $p = r_1$. If $\tau \in (\beta, L]$, then $v(t, \tau_1, \varepsilon)$ also satisfies an estimate of the form (2.25) with r_1 replaced by r_2 and, possibly, the constant σ_9 replaced by another constant $\widetilde{\sigma}_9$. Taking this fact and inequalities (2.24) and (2.25) into account, for all $\tau_1 \in [\alpha, \beta]$ and $\tau \in [0, L]$ we obtain estimate (2.22), where

$$\sigma_6 = 3\sigma_1(1+\sigma_1)(\sigma_8 + \sigma_9 + \widetilde{\sigma}_9)\|\widetilde{K}\| e^{\sigma_7 L}.$$

The smallness of $\varepsilon_0 > 0$ is determined by conditions for the validity of inequalities (1.20) and (1.21) and by the estimate $\sigma_6 \varepsilon_0^{\frac{1}{m+r}} \leq \frac{1}{2}\rho$, which guarantees that the solution of system (2.19) under investigation does not leave the domain $\mathcal{D} \times R^m$ $\forall (\tau, \varepsilon) \in [0, L] \times (0, \varepsilon_0]$. For $\tau_1 \in [0, L] \backslash [\alpha, \beta]$, the proof of the theorem is analogous.

Remark 3. The first two inequalities in (2.17) mean that, at certain times, resonance occurs in the multifrequency system (2.1), but the system quickly leaves the resonance state. If identities (2.18) are satisfied, then the system remains in the resonance state for a sufficiently long time period $\Delta\tau = \beta - \alpha$. In this connection, there arises the necessity of using the method of averaging over a part of angular variables. The averaging scheme proposed above is not unique. Efficient estimates for the norm of the difference of solutions of perturbed and averaged equations can also be obtained by averaging over all angular variables on the intervals $[0, \alpha)$ and $(\beta, L]$ and over a part of these variables on $[\alpha, \beta]$ and then "glueing" the integral curves in a proper way. Note that the order of the estimates obtained with respect to ε is the same as in inequality (2.22).

We give an example of frequencies satisfying conditions (2.17), (2.18), and (2.21). Let

$$\omega_1(\tau) = \tau + 1 \quad \forall \tau \in [0, 3],$$

$$\omega_2(\tau) = \begin{cases} \tau^2 + 3, & \tau \in [0, 1), \\ 2(\tau+1), & \tau \in [1, 2], \\ \frac{1}{2}\tau^2 + 4, & \tau \in (2, 3]. \end{cases}$$

It is clear that $\omega_1(\tau)$ and $\omega_2(\tau)$ are continuously differentiable on $[0,3]$ and

$$\Delta(\tau) = \begin{vmatrix} \omega_1(\tau) & \omega_2(\tau) \\ \omega_1^{(1)}(\tau) & \omega_2^{(1)}(\tau) \end{vmatrix} = \begin{cases} -(\tau+3)(1-\tau), & \tau \in [0,1), \\ 0, & \tau \in [1,2], \\ \frac{1}{2}(\tau-2)(\tau+4), & \tau \in (2,3], \end{cases}$$

$$(k^{(1)}, \omega(\tau)) = 2\omega_1(\tau) - 1 \cdot \omega_2(\tau) \equiv 0 \quad \forall \tau \in [1,2].$$

The functions $\omega_1(\tau)$ and $\omega_2(\tau)$ thus chosen satisfy conditions (2.17) for $r_1 = r_2 = 1$, $\alpha = 1$, $\beta = 2$, $L = 3$, and $\sigma_5 = 3$ and identity (2.18) for $h = 1$ and $k^{(1)} = (2,-1)$. In this case, inequality (2.21) takes the form $\omega_1(\tau) \geq 1$ or $\omega_2(\tau) \geq 3$ $\forall \tau \in [0,3]$.

At the end of this section, we justify the averaging method on the semiaxis $[0,\infty) = R_+$. Note that, in Chapter 2, we establish an efficient estimate for the error of the averaging method on the entire axis.

We assume that

$$\|\overline{a}(x,\tau,\varepsilon) - \overline{a}(x,\tau,0)\| \leq \sigma_{10}\varepsilon^{\delta} \quad \forall (x,\tau,\varepsilon) \in \mathcal{D} \times R_+ \times [0,\varepsilon_0], \tag{2.26}$$

$$\overline{a}(x,\tau,0) \in C_x^2(\mathcal{D} \times R_+, \sigma_{10}),$$

and consider the averaged equations of the first approximation for slow variables

$$\frac{d\overline{x}}{d\tau} = \overline{a}\left(\overline{x},\tau,0\right). \tag{2.27}$$

Theorem 2.4. *Suppose that the following conditions are satisfied:*

(a) $\|(W_p^T(\tau)W_p(\tau))^{-1}W_p^T(\tau)\|$ *is uniformly bounded for certain* $p \geq m$ *and all* $\tau \in R_+$, *and the functions* $\omega_\nu^{(j)}(\tau)$, $\nu = \overline{1,m}$, $j = \overline{0,p-1}$, *are uniformly continuous on* R_+;

(b) *there exists a solution* $\overline{x} = \overline{x}(\tau)$ *of Eq. (2.27) that lies in* $\mathcal{D}$ *together with its* ρ*-neighborhood for all* $\tau \in R_+$;

(c) *the normal fundamental matrix* $Q(\tau,t)$, $Q(t,t) = E_n$, *of solutions of the variational equation* $\dfrac{dz}{d\tau} = \dfrac{\partial\overline{a}(\overline{x}(\tau),\tau,0)}{\partial x}z$ *satisfies the estimate*

$$\|Q(\tau,t)\| \leq Ke^{-\gamma(\tau-t)} \quad \forall \tau \geq t \geq 0,$$

$$K = \text{const} \geq 1, \quad \gamma = \text{const} \geq 0; \tag{2.28}$$

(d) *conditions (2.2) are satisfied for* $q = 0$ *and* $\tau \in R_+$, *and relation (2.26) holds for* $\delta \geq \dfrac{1}{p}$.

Then there exist positive constants σ_{11}, ε_2, *and* $\rho_1 < \rho$ *such that the following assertions are true:*

(i) *for all* $\tau \in R_+$, $\psi \in R^m$, *and* $\varepsilon \in (0, \varepsilon_0]$, $\varepsilon_0 \leq \varepsilon_2$, *the following estimate is true:*

$$\|x(\tau, \overline{x}(0), \psi, \varepsilon) - \overline{x}(\tau)\| \leq \sigma_{11}\varepsilon^{\frac{1}{p}}; \tag{2.29}$$

(ii) *the slow variables* $x(\tau, y, \psi, \varepsilon)$ *of any solution* $(x(\tau, y, \psi, \varepsilon); \varphi(\tau, y, \psi, \varepsilon))$ *of system (2.1) such that*

$$\psi \in R^m, \quad \varepsilon \in (0, \varepsilon_0],$$

$$y \in \mathcal{D}_{\rho_1}(\overline{x}(0)) \equiv \{y \colon y \in R^n, \|y - \overline{x}(0)\| < \rho_1\}$$

are uniformly bounded for any $\tau \in R_+$.

Proof. It follows from the smoothness conditions for the right-hand sides of Eqs. (2.1) that, for

$$y \in \mathcal{D}_{\rho_1}(\overline{x}(0)), \quad \rho_1 \leq \frac{1}{2K}\rho, \quad \psi \in R^m, \quad \varepsilon \in (0, \varepsilon_0],$$

the curve $x = x(\tau, y, \psi, \varepsilon)$ lies in the domain $\mathcal{D}_{2K\rho_1}(\overline{x}(\tau))$ for all τ from a certain maximum half-interval $[0, T)$. For such τ, the function $\xi(\tau, y, \psi, \varepsilon) = x(\tau, y, \psi, \varepsilon) - \overline{x}(\tau)$ satisfies the equation

$$\xi(\tau, y, \psi, \varepsilon) = Q(\tau, 0)\xi(0, y, \psi, \varepsilon) + \int_0^\tau Q(\tau, t)[F(\xi(t, y, \psi, \varepsilon), t, \varepsilon) + \widetilde{a}(x(t, y, \psi, \varepsilon), \varphi(t, y, \psi, \varepsilon), t, \varepsilon)]dt, \tag{2.30}$$

where

$$F(\xi, t, \varepsilon) = \overline{a}(\xi + \overline{x}(t), t, \varepsilon) - \overline{a}(\overline{x}(t), t, 0) - \frac{\partial}{\partial x}\overline{a}(\overline{x}(t), t, 0)\xi,$$

$$\widetilde{a}(x, \varphi, t, \varepsilon) = a(x, \varphi, t, \varepsilon) - \overline{a}(x, t, \varepsilon), \quad \|F\| \leq \sigma_{10}(\varepsilon^\delta + n^2\|\xi\|^2).$$

Using the inequality $\|\xi(\tau, y, \psi, \varepsilon)\| \leq 2K\rho_1$ and relations (2.28) and (2.30), we get

$$\begin{aligned}\sup_{\tau\in[0,T)} &\|\xi(\tau, y, \psi, \varepsilon)\| \\ &\leq K\|\xi(0, y, \psi, \varepsilon)\| \\ &\quad + \varepsilon^{\delta}\sigma_{10}\,\frac{1}{\gamma}\,K + n^2\sigma_{10}\frac{2}{\gamma}\,K^2\rho_1 \sup_{\tau\in[0,T)} \|\xi(\tau, y, \psi, \varepsilon)\| \\ &\quad + \sup_{\tau\in[0,T)}\left\|\int_0^{\tau} Q(\tau, t)\tilde{a}(x(t, y, \psi, \varepsilon), \varphi(t, y, \psi, \varepsilon), t, \varepsilon)\, dt\right\|, \quad (2.31)\end{aligned}$$

which, for $\rho_1 = \min\left\{\frac{\rho}{2K}; \frac{\gamma}{6n^2\sigma_{10}K^2}\right\}$, yields

$$\begin{aligned}\sup_{\tau\in[0,T)} &\|\xi(\tau, y, \psi, \varepsilon)\| \\ &\leq \frac{3}{2}K\rho_1 + \frac{3}{2\gamma}K\sigma_{10}\varepsilon^{\delta} \\ &\quad + \frac{3}{2}\sup_{\tau\in[0,T)}\left\|\int_0^{\tau} Q(\tau, t)\,\tilde{a}\,(x(t, y, \psi, \varepsilon), \varphi\,(t, y, \psi, \varepsilon), t, \varepsilon)\, dt\right\|. \quad (2.32)\end{aligned}$$

We represent the last term on the right-hand side of (2.32) in the form

$$\begin{aligned}\sup_{\tau\in[0,T)} &\left\|\int_0^{\tau} Q(\tau, t)\tilde{a}\,(x, \varphi, t, \varepsilon)dt\right\| \\ &\leq \sum_{k\neq 0}\sup_{\tau\in[0,T)}\left[\sum_{r=0}^{s-1}\left\|\int_r^{r+1} Q(\tau, t)a_k(x, t, \varepsilon)\,\exp\{i(k, \widetilde{\varphi})\}\right.\right. \\ &\quad \left.\left.\times \exp\left\{\frac{i}{\varepsilon}\int_0^t (k, \omega(z))\, dz\right\} dt\right\|\right] \\ &\quad + \left\|\int_s^{\tau} Q(\tau, t)a_k(x, t, \varepsilon)\exp\{i(k, \widetilde{\varphi})\}\,\exp\left\{\frac{i}{\varepsilon}\int_0^t (k, \omega(z))dz\right\} dt\right\|,\end{aligned}$$

where s is the integer part of τ, $x = x(t, y, \psi, \varepsilon)$, $\varphi = \varphi(t, y, \psi, \varepsilon)$, and $\widetilde{\varphi} = \varphi - \frac{1}{\varepsilon}\int\limits_0^t \omega(z)dz$, and estimate each of the integrals over the segments $[r, r+1]$ of unit length using inequalities (1.20) and (2.28) as follows:

$$\Big\| \int\limits_r^{r+1} Q(\tau, t) a_k(x, t, \varepsilon) \exp\{i(k, \widetilde{\varphi})\} \exp\Big\{\frac{i}{\varepsilon}\int\limits_0^t (k, \omega(z))\, dz\Big\}\, dt \Big\|$$

$$\leq \sigma_{12}\varepsilon^{\frac{1}{p}}(2 + \sigma_1 + \sigma_{10}n^2) K e^{-\gamma(\tau - r - 1)}$$

$$\times \left[\sup_G \|a_k\| + \left(\sup_G \Big\|\frac{\partial a_k}{\partial \tau}\Big\| + \sup_G \Big\|\frac{\partial a_k}{\partial x}\Big\|\right) \frac{1}{\|k\|}\right].$$

Here, σ_{12} is a constant corresponding to the constant σ_3 in estimate (1.20). Since $\tau - s < 1$, the integral over the segment $[s, \tau]$ satisfies the same inequality with the factor $e^{-\gamma(\tau - r - 1)}$ replaced by 1. Then, taking into account condition (2.2) for $q = 0$ and the inequality

$$\sum_{r=0}^{s-1} e^{-\gamma(\tau - r - 1)} < \frac{e^\gamma}{e^\gamma - 1},$$

we get

$$\sup_{\tau \in [0, T)} \Big\| \int\limits_0^\tau Q(\tau, t) \widetilde{a}(x, \varphi, t, \varepsilon)\, dt \Big\|$$

$$\leq K\left(1 + \frac{e^\gamma}{e^\gamma - 1}\right) \sigma_1 (2 + \sigma_1 + n^2\sigma_{10})\, \sigma_{12}\varepsilon^{\frac{1}{p}} \equiv \sigma_{13}\varepsilon^{\frac{1}{p}}.$$

Using the last inequality, we can rewrite estimate (2.32) in the form

$$\sup_{\tau \in [0, T)} \|\xi(\tau, y, \psi, \varepsilon)\| \leq \frac{3}{2} K\rho_1 + \frac{3}{2\gamma} K\sigma_{10}\varepsilon^\delta + \frac{3}{2}\sigma_{13}\varepsilon^{\frac{1}{p}} \leq \frac{3}{2} K\rho_1 + \sigma_{11}\varepsilon^{\frac{1}{p}},$$

$$\sigma_{11} = \frac{3}{2}\left(\sigma_{13} + \frac{1}{\gamma} K\sigma_{10}\right). \tag{2.33}$$

Further, setting $\sigma_{11}\varepsilon^{1/p} \leq \frac{1}{4} K\rho_1$, we obtain

$$\sup_{\tau \in [0, T)} \|\xi(\tau, y, \psi, \varepsilon)\| \leq \frac{7}{4} K\rho_1 < 2K\rho_1, \tag{2.34}$$

i.e., the curve $x = x(\tau, y, \psi, \varepsilon)$ does not leave the $\frac{7}{4}K\rho_1$-neighborhood of the curve $x = \overline{x}(\tau)$. Therefore, the solution $(x(\tau, y, \psi, \varepsilon); \varphi(\tau, y, \psi, \varepsilon))$ of system (2.1) can be extended to all $\tau \in R_+$. Inequality (2.34) does not change for $T = \infty$. Thus, relation (2.33) yields the uniform estimate

$$\|x(\tau, y, \psi, \varepsilon)\| < 2K\rho_1 + \sup_{\tau \in R_+} \|\overline{x}(\tau)\| \equiv \sigma_{14}$$

for all $\tau \in R_+$, $y \in \mathcal{D}_{\rho_1}(\overline{x}(0))$, $\psi \in R^m$, and $\varepsilon \in (0, \varepsilon_0]$. Inequality (2.29) can be obtained from (2.31) and (2.33) for $\xi(0, y, \psi, \varepsilon) = 0$. Theorem 2.4 is proved.

3. Investigation of Two-Frequency Systems

In this section, we consider the case where system (2.1) is a two-frequency system, i.e., $\varphi = (\varphi_1, \varphi_2)$ and $\omega(\tau) = (\omega_1(\tau), \omega_2(\tau))$, and study the problem of the justification of the averaging method on an asymptotically large time interval $[0, T(\varepsilon)]$, where $T(\varepsilon) \to \infty$ as $\varepsilon \to 0$, and on the infinite time interval $[0, \infty) = R_+$ under assumptions for $\omega(\tau)$ weaker than those in Section 2.

Assume that $\omega(\tau) \in C^1_{[0,\infty)}$ and

$$\omega_2(\tau) \geq d_1, \quad \left| \frac{d}{d\tau} \left(\frac{\omega_1(\tau)}{\omega_2(\tau)} \right) \right| \geq d_1 \quad \forall \tau \in R_+, \tag{3.1}$$

where d_1 is a positive constant. For $\tau \in [0, L]$, condition (3.1) is the Arnol'd condition [Arn2], by using which Arnol'd obtained an estimate for the error of the averaging method on a finite time interval.

We also require that the function $\omega_2(\tau)$ satisfy at least one of the following conditions:

(i) $\left| \omega_2^{-2}(\tau) \dfrac{d\omega_2(\tau)}{d\tau} \right| \leq d_2 = \text{const} \quad \forall \tau \in R_+;$

(ii) $\omega_2(\tau)$ is nondecreasing or nonincreasing on R_+.

Denote by $h_d(\tau)$, $d = \text{const} > 0$, the following even function continuously differentiable on $\forall \tau \in R$:

$$h_d(\tau) = \begin{cases} 1, & \tau \in [0, d], \\ d^{-4}\tau^2(2d - \tau)^2, & \tau \in (d, 2d), \\ 0, & \tau \in [2d, \infty). \end{cases}$$

It is easy to verify that, for all $\tau \in R$, the function $h_d(\tau)$ satisfies the inequalities

$$0 \le h_d(\tau) \le 1, \qquad \left| \frac{d}{d\tau} h_d(\tau) \right| \le \frac{16}{d} f_d(\tau),$$

where $f_d(\tau) = 1$ for $d < |\tau| < 2d$ and $f_d(\tau) = 0$ for $|\tau| \le d$ and $|\tau| \ge 2d$.

The statement below gives an estimate of the time for which the two-frequency system (2.1) passes through the resonance zone.

Lemma 3.1. *Let conditions (3.1) be satisfied and let* $k = (k_1, k_2) \neq 0$ *be an arbitrary vector with integer-valued coordinates. Then, for all* $\tau \in R_+$ *except, possibly, a time interval whose length does not exceed* 2μ, $\mu \le d_1^{-1}$, *the function* $(k, \omega(\tau)) = k_1\omega_1(\tau) + k_2\omega_2(\tau)$ *satisfies the inequality* $|k, \omega(\tau)| \ge d_1^2\mu$.

Proof. If $(k, \omega(\tau_k)) = 0$, then it follows from (3.1) that $k_1 \neq 0$ and the function $\omega(\tau, k) \equiv \dfrac{k_2}{k_1} + \dfrac{\omega_1(\tau)}{\omega_2(\tau)}$ is monotone. Hence,

$$|(k, \omega(\tau))| = |k_1\omega_2(\tau)||\omega(\tau_k, k) - \omega(\tau, k)| \ge d_1^2\mu$$

for $|\tau - \tau_k| \ge \mu$. Now let $(k, \omega(\tau)) \neq 0 \ \forall \tau \in R_+$. If $k_1 = 0$, then $|(k, \omega(\tau))| = |k_2\omega_2(\tau)| \ge d_1 \ge d_1^2\mu$. If $k_1 \neq 0$, then we obtain the estimate $|(k, \omega(\tau))| \ge d_1^2\mu$ for all $\tau \in [\mu, \infty)$. Lemma 3.1 is proved.

Lemma 3.2. *Suppose that inequalities (3.1) and at least one of conditions (i) and (ii) are satisfied. Then, for every* $\gamma > 0$, *one can find a constant* $d_3 = d_3(\gamma) > 0$ *such that, for any* $T \in R_+$, *the following estimate is true:*

$$A(T) \equiv \int_0^T e^{-\gamma(T-\tau)} \left| \frac{d}{d\tau} \frac{1 - h_\mu(\tau - \tau_k)}{(k, \omega(\tau))} \right| d\tau \le \frac{d_3}{\mu}, \qquad k \neq 0, \tag{3.2}$$

where $\tau_k \in R_+$ *is a point at which* $(k, \omega(\tau))$ *turns into zero; if* $(k, \omega(\tau)) \neq 0$ $\forall \tau \in R_+$, *then* $\tau_k = 0$.

Proof. Let $k_1 = 0$. Then $|k_2| \geq 1, \ |(k, \omega(\tau))| \geq d_1$, and

$$A(T) \leq \int_0^T e^{-\gamma(T-\tau)} \Big| \frac{d}{d\tau} h_\mu(\tau) \Big| \frac{1}{\omega_2(\tau))} \, d\tau$$

$$+ \int_0^T e^{-\gamma(T-\tau)} \Big| \frac{d}{d\tau} \Big(\frac{1}{\omega_2(\tau))} \Big) \Big| (1 - h_\mu(\tau)) \, d\tau$$

$$\leq \frac{16}{d_1 \mu} \int_0^T f_\mu(\tau) d\tau + \int_0^T e^{-\gamma(T-\tau)} \Big| \frac{d\omega_2(\tau)}{d\tau} \Big| \frac{1}{\omega_2^2(\tau)} \, d\tau.$$

If condition (i) is satisfied, then

$$A(T) \leq \frac{16}{d_1 \mu} + d_2 \int_0^T e^{-\gamma(T-\tau)} d\tau \leq \frac{16}{d_1} + \frac{d_2}{\gamma},$$

and if condition (ii) is satisfied, then $A(T)$ can be estimated as follows:

$$A(T) \leq \frac{16}{d_1} + \int_0^T \Big| \frac{d\omega_2}{d\tau} \frac{1}{\omega_2^2(\tau)} \Big| \, d\tau = \frac{16}{d_1} + \Big| \int_0^t \frac{d}{d\tau} \Big(\frac{1}{\omega_2(\tau)} \Big) \, d\tau \Big| \leq \frac{18}{d_1}.$$

Thus, in the case $k_1 = 0$, we get $A(T) < \dfrac{18}{d_1} + \dfrac{1}{\gamma} d_2$.

Now let $k_1 \neq 0$. Then

$$A(T) \leq \int_0^T e^{-\gamma(T-\tau)} \frac{16}{\mu} f_\mu(\tau - \tau_k) \frac{1}{|(k, \omega(\tau))|} \, d\tau$$

$$+ \frac{1}{|k_1|} \int_0^T e^{-\gamma(T-\tau)} (1 - h_\mu(\tau - \tau_k)) \Big| \frac{d}{d\tau} \Big(\frac{1}{\omega_2(\tau)} \Big) \Big| \frac{1}{|\omega(\tau, k)|} \, d\tau$$

$$+ \int_0^T e^{-\gamma(T-\tau)} (1 - h_\mu(\tau - \tau_k)) \frac{1}{|k_1| \omega_2(\tau)} \Big| \frac{d}{d\tau} \Big(\frac{1}{\omega(\tau, k)} \Big) \Big| \, d\tau. \quad (3.3)$$

According to the definition of the function $f_\mu(\tau)$ and Lemma 3.1, the first integral on the right-hand side of (3.3) can be estimated from above by the constant

$32d_1^{-2}\mu^{-1}$. In view of (3.1), the function $\omega(\tau, k)$ is monotone. Therefore, the last integral on the right-hand side of (3.3) is estimated as follows:

$$\frac{1}{d_1}\Big|\int_0^T \frac{d}{d\tau}\Big(\frac{1}{\omega(\tau,k)}\Big)(1 - h_\mu(\tau - \tau_k))\,d\tau\Big|$$
$$\leq \frac{1}{d_1}\Big(\Big|\int_0^{\tau_k-\mu} \frac{d}{d\tau}\Big(\frac{1}{\omega(\tau,k)}\Big)\,d\tau\Big| + \Big|\int_{\tau_k+\mu}^{T} \frac{d}{d\tau}\Big(\frac{1}{\omega(\tau,k)}\Big)\,d\tau\Big|\Big)$$
$$\leq \frac{4}{d_1^2\mu},$$

where $\tau_k - \mu \geq 0$ and $\tau_k + \mu \leq T$. If $\tau_k < \mu$ or $\tau_k > T - \mu$, then the last inequality remains the same.

Consider the second integral on the right-hand side of (3.3). In case (i), it can be estimated by the value

$$\frac{d_2}{d_1\mu}\int_0^T e^{-\gamma(T-\tau)}d\tau \leq \frac{d_2}{d_1\gamma\mu};$$

in case (ii), it can be estimated by the value

$$\frac{1}{d_1\mu}\int_0^T\Big|\frac{d}{d\tau}\Big(\frac{1}{\omega_2(\tau)}\Big)\Big|\,d\tau = \frac{1}{d_1\mu}\Big|\int_0^T \frac{d}{d\tau}\Big(\frac{1}{\omega_2(\tau)}\Big)\,d\tau\Big| \leq \frac{2}{d_1^2\mu}.$$

Thus, it follows from (3.3) for $k_1 \neq 0$ that

$$A\,(T) \leq \frac{36}{d_1^2\mu} + \frac{d_2}{d_1\gamma\mu}.$$

Combining the estimates for $A(T)$ in the cases $k_1 = 0$ and $k_1 \neq 0$, we obtain inequality (3.2) with the constant $d_2 = 36d_1^{-2} + d_2(d_1\gamma)^{-1}$ for $d_1\mu \leq 1$. Lemma 3.2 is proved.

Consider a two-frequency system of the form

$$\frac{dx}{d\tau} = a\,(x, \varphi, \tau, \varepsilon),$$
$$\frac{d\varphi}{d\tau} = \frac{\omega(\tau)}{\varepsilon} + b\,(x, \varphi, \tau, \varepsilon), \tag{3.4}$$

where the functions a, b, and ω are defined for $(x, \varphi, \tau, \varepsilon) \in \mathcal{D} \times R^2 \times R_+ \times [0, \varepsilon_0]$, 2π-periodic in φ_ν, $\nu = 1, 2$, continuously differentiable with respect to x, φ, and τ for every fixed ε, and such that

$$\|b(x, \varphi, \tau, \varepsilon)\| + \left\|\frac{\partial}{\partial x} b(x, \varphi, \tau, \varepsilon)\right\| + \left\|\frac{\partial}{\partial \varphi} b(x, \varphi, \tau, \varepsilon)\right\| \leq d_4$$

$$\forall (x, \varphi, \tau, \varepsilon) \in \overline{G},$$

$$\sum_k \left[(1 + \|k\|) \sup_G \|a_k\| + \sup_G \left\|\frac{\partial a_k}{\partial x}\right\| + \sup_G \left\|\frac{\partial a_k}{\partial \tau}\right\|\right] \leq d_4. \tag{3.5}$$

Here, $G = \mathcal{D} \times R_+ \times [0, \varepsilon_0]$, $\mathcal{D}$ is a bounded domain from R^n, and $a_k = a_k(x, \tau, \varepsilon)$ are the Fourier coefficients of the function $a(x, \varphi, \tau, \varepsilon)$.

As in the previous section, by $(x(\tau, y, \psi, \varepsilon); \varphi(\tau, y, \psi, \varepsilon))$, $x(0, y, \psi, \varepsilon) = y$, $\varphi(0, y, \psi, \varepsilon) = \psi$, we denote a solution of system (3.4) and by $\overline{x} = \overline{x}(\tau)$ a solution of the averaged system of the first approximation (2.27), which is defined for all $\tau \in R_+$.

Theorem 3.1. *Suppose that the following conditions are satisfied:*

(I) inequalities (3.1) and (3.5) and at least one of restrictions (i) and (ii) are satisfied;

(II) conditions (b) and (c) of Theorem 2.4 and relation (2.26) for $\delta \geq \frac{1}{2}$ *are satisfied.*

Then, for all $\tau \in R_+$, $\psi \in R^2$, *and* $\varepsilon \in (0, \varepsilon_0]$ *(*ε_0 *is sufficiently small), the following inequality is true:*

$$\|x(\tau, \bar{x}(0), \psi, \varepsilon) - \bar{x}(\tau)\| \leq d_5 \sqrt{\varepsilon}, \tag{3.6}$$

where the constant d_5 *is independent of* τ, ψ, *and* ε.

Prior to the proof of Theorem 3.1, we indicate its difference from Theorem 2.4. In Theorem 2.4, the condition of the uniform continuity of the functions $\omega_\nu^{(j)}(\tau)$, $\nu = \overline{1, m}$, $j = \overline{0, p-1}$, on the semiaxis R_+ is imposed. This condition substantially restricts the growth of the functions $\omega_\nu(\tau)$; in particular, $\omega_\nu(\tau) = \tau^l$ for $l > 1$ is not uniformly continuous on R_+. Moreover, in Theorem 2.4, the boundedness of the norm of the matrix $(W_p^T(\tau) W_p(\tau))^{-1} W_p^T(\tau)$ is an essential assumption. In Theorem 3.1, conditions (3.1), (i), and (ii) do not require such strong restrictions on the components $\omega_1(\tau)$ and $\omega_2(\tau)$ of the vector $\omega(\tau)$. For example, if $\omega_1(\tau) = \tau^2 + \tau$ and $\omega_2(\tau) = 1$, then all conditions (3.1), (i), and (ii) are satisfied for $\tau \in R_+$. It is easy to see that $\omega_1(\tau) = \tau^2 + \tau$ is not a uniformly continuous function on R_+. Moreover, for these frequencies, we have

$$\|(W_2^T(\tau)W_2(\tau))^{-1}W_2^T(\tau)\| = \|W_2^{-1}(\tau)\| = \frac{\tau^2 + 3\tau + 2}{2\tau + 1} > 1 + \frac{1}{2}\tau,$$

i.e., the indicated norm is not bounded on the semiaxis. It is easy to verify that, for $p > 2$, the value $\|(W_2^T(\tau)W_2(\tau))^{-1}W_2^T(\tau)\|$ is also not bounded for all $\tau \in R_+$. Therefore, for this collection of frequencies $\omega_1(\tau)$ and $\omega_2(\tau)$, we cannot use Theorem 2.4.

Proof of Theorem 3.1. Assume that, for $t \in [0, T)$, $T = T(\psi, \varepsilon)$, the curve $x = x(\tau, \bar{x}(0), \psi, \varepsilon)$ does not leave a ρ_1-neighborhood of the curve $x = \overline{x}(\tau)$. We fix the constant $\rho_1 \in (0, \rho)$ below. For $\tau \in [0, T)$, $\psi \in R^2$, and $\varepsilon \in (0, \varepsilon_0]$, we consider the function

$$y(\tau, \psi, \varepsilon) = x(\tau, \overline{x}(0), \psi, \varepsilon) - \overline{x}\,(\tau)$$
$$- \varepsilon U(x(\tau, \overline{x}(0), \psi, \varepsilon), \varphi(\tau, \overline{x}(0), \psi, \varepsilon), \tau, \varepsilon), \qquad (3.7)$$

where

$$U(x, \varphi, \tau, \varepsilon) = \sum_{k \neq 0} a_k(x, \tau, \varepsilon) \frac{1 - h_\mu(\tau - \tau_k)}{i(k, \omega(\tau))} \exp\{i(k, \varphi)\}$$

and $\tau_k \in R_+$ is a point at which $(k, \omega(\tau))$ turns into zero; if $(k, \omega(\tau)) \neq 0$ $\forall \tau \in R_+$, then $\tau_k = 0$.

By direct differentiation, one can easily verify that $y = y(\tau, \psi, \varepsilon)$ satisfies the equation

$$\frac{dy}{d\tau} = H(\tau)y + F(y, \tau) + \delta(x, \varphi, \tau, \varepsilon)$$
$$- \varepsilon \frac{\partial}{\partial \tau} U(x, \varphi, \tau, \varepsilon) + Y\,(x, y, \varphi, \tau, \varepsilon), \qquad (3.8)$$

where $x = x(\tau, \overline{x}(0), \psi, \varepsilon)$, $\varphi = \varphi(\tau, \overline{x}(0), \psi, \varepsilon)$,

$$\delta(x, \varphi, \tau, \varepsilon) = \sum_{k \neq 0} a_k(x, \tau, \varepsilon) h_\mu(\tau - \tau_k) \exp\{i(k, \varphi)\},$$

$$F(y, \tau) = \overline{a}(\overline{x}(\tau) + y, \tau, 0) - \overline{a}(\overline{x}(\tau), \tau, 0) - H(\tau) y,$$

$$H(\tau) = \frac{\partial}{\partial x} \overline{a}(\overline{x}(\tau), \tau, 0),$$

$$Y(x, y, \varphi, \tau, \varepsilon) = \overline{a}(x, \tau, \varepsilon) - \overline{a}(\overline{x}(\tau) + y, \tau, 0) - \varepsilon \frac{\partial}{\partial \varphi} U(x, \varphi, \tau, \varepsilon) b(x, \varphi, \tau, \varepsilon)$$
$$- \varepsilon \frac{\partial}{\partial x} U(x, \varphi, \tau, \varepsilon) a(x, \varphi, \tau, \varepsilon).$$

Using the definition of the function $h_\mu(\tau)$, Lemma 3.1, and inequalities (2.26) and (3.5), we obtain

$$\|F(y, \tau)\| \leq n^2 \sigma_{10} \|y\|^2, \quad \|Y(x, y, \varphi, \tau, \varepsilon)\| \leq \sigma_{10} \varepsilon^\delta + \frac{\varepsilon}{\mu} d_6,$$

$$d_6 = n^2 (1 + 2d_4)\, d_4^2 d_1^{-2}.$$

Equation (3.8) yields the following representation of the function $y = y(\tau, \psi, \varepsilon)$ for all $\tau \in [0, T)$:

$$y = Q(\tau, 0) y(0, \psi, \varepsilon) + \int_0^\tau Q(\tau, t) \Big[F(y, t) + \delta(x, \varphi, t, \varepsilon)$$
$$- \varepsilon \frac{\partial}{\partial t} U(x, \varphi, t, \varepsilon) + Y(x, y, \varphi, t, \varepsilon) \Big]\, dt.$$

Using this representation, we get

$$\sup_{[0,T)} \|y(\tau, \psi, \varepsilon)\|$$
$$\leq K \|y(0, \psi, \varepsilon)\| + \Big[\sigma_{10} \varepsilon^\delta + d_6 \frac{\varepsilon}{\mu} + n^2 \sigma_{10} \sup_{[0,T)} \|y(\psi, t, \varepsilon)\|^2 \Big] \frac{K}{\gamma}$$

$$+ K \sum_{k \neq 0} \Big\{ \Big(\sup_G \|a_k(x,\tau,\varepsilon)\| \sup_G \Big\| \frac{\partial}{\partial \tau} a_k(x,\tau,\varepsilon) \Big\| \Big) \sup_{[0,T)} \int_0^\tau \Big[h_\mu(t-\tau_k)$$

$$+ \varepsilon \frac{1 - h_\mu(t-\tau_k)}{|(k,\omega(t))|} + \varepsilon \Big| \frac{\partial}{\partial t} \frac{1 - h_\mu(t-\tau_k)}{(k,\omega(t))} \Big| \Big] e^{-\gamma(\tau - t)} dt \Big\}. \tag{3.9}$$

Since

$$\|y(\tau,\psi,\varepsilon)\| \leq \|x(\tau,\overline{x}(0),\psi,\varepsilon) - \overline{x}(\tau)\| + \varepsilon \sup_{\overline{G}} \|U(x,\varphi,\tau,\varepsilon)\|$$

$$< \rho_1 + \frac{\varepsilon}{\mu} d_4 d_1^{-2},$$

$$\|y(0,\psi,\varepsilon)\| \leq \frac{\varepsilon}{\mu} d_4 d_1^{-2},$$

for

$$\rho_1 \leq \Big(\frac{4}{\gamma} K n^2 \sigma_{10} \Big)^{-1} \quad \text{and} \quad \frac{\varepsilon}{\mu} \leq \frac{\gamma d_1^2}{4Kn^2 \sigma_{10} d_4}$$

relation (3.9) yields

$$\sup_{[0,T)} \|y(\tau,\psi,\varepsilon)\| \leq \frac{2}{d_1^2} K \Big(d_4 + \frac{1}{\gamma} d_1^2 d_6 \Big) \frac{\varepsilon}{\mu} + \frac{2\sigma_{10}}{\gamma} K \varepsilon^\delta$$

$$+ 2K \sum_{k \neq 0} \Big\{ \Big(\sup_G \|a_k\| + \sup_G \Big\| \frac{\partial a_k}{\partial \tau} \Big\| \Big) \sup_{[0,T)} \int_0^\tau \Big[h_\mu(t-\tau_k)$$

$$+ \varepsilon \frac{1 - h_\mu(t-\tau_k)}{|(k,\omega(t))|} + \varepsilon \Big| \frac{d}{dt} \frac{1 - h_\mu(t-\tau_k)}{(k,\omega(t))} \Big| \Big] e^{-\gamma(\tau - t)} dt \Big\}.$$

To estimate the integral on the right-hand side of the last inequality, we use Lemmas 3.1 and 3.2. Then, taking (3.5) into account, we get

$$\sup_{[0,T)} \|y(\tau,\psi,\varepsilon)\|$$

$$\leq \frac{2}{\gamma} K d_1^{-2} d_4 [\gamma + 1 + d_1^2 d_4^{-1} d_6 + \gamma d_1^2 d_3] \frac{\varepsilon}{\mu} + 8K d_4 \mu + \frac{2}{\gamma} K \sigma_{10} \varepsilon^\delta,$$

which (for $\delta \geq \frac{1}{2}$ and $\mu = \sqrt{\varepsilon}$) yields

$$\|x(\tau,\overline{x}(0),\psi,\varepsilon) - \overline{x}(\tau)\| \leq \Big[\frac{2}{\gamma} d_1^{-2} d_4 K (\gamma + 1 + \gamma d_1^2 d_3$$

$$+ d_1^2 d_4^{-1} d_6) + \frac{2}{\gamma} K \sigma_{10} + 8K d_4 + d_1^{-2} d_4 \Big] \sqrt{\varepsilon}$$

$$\equiv d_5 \sqrt{\varepsilon} \tag{3.10}$$

for any $\tau \in [0, T)$, $\psi \in R^2$, and $\varepsilon \in (0, \varepsilon_0]$. We now choose $\varepsilon_0 > 0$ so small that

$$\varepsilon_0^{1/2} \le \min\Big\{\frac{1}{2\sigma_5}\rho_1; \frac{\gamma d_1^2}{4Kn^2 d_4 \sigma_{10}}\Big\}, \quad \rho_1 = \min\Big\{\frac{\rho}{2}; \Big(\frac{4}{\gamma}Kn^2\sigma_{10}\Big)^{-1}\Big\}.$$

According to estimate (3.10), the curve $x = x(\tau, \overline{x}(0), \psi, \varepsilon)$ does not leave the $\frac{1}{2}\rho_1$-neighborhood of the curve $x = \overline{x}(\tau)$ for all $\tau \in [0, T)$. This implies that $T = \infty$ and inequality (3.10) holds for any $\tau \in R_+$. Theorem 3.1 is proved.

Now assume that the function $a(x, \varphi, \tau, \varepsilon)$ averaged with respect to φ over the cube of periods is identically equal to zero, i.e.,

$$\overline{a}\,(x, \tau, \varepsilon) \equiv 0 \quad \forall (x, \tau, \varepsilon) \in \mathcal{D} \times R_+ \times [0, \varepsilon_0].$$

In this case, the solutions $\overline{x}(\tau) \equiv x^0 = \text{const}$ $\forall \tau \in R_+$ of the averaged equations for slow variables are stationary, and, therefore, condition (II) of Theorem 3.1 is not satisfied. This is the case, in particular, for Hamiltonian systems [Arn4]. Nekhoroshev [Nek1, Nek2] established that, for time $\exp\{\varepsilon^{-\alpha}\}$, $\alpha = \text{const} > 0$, the slow variable x of the solution $(x; \varphi)$ of a Hamiltonian system deviates from its initial value by a value not greater than $c\varepsilon^{\beta}$, $c = \text{const} > 0$, $\beta = \text{const} > 0$. In what follows, we obtain an analogous result for a two-frequency system under certain additional assumptions concerning the frequency vector $\omega(\tau)$. The following statement can be proved by analogy with Lemma 3.2:

Lemma 3.3. *Suppose that conditions (3.1) are satisfied and the following inequality is true:*

$$\int_0^T \Big[\frac{1}{\omega_2(\tau)} + \Big|\frac{d}{d\tau}\Big(\frac{1}{\omega_2(\tau)}\Big)\Big|\Big]\, d\tau \le d_7 \ln T + d_8 \qquad \forall T \ge 1, \tag{3.11}$$

where d_7 *and* d_8 *are nonnegative constants. Then, for* $k \ne 0$, $0 < \mu < \min\Big\{\dfrac{1}{d_1}; \dfrac{1}{3}\Big\}$, *and* $T > 1$, *the following estimate is true:*

$$\begin{aligned} B(T) &\equiv \int_0^T \Big[\frac{1 - h_\mu(\tau - \tau_k)}{|(k, \omega(\tau))|} + \Big|\frac{d}{d\tau}\,\frac{1 - h_\mu(\tau - \tau_k)}{(k, \omega(\tau))}\Big|\Big]\, d\tau \\ &\le \frac{36}{d_1^2 \mu} + \frac{1}{d_1 \mu}(d_7 \ln T + d_8); \end{aligned} \tag{3.12}$$

for $T \in [0, 1]$, *the following estimate is true:*

$$B(T) \le \frac{d_9}{\mu}, \quad d_9 = d_1^{-2}\Big[37 + d_1^{-1} \max_{[0,1]}\Big|\frac{d}{d\tau}\Big(\frac{1}{\omega_2(\tau)}\Big)\Big|\Big]. \tag{3.13}$$

We denote by $\mathcal{D}_\rho$ the set of points belonging to $\mathcal{D}$ together with their ρ-neighborhoods and choose $\rho > 0$ so small that $\mathcal{D}_\rho \ne \emptyset$.

Theorem 3.2. *Suppose that* $\overline{a}(x, \tau, \varepsilon) \equiv 0 \;\; \forall (x, \tau, \varepsilon) \in G$ *and conditions (3.1), (3.5), and (3.11) are satisfied. Then, for all* $x^0 \in \mathcal{D}_\rho$, $\psi \in R^2$, *and* $\varepsilon \in (0, \varepsilon_0]$ (ε_0 *is sufficiently small), the following estimates are true:*

(a) if $d_7 > 0$, *then*

$$\|x(\tau, x^0, \psi, \varepsilon) - x^0\| < d_{10}\varepsilon^\beta \quad \forall \tau \in [0, \exp\{\varepsilon^{-(1-2\beta)}\}], \tag{3.14}$$

where β *is an arbitrary number from the interval* $\left(0, \frac{1}{2}\right)$;

(b) if $d_7 = 0$, *then*

$$\|x(\tau, x^0, \psi, \varepsilon) - x^0\| < d_{10}\sqrt{\varepsilon} \quad \forall \tau \in [0, \infty), \tag{3.15}$$

where d_{10} *is a constant independent of* ε, x^0, *and* ψ.

Proof. We use the method proposed in the proof of Theorem 3.1. The function $y(\tau, \psi, \varepsilon)$ defined by equality (3.7) for $\overline{x}(\tau) \equiv x^0$ admits the representation

$$y(\tau, \psi, \varepsilon) = y(0, \psi, \varepsilon) + \int_0^\tau \Big[\delta(x, \varphi, t, \varepsilon) - \varepsilon \frac{\partial}{\partial t} U(x, \varphi, t, \varepsilon)$$
$$- \varepsilon \frac{\partial}{\partial x} U(x, \varphi, t, \varepsilon) a(x, \varphi, t, \varepsilon) - \varepsilon \frac{\partial}{\partial \varphi} U(x, \varphi, t, \varepsilon) b(x, \varphi, t, \varepsilon)\Big]\, dt,$$

which yields

$$\|y(\tau, \psi, \varepsilon)\| \le \|y(0, \psi, \varepsilon)\| + \Big(1 + \sup_{\overline{G}} \|a\| + \sup_{\overline{G}} \|b\|\Big)$$
$$\times \sum_{k \ne 0} \Big[\|k\| \sup_G \|a_k\| + \sup_G \Big\|\frac{\partial a_k}{\partial \tau}\Big\| + \sup_G \Big\|\frac{\partial a_k}{\partial x}\Big\|\Big] \Big[\int_0^\tau [h_\mu(t - \tau_k)] dt$$
$$+ \varepsilon \int_0^\tau \Big(\frac{1 - h_\mu(t - \tau_k)}{|(k, \omega(t))|} + \Big|\frac{d}{dt} \frac{1 - h_\mu(t - \tau_k)}{(k, \omega(t))}\Big|\Big)\, dt\Big]. \tag{3.16}$$

If $\tau \in [0,1]$, then inequalities (3.5), (3.13), and (3.16) for $\mu = \sqrt{\varepsilon}$ yield

$$\|y(\tau,\psi,\varepsilon)\| \le d_{11}\mu, \quad d_{11} = d_4 d_1^{-2} + (1+2d_4)d_4(4+d_9). \tag{3.17}$$

If $\tau > 1$, then, in view of (3.12), estimate (3.16) takes the form

$$\|y(\tau,\psi,\varepsilon)\|$$
$$\le \frac{\varepsilon}{\mu} d_1^{-2} d_4 + (1+2d_4)d_4\left[4\mu + (36d_1^{-2} + d_1^{-1}(d_8 + d_7 \ln \tau))\frac{\varepsilon}{\mu}\right]. \tag{3.18}$$

Let $d_7 = 0$. Then relations (3.17) and (3.18) for $\mu = \sqrt{\varepsilon}$ yield

$$\|y(\tau,\psi,\varepsilon)\| \le d_{12}\sqrt{\varepsilon},$$
$$d_{12} = \max\{d_{11}; d_1^{-2}d_4 + (1+2d_4)d_4(4+36d_1^{-2} + d_1^{-1}d_8)\},$$
$$\|x(\tau,x^0,\psi,\varepsilon) - x^0\| \le \|y(\tau,\psi,\varepsilon)\| + \varepsilon \sup_{\overline{G}} \|U\| \le (d_1^{-2}d_4 + d_{12})\sqrt{\varepsilon}$$

for all $\tau \in [0,\infty)$, $\psi \in R^2$, $\varepsilon \in (0,\varepsilon_0]$, and $x^0 \in \mathcal{D}_\rho$. Thus, estimate (3.15) is proved.

Consider the case $d_7 > 0$. We fix an arbitrary $\beta \in \left(0; \frac{1}{2}\right)$ and set $\varepsilon^\beta = \mu$. Analyzing inequality (3.18), we establish that $y(\tau,\psi,\varepsilon)$ satisfies an estimate of the form $\|y(\tau,\psi,\varepsilon)\| \le c\varepsilon^\beta$ on the maximum (in order with respect to ε) time interval $[1, T(\varepsilon)]$ if $\varepsilon\mu^{-1}\ln T(\varepsilon) = \mu$, i.e., $T(\varepsilon) = \exp\{\varepsilon^{-(1-2\beta)}\}$. Hence, for all $\tau \in [0,+\infty)$, $\psi \in R^2$, $\varepsilon \in (0,\varepsilon_0]$, and $x^0 \in \mathcal{D}_\rho$, relations (3.17) and (3.18) yield

$$\|y(\tau,\psi,\varepsilon)\| \le d_{13}\varepsilon^\beta, \quad \|x(\tau,x^0,\psi,\varepsilon) - x^0\| \le (d_{13} + d_1^{-2}d_4)\varepsilon^\beta,$$
$$d_{13} = \max\{d_{11}; d_1^{-2}d_4 + (1+2d_4)d_4(4 + 36d_1^{-2} + d_1^{-1}(d_7 + d_8))\}.$$

To complete the proof of the theorem, we set $d_{10} = d_{13} + d_1^{-2}d_4$ and choose ε_0 so small that

$$d_{10}\varepsilon_0^\beta \le \frac{1}{2}\rho, \quad \varepsilon_0^\beta \le \min\left\{\frac{1}{3}; \frac{1}{d_1}\right\}.$$

The first of these inequalities guarantees that the curve $x = x(\tau,x^0,\psi,\varepsilon)$ lies in $\mathcal{D}$ together with its $\frac{1}{2}\rho$-neighborhood for any $\tau \in R_+$ if $d_7 = 0$ and for any $\tau \in [0, \exp\{\varepsilon^{-(1-2\beta)}\}]$ if $d_7 > 0$. The second inequality follows from Lemma 3.3. Theorem 3.2 is proved.

Remark 4. Restrictions (3.1), (i), (ii), and (3.11) imposed on the frequencies $\omega_1(\tau)$ and $\omega_2(\tau)$ of system (3.4) are sufficient and do not exhaust all possibilities of establishing the results presented in this section. For example, Theorem 3.1 remains true if, instead of condition (ii), one assumes that $\dfrac{d}{d\tau}\omega_2(\tau)$ does not change its sign on finitely many intervals that cover $[0, \infty)$, and Theorem 3.2 remains true if, on the left-hand side of (3.11), the integral over the segment $[0, T]$ is replaced by the integral over $[\tau_0, T]$, where τ_0 is positive and fixed. However, as follows from the example presented below, the restrictions indicated above are essential.

Consider the problem

$$\frac{dx}{d\tau} = x\cos\varphi_2 + \sin\varphi_2, \qquad \frac{d\varphi_1}{d\tau} = \frac{\tau}{\varepsilon}, \qquad \frac{d\varphi_2}{d\tau} = \frac{1}{\varepsilon},$$
$$x(0) = \varphi_1(0) = \varphi_2(0) = 0,$$

for which all conditions of Theorem 3.2 except inequality (3.11) are satisfied. Below, we show that estimate (3.14) is not true for $\tau \sim \dfrac{1}{\varepsilon}$. Indeed, the x-component of the solution of this problem is determined by the relation

$$x = x(\tau, \varepsilon) = \varepsilon e^{\varepsilon \sin(\tau/\varepsilon)} \int\limits_0^{\tau/\varepsilon} e^{-\varepsilon \sin\tau} \sin\tau d\tau.$$

The integrand $e^{-\varepsilon\sin\tau}\sin\tau$ is 2π-periodic. Therefore, we first estimate the integral over the segment $[0, 2\pi]$. We have

$$\int\limits_0^{2\pi} e^{-\varepsilon\sin\tau}\sin\tau d\tau = \int\limits_0^{\pi} (e^{-\varepsilon\sin\tau} - e^{\varepsilon\sin\tau})\sin\tau d\tau$$
$$\le -\int\limits_0^{\pi} \frac{2}{e}\varepsilon\sin^2\tau d\tau = -\frac{\varepsilon\pi}{e}.$$

We set $\tau = 2\pi\varepsilon E\{\varepsilon^{-2}\}$, where $E\{\alpha\}$ is the integer part of the number α. Then

$$x(2\pi\varepsilon E\{\varepsilon^{-2}\},\varepsilon) = \varepsilon \int\limits_0^{2\pi E\{\varepsilon^{-2}\}} e^{-\varepsilon\sin\tau}\sin\tau d\tau$$

$$= \varepsilon E\{\varepsilon^{-2}\}\int\limits_0^{2\pi} e^{-\varepsilon\sin\tau}\sin\tau d\tau \le -\frac{\pi}{e}(1-\varepsilon^2) \le -\frac{\pi}{2e}$$

for $\varepsilon^2 \le \frac{1}{2}$. Hence,

$$|x(\tau,\varepsilon) - 0| \ge \frac{\pi}{2e} \qquad \text{for} \quad \tau = 2\pi\varepsilon E\{\varepsilon^{-2}\} \sim \frac{1}{\varepsilon}.$$

4. Justification of Averaging Method for Oscillation Systems with $\omega = \omega(x,\tau)$

Consider a multifrequency system of the form

$$\begin{aligned}\frac{dx}{d\tau} &= a(x,\varphi,\tau) + \varepsilon A(x,\varphi,\tau,\varepsilon),\\ \frac{d\varphi}{d\tau} &= \frac{\omega(x,\tau)}{\varepsilon} + b(x,\varphi,\tau,\varepsilon),\end{aligned} \tag{4.1}$$

where $(x,\varphi,\tau,\varepsilon) \in \mathcal{D}\times R^m \times R_+ \times (0,\varepsilon_0] \equiv \overline{G}$ and the real vector functions a, A, ω, and b belong to certain classes of smooth functions 2π-periodic in φ_ν, $\nu = \overline{1,m}$, $m \ge 2$. We also consider the corresponding averaged (with respect to φ) system of equations of the first approximation for slow variables:

$$\frac{d\overline{x}}{d\tau} = \overline{a}(\overline{x},\tau) \equiv (2\pi)^{-m}\int\limits_0^{2\pi}\ldots\int\limits_0^{2\pi} a(\overline{x},\varphi,\tau)\,d\varphi_1\ldots d\varphi_m. \tag{4.2}$$

The main result on this section is the following: we establish an estimate for $\|x-\overline{x}\|$ on a finite time interval and prove an analog of the Banfi–Filatov theorem [Fil, Ban] for systems of the standard form in the case of an infinite time interval.

For a two-frequency system, the estimate

$$\|x-\overline{x}\| \le c\sqrt{\varepsilon}\,\ln^2\frac{1}{\varepsilon} \quad \forall\tau\in[0,L]$$

was first established by Arnol'd [Arn2]. Later, Neishtadt improved this estimate as follows: $\|x - \overline{x}\| \leq c\sqrt{\varepsilon}$ [Arn4]. The main assumption was the following: the rate of the variation of the ratio of frequencies $\dfrac{\omega_1}{\omega_2}$ along integral curves of system (4.1) is nonzero. This assumption becomes obvious if we represent the equation $(k, \omega(x, \tau)) = 0,\ \ k \neq 0,$ of the resonance surface in the form

$$\frac{\omega_1(x, \tau)}{\omega_2(x, \tau)} = -\frac{k_2}{k_1}.$$

In other words, the resonance surfaces in the two-frequency case are level surfaces. If the number of frequencies m is greater than 3, then the structure of such surfaces is often fairly complex, which significantly complicates the investigation of multifrequency systems. Therefore, it is necessary to impose certain restrictions on resonance surfaces [Bak1, Gre, GrR1–GrR3, Neis2, Sam5]. One should also note Khapaev's paper [Kha1], in which the restrictions considered are related only to resonance harmonics of the function $a(x, \varphi, \tau)$; however, in this case, the estimate of the error of the averaging method is not expressed explicitly in terms of ε. Below, we establish the estimate $\|x - \overline{x}\| \leq c\sqrt{\varepsilon}$ under analogous assumptions for system (4.1).

Assume that $\overline{x} = \overline{x}(\tau)$ is a certain solution of Eqs. (4.2) defined on the semiaxis R_+ and lying in $\mathcal{D}$ together with its ρ-neighborhood. Denote by P the set of m-dimensional vectors $k = (k_1, \ldots, k_m)$ with integer-valued coordinates for which the Fourier coefficients $a_k(x, \tau)$ of the function $a(x, \varphi, \tau)$ are not identically equal to zero on the set $\mathcal{D}_{\rho_1}(\overline{x}(\tau)) \times R_+$, where

$$\mathcal{D}_{\rho_1}(\overline{x}(\tau)) = \{x : x \in \mathcal{D}, \|x - \overline{x}(\tau)\| < \rho_1 \quad \forall\ \tau \in R_+\}$$

and $\rho_1 \in (0, \rho]$ is a fixed constant.

Assume that the functions $\omega(x, \tau)$, $\dfrac{\partial}{\partial x}\omega(x, \tau)$, and $\dfrac{\partial}{\partial \tau}\omega(x, \tau)$ are continuous on $\mathcal{D} \times R_+$ and, for all $k \in P$ and $(x, \varphi, \tau, \varepsilon) \in \overline{G}$, the following inequality is true:

$$|(k, \omega(x, \tau))| + |(k, \Omega(x, \varphi, \tau, \varepsilon))| \geq \sigma_1 \|k\|^{-s}, \tag{4.3}$$

where $\sigma_1 > 0$ and $s \geq -1$ are constants, (k, ω) and (k, Ω) are the scalar products of vectors,

$$\Omega(x, \varphi, \tau, \varepsilon) = \frac{\partial \omega(x, \tau)}{\partial \tau} + \frac{\partial \omega(x, \tau)}{\partial x}\,\delta\,(x, \varphi, \tau, \varepsilon),$$

$$\delta(x, \varphi, \tau, \varepsilon) = \overline{a}(x, \tau) + \sum_{k \in P} a_k(x, \tau) h_{\varepsilon^\alpha}((k, \omega(x, \tau))) \exp\{i(k, \varphi)\},$$

$h_\alpha(\tau)$ for $d = \varepsilon^\alpha$ is the function defined in Section 3, and $\alpha \in \left[0, \frac{1}{2}\right)$ is a fixed constant.

For $s = -1$, condition (4.3) is an analog of condition (23) in [Sam5], which allows one to obtain a uniform estimate for the oscillation integral. Also note that, by virtue of the finiteness of the function $h_{\varepsilon^\alpha}((k,\omega))$, restriction (4.3) deals only with the resonance harmonics of the function a. This restriction was analyzed in [Kha2].

Assume that the following conditions are satisfied:

$$[\omega(x,\tau); a(x,\varphi,\tau); A(x,\varphi,\tau,\varepsilon); b(x,\varphi,\tau,\varepsilon)] \in C^1_{x,\varphi,\tau}(\overline{G},\sigma_2),$$

$$a \in C^{l_1}_\varphi(\overline{G},\sigma_2), \quad \frac{\partial a}{\partial \tau} \in C^{l_2}_\varphi(\overline{G},\sigma_2), \quad \frac{\partial a}{\partial x} \in C^{l_3}_\varphi(\overline{G},\sigma_2), \tag{4.4}$$

$$l_1 > m + 1 + \max\left\{0; 2s; \frac{s+1}{1-2\alpha} - 1\right\}, \quad \min\{l_2; l_3\} \geq m + \max\{0; s\},$$

where σ_2 is a certain constant. We also assume that there exists a solution $(x(\tau,\varepsilon); \varphi(\tau,\varepsilon))$ of system (4.1) defined for any $(\tau,\varepsilon) \in R_+ \times (0,\varepsilon_0]$ and lying in $\mathcal{D}_{\frac{1}{2}\rho_1}(\overline{x}(\tau)) \times R^m$.

Lemma 4.1. *If* $f(\tau) = (f_1(\tau),\ldots,f_n(\tau)) \in C^1_{R_+}$, $L > 0$ *is a constant, and conditions (4.3) and (4.4) are satisfied, then, for* $s > -1$, *one can find a sufficiently large number* $N \sim \varepsilon^{\frac{2\alpha-1}{2(s+1)}}$ *such that, for all* $(\tau,\overline{\tau},\overline{t},\varepsilon) \in R_+ \times R_+ \times [0,L] \times (0,\varepsilon_1]$ *and* $k \in P_N = \{k\colon k \in P, \|k\| \leq N\}$, *the oscillation integral*

$$I_k(\tau,\overline{\tau},\overline{t},\varepsilon) = \int\limits_\tau^{\tau+\overline{t}} f(t)\exp\left\{\frac{i}{\varepsilon}\int\limits_{\overline{\tau}}^{t}(k,\omega(x(z,\varepsilon),z))\,dz\right\}dt \tag{4.5}$$

satisfies the inequality

$$\|I_k\| \leq \sigma_3\sqrt{\varepsilon}\left[(1+\|k\|^s)\|k\|^{s+1}\max_{[\tau,\tau+L]}\|f(t)\| + \|k\|^s\max_{[\tau,\tau+L]}\left\|\frac{d}{dt}f(t)\right\|\right], \tag{4.6}$$

where σ_3 *and* $\varepsilon_1 \in (0,\varepsilon_0]$ *are constants independent of* τ, $\overline{\tau}$, $\overline{t}$, ε, *and* k. *If* $s = -1$, *then (4.6) holds for all* $k \in P$.

Proof. For $(\tau, \varepsilon) \in R_+ \times (0, \varepsilon_0]$, we consider the functions

$$y(\tau, \varepsilon) = x(\tau, \varepsilon) + U(\tau, \varepsilon),$$

$$U(\tau,\varepsilon) = \varepsilon \sum_{k\in P} \frac{a_k(x(\tau,\varepsilon),\tau)}{i(k,\omega(x(\tau,\varepsilon),\tau))} \times [1 - h_\mu((k,\omega(x(\tau,\varepsilon),\tau)))] \exp\{i(k,\varphi(\tau,\varepsilon))\}. \tag{4.7}$$

The smoothness conditions (4.4) and the estimates for the Fourier coefficients $\forall(\tau,\varepsilon) \in R_+ \times (0,\varepsilon_0]$ presented in [BMS] yield

$$\|U(\tau,\varepsilon)\| \le \varepsilon^{1-\alpha} \sum_{k\neq 0} \sup_{\overline{G}} \|a_k\| \le \varepsilon^{1-\alpha} \sum_{k\neq 0} m^{l_1} \sigma_2 \|k\|^{-l_1}$$
$$\le \varepsilon^{1-\alpha} m^{l_1} 2^m \sigma_2 \left(1 + \frac{1}{l_1 - m}\right) \equiv \sigma_4 \varepsilon^{1-\alpha}. \tag{4.8}$$

We set

$$\varepsilon_2 = \min\left\{ \left(\frac{1}{2}\,\rho_1 \sigma_4^{-1}\right)^{\frac{1}{1-\alpha}} ; \varepsilon_0 \right\}.$$

Then, for $\tau \in R_+$ and $\varepsilon \in (0, \varepsilon_2]$, the curve $y = y(\tau,\varepsilon)$ lies in $\mathcal{D}_{\rho_1}(\overline{x}(\tau))$. By direct differentiation, one can easily verify that

$$\frac{dy(\tau,\varepsilon)}{d\tau} = \delta(x(\tau,\varepsilon),\varphi(\tau,\varepsilon),\tau,\varepsilon) + B(\tau,\varepsilon),$$

where the function $B(\tau,\varepsilon)$ is continuous in $\tau \in R_+$ for every fixed ε and satisfies the inequality

$$\|B(\tau,\varepsilon)\| \le \varepsilon^{1-2\alpha} \sum_{k\neq 0} \Big\{ \sup \|a_k\| \|k\| \times \Big[\sup \|b\| + 17 \Big(\sup \|a + \varepsilon A\| \sup \Big\| \frac{\partial \omega}{\partial x} \Big\| + \sup \Big\| \frac{\partial \omega}{\partial \tau} \Big\| \Big) \Big] + \sup \Big\| \frac{\partial a_k}{\partial \tau} \Big\| + \sup \Big\| \frac{\partial a_k}{\partial x} \Big\| \sup \|a + \varepsilon A\| \Big\} + \varepsilon \sup \|A\|$$
$$\le \sigma_5 \varepsilon^{1-2\alpha}, \tag{4.9}$$

$$\sigma_5 = 2^m \sigma_2 (2 + 18\sigma_2 + 17 n \sigma_2) m^{\max\{l_1; l_2; l_3\}} \times \left(3 + \frac{1}{l_1 - m - 1} + \frac{1}{l_2 - m} + \frac{1}{l_3 - m}\right).$$

In the last inequality, the supremum is taken over all $(x, \varphi, \tau, \varepsilon) \in \overline{G}$, and this inequality can be established with the use of (4.4) by analogy with (4.8).

We now consider an arbitrary vector $k \in P$. Assuming that $x = x(\tau, \varepsilon)$, $y = y(\tau, \varepsilon)$, and $\varphi = \varphi(\tau, \varepsilon)$, we obtain

$$
\begin{aligned}
&|(k, \omega(y, \tau))| + \left|\frac{d}{d\tau}(k, \omega(y, \tau))\right| \\
&\quad \geq |(k, \omega(y, \tau))| + |(k, \Omega(y, \varphi, \tau, \varepsilon))| - \|k\| \sup_{\overline{G}} \left\|\frac{\partial \omega}{\partial x}\right\| \\
&\quad \times \Big(\sup_{\tau \in R_+} \|B\| + \sup_{\tau \in R_+} \|\delta(x, \varphi, \tau, \varepsilon) - \delta(y, \varphi, \tau, \varepsilon)\|\Big). \qquad (4.10)
\end{aligned}
$$

Using inequalities (4.8) and (4.9) and conditions (4.4), we get

$$
\|B(\tau, \varepsilon)\| + \|\delta(x, \varphi, \tau, \varepsilon) - \delta(y, \varphi, \tau, \varepsilon)\| \leq \sigma_6 \varepsilon^{1-2\alpha},
$$

$$
\sigma_6 = \sigma_5 + \sigma_2\sigma_4[n + (16n\sigma_2 + 1)]2^m \Big(\frac{m^{l_1}}{l_1 - m - 1} + \frac{m^{l_3}}{l_3 - m} + 2\Big),
$$

which, together with (4.3) and (4.10), yields

$$
|(k, \omega(y(\tau, \varepsilon), \tau))| + \left|\frac{d}{d\tau}(k, \omega(y(\tau, \varepsilon), \tau))\right| \geq \frac{\sigma_1}{2}\|k\|^{-s} \qquad (4.11)
$$

for $s > -1$, $\varepsilon \in (0, \varepsilon_2]$, $\tau \in R_+$, and $k \in P_{N_1}$; here,

$$
N_1 = E\Big\{\varepsilon^{\frac{2\alpha-1}{s+1}} \Big(\frac{\sigma_1}{2n\sigma_2\sigma_6}\Big)^{\frac{1}{s+1}}\Big\}
$$

and $E\{t\}$ is the integer part of the number t. If $s = -1$, then inequality (4.11) is satisfied for all $k \in P$ and

$$
\varepsilon \leq \varepsilon_3 = \min\Big\{\varepsilon_2; (2n\sigma_1^{-1}\sigma_2\sigma_6)^{\frac{1}{2\alpha-1}}\Big\}.
$$

It follows from (4.11) that, for all $k \in P_{N_1}$ and $s > -1$ or for $k \in P$, $s = -1$, $\varepsilon \in (0, \varepsilon_3]$, and $\tau_0 \in [\tau, \tau + \overline{t}]$, at least one of the following inequalities is satisfied:

$$
|(k, \omega(y_0, \tau_0))| \geq \frac{1}{4}\sigma_1\|k\|^{-s}, \quad \left|\frac{d}{d\tau}(k, \omega(y_0, \tau_0))\right| \geq \frac{1}{4}\sigma_1\|k\|^{-s}, \qquad (4.12)
$$

where $y_0 = y(\tau_0, \varepsilon)$. Let $|(k, \omega(y_0, \tau_0))| \geq \frac{1}{4}\sigma_1 \|k\|^{-s}$. Then, according to the Lagrange mean-value theorem and the condition of the boundedness of $\left\| \frac{d}{d\tau}\omega \right\|$ on the segment

$$\tau \in [\tau_0, \tau_0 + \delta_k], \quad \delta_k = \frac{\sigma_1}{8\sigma_2}[1 + n(\sigma_4 + \sigma_5)]^{-1}\|k\|^{-s-1},$$

we have

$$|(k, \omega(y(\tau, \varepsilon), \tau))| \geq \frac{1}{8}\sigma_1 \|k\|^{-s}.$$

In view of (4.7) and (4.8), this estimate yields

$$|(k, \omega(x(\tau, \varepsilon), \tau))| \geq \frac{1}{16}\sigma_1 \|k\|^{-s} \tag{4.13}$$

for all $\tau \in [\tau_0, \tau_0 + \delta_k]$, $\varepsilon \in (0, \varepsilon_3]$, $s > -1$, and $k \in P_{N_2}$, where

$$N_2 = \min\Big\{N_1; E\Big\{\sigma_7^{\frac{1}{s+1}} \varepsilon^{\frac{\alpha-1}{s+1}}\Big\}\Big\}, \quad \sigma_7 = \Big(\frac{16}{\sigma_1} n\sigma_2\sigma_4\Big)^{-1}.$$

For $s = -1$ and $\varepsilon \leq \varepsilon_4 = \min\{\varepsilon_3; \sigma_7^{\frac{1}{\alpha-1}}\}$, estimate (4.13) holds for all $k \in P$.

If the first inequality in (4.12) is not satisfied, then, by virtue of the continuity of the function $\frac{d}{d\tau}(k, \omega(y(\tau, \varepsilon), \tau))$ in τ, the inequality

$$\left| \frac{d}{d\tau}(k, \omega(y(\tau, \varepsilon), \tau)) \right| \geq \frac{1}{8}\sigma_1 \|k\|^{-s} \tag{4.14}$$

holds on a certain segment $[\tau_0, \alpha_k]$ of maximum length that does not exceed δ_k. Let τ_k^* denote the minimum point of the function $|(k, \omega(y(\tau, \varepsilon), \tau))|$ on this segment. It follows from (4.14) that

$$|(k, \omega(y(\tau, \varepsilon), \tau))| \geq \frac{1}{8}\sigma_1 \|k\|^{-s} |\tau - \tau_k^*| \quad \forall \tau \in [\tau_0, \alpha_k].$$

Therefore,

$$|(k, \omega(x(\tau, \varepsilon), \tau))| \geq \frac{1}{16}\sigma_1 \|k\|^{-s}\sqrt{\varepsilon} \tag{4.15}$$

$$\forall \tau \in [\tau_0, \alpha_k] \backslash [\tau_k^* - \sqrt{\varepsilon}, \tau_k^* + \sqrt{\varepsilon}]$$

for

$$s > -1, \;\; \varepsilon \in (0, \varepsilon_4], \;\; k \in P_N, \;\; N = \min\Big\{E\{\sigma_7^{\frac{1}{s+1}} \varepsilon^{\frac{2\alpha-1}{2(s+1)}}\}; N_2\Big\}.$$

If $s = -1$, then estimate (4.15) holds for every $k \in P$ and $\varepsilon \leq \varepsilon_1 = \min\{\varepsilon_4; \sigma_7^{\frac{2}{1-2\alpha}}\}$.

We now represent $I_k(\tau, \overline{\tau}, \overline{t}, \varepsilon)$ in the form of the sum of the integrals

$$I_k = \sum_{r=0}^{q_k-1} \int\limits_{\tau+\delta_k r}^{\tau+\delta_k(r+1)} F dt + \int\limits_{\tau+\delta_k q_k}^{\tau+\overline{t}} F dt, \tag{4.16}$$

where F is the integrand of integral (4.5) and q_k is the integer part of the number $\overline{t}\delta_k^{-1}$,

$$q_k \leq \frac{1}{\sigma_1} 8\sigma_2 L[1 + n(\sigma_4 + \sigma_5)]\|k\|^{s+1} \equiv \sigma_8 \|k\|^{s+1}.$$

Let us estimate each integral on the right-hand side of (4.16). If the first inequality in (4.12) is satisfied at the point $\tau_0 = \tau + \delta_k r$, then, integrating by parts and using (4.13), we get

$$\left\| \int\limits_{\tau+\delta_k r}^{\tau+\delta_k(r+1)} F dt \right\| \leq \varepsilon \Bigg\{ \frac{16}{\sigma_1} \delta_k \|k\|^s \max_{[\tau,\tau+L]} \left\| \frac{d}{dt} f(t) \right\|$$
$$+ \left[(n\sigma_2 + 1)\sigma_2 \left(\frac{16}{\sigma_1} \right)^2 \delta_k \|k\|^{1+2s} \right.$$
$$\left. + \frac{32}{\sigma_1} \|k\|^s \right] \max_{[\tau,\tau+L]} \|f(t)\| \Bigg\}. \tag{4.17}$$

Now assume that, at the point $\tau_0 = \tau + \delta_k r$, the first inequality in (4.12) is not satisfied, but the second inequality in (4.12) is true. Then, on a segment of length $2\sqrt{\varepsilon}$, the integral under investigation can be estimated by the value $2\sqrt{\varepsilon} \max \|f(t)\|$, and, outside this segment, inequality (4.15) holds. Therefore, the following estimate holds for $\alpha_k \leq \tau + \delta_k(r+1)$:

$$\left\| \int\limits_{\tau+\delta_k r}^{\alpha_k} F dt \right\|$$
$$\leq 2\sqrt{\varepsilon} \Bigg\{ \max_{[\tau,\tau+L]} \|f(t)\| \left(1 + \frac{64}{\sigma_1} \|k\|^s \right)$$
$$+ \max_{[\tau,\tau+L]} \left\| \frac{d}{dt} f(t) \right\| \frac{16}{\sigma_1} \delta_k \|k\|^s \Bigg\}. \tag{4.18}$$

Note that if $\alpha_k < \tau + \delta_k(r+1)$, then the definition of the number α_k yields

$$|(k, \omega(y(\alpha_k, \varepsilon), \alpha_k))| \geq \frac{\sigma_1}{4}\|k\|^{-s}.$$

In view of this inequality, the integral of the function F over the segment $[\alpha_k, \tau + \delta_k(r+1)]$ can be estimated from above by the value presented on the right-hand side of (4.17). Combining (4.16)–(4.18), we obtain estimate (4.6) $\forall k \in P_N$ for $s > -1$ or $\forall k \in P$ for $s = -1$ with the constant

$$\sigma_3 = \frac{64}{\sigma_1}L + 2\,(1+\sigma_8)\left(1 + \frac{96}{\sigma_1}\right) + \left(\frac{16}{\sigma_1}\right)^2 (2 + 2n\sigma_2)\,\sigma_2 L.$$

Lemma 4.1 is proved.

Theorem 4.1. *Suppose that there exists a solution $\overline{x} = \overline{x}(\tau)$ of the averaged system (4.2) that lies in $\mathcal{D}$ together with its ρ-neighborhood $\forall \tau \in [0, L]$ and conditions (4.3) and (4.4) are satisfied for $\tau \in [0, L]$. Then one can find constants $\overline{\varepsilon}_0 \in (0, \varepsilon_0]$ and $\sigma_9 > 0$ such that*

$$\|x\,(\tau, \overline{x}\,(0), \psi, \varepsilon) - \overline{x}\,(\tau)\| \leq \sigma_9\sqrt{\varepsilon} \tag{4.19}$$

for all $\tau \in [0, L]$, $\psi \in R^m$, and $\varepsilon \in (0, \overline{\varepsilon}_0]$.

Proof. It follows from the conditions of the smoothness of the right-hand side of system (4.1) that the slow variables $x(\tau, \overline{x}(0), \psi, \varepsilon)$ of every solution $(x(\tau, \overline{x}(0), \psi, \varepsilon); \varphi(\tau, \overline{x}(0), \psi, \varepsilon))$, $\psi \in R^m$, $\varepsilon \leq \varepsilon_0$, lie in the $\frac{1}{2}\rho_1$-neighborhood of the curve $\overline{x} = \overline{x}(\tau)$ for all τ from a certain segment $[0, L_1] \subset [0, L]$, $L_1 = L_1(\psi, \varepsilon)$. Here, $\rho_1 \in (0, \rho]$ is the constant determined by condition (4.3). Then it follows from Eqs. (4.1) and (4.2) and the Gronwall–Bellman lemma that, for any $\tau \in [0, L_1]$,

$$\|x(\tau, \overline{x}(0), \psi, \varepsilon) - \overline{x}(\tau)\| \leq e^{Ln\sigma_2}\Big[\varepsilon\sigma_2 L + L \sup_{\overline{G}} \|R_N a(x, \varphi, \tau)\|\Big]$$

$$+ \sum_{k \in P_N} \sup_{\tau} \Big\| \int_0^{\tau} a_k(x(t, \overline{x}(0), \psi, \varepsilon), t) \Big\| \exp\{i(k, \theta)\}$$

$$\times \exp\Big\{\frac{i}{\varepsilon}\int_0^t (k, \omega(x(z, \overline{x}(0), \psi, \varepsilon), z))dz\Big\}\,dt, \tag{4.20}$$

where

$$\theta = \varphi(t, \overline{x}(0), \psi, \varepsilon) - \frac{1}{\varepsilon}\int_0^t \omega(x(z, \overline{x}(0), \psi, \varepsilon), z)dz$$

and

$$R_N a(x, \varphi, \tau) = \sum_{\|k\|>N} a_k(x, \tau)\exp\{i(k, \varphi)\}$$

for $s > -1$. If $s = -1$, then we set $P_N = P$ and $R_N a(x, \varphi, \tau) \equiv 0$. Here, N is the number defined in Lemma 4.1. It is obvious that

$$N > \frac{1}{2}\sigma_{10}\,\varepsilon^{\frac{2\alpha-1}{2(s+1)}} \quad \forall\varepsilon \in (0, \varepsilon_0], \qquad \varepsilon_0 \le \sigma_{10}^{\frac{2(s+1)}{1-2\alpha}},$$

$$\sigma_{10} = \min\Big\{\sigma_7^{\frac{1}{s+1}};\Big(\frac{1}{\sigma_1}2n\sigma_2\sigma_6\Big)^{-\frac{1}{s+1}}\Big\}.$$

The smoothness conditions (4.4) for the function $a(x, \varphi, \tau)$ guarantee that

$$\begin{aligned}
\sup_{\overline{G}}\|R_N a(x, \varphi, \tau)\| &\le \sum_{\|k\|>N}\sup_{\overline{G}}\|a_k(x, \tau)\| \le \sum_{\|k\|>N}\sigma_2 m^{l_1}\|k\|^{-l_1} \\
&\le 2^m m^{l_1}\sigma_2\sum_{r=N+1}^{\infty} r^{m-1-l_1} \le 2^m m^{l_1}\sigma_2\int_N^{\infty} r^{m-1-l_1}dr \\
&= 2^m m^{l_1}\sigma_2\frac{1}{l_1-m}N^{m-l_1} \le \sigma_{11}\varepsilon^{\frac{(l_1-m)(1-2\alpha)}{2(s+1)}}, \qquad (4.21)
\end{aligned}$$

$$\sigma_{11} = \frac{2^m m^{l_1}\sigma_2}{l_1-m}\Big(\frac{2}{\sigma_{10}}\Big)^{l_1-m},$$

for $s > -1$. Using Lemma 4.1, conditions (4.4), and inequality (4.21), we obtain the following estimate for $s > -1$:

$$\|x(\tau, \overline{x}(0), \psi, \varepsilon) - \overline{x}(\tau)\| \le \sigma_{12}(\sqrt{\varepsilon} + \varepsilon^{\frac{(l_1-m)(1-2\alpha)}{2(s+1)}}) \le 2\sigma_{12}\sqrt{\varepsilon} \equiv \sigma_9\sqrt{\varepsilon},$$

where $\tau \in [0, L_1]$, $\psi \in R^m$, $\varepsilon \le \min\{\varepsilon_0; \sigma_{10}^{\frac{2(s+1)}{1-2\alpha}}; \varepsilon_1\} \equiv \underline{\varepsilon}_0$, and

$$\sigma_{12} = e^{Ln\sigma_2}\Big[\sigma_2 L + \sigma_{11} L + 2^m \sigma_2\sigma_3(1+\sigma_2)m^{\max\{l_1;l_2;l_3\}}$$
$$\times\Big(4 + \frac{1}{l_1 - 2s - 1 - m} + \frac{1}{l_1 - m - s - 1}$$
$$+ \frac{1}{l_2 - s - m} + \frac{1}{l_3 - s - m}\Big)\Big].$$

Here, ε_1 is the constant defined in Lemma 4.1. It is easy to see that the last estimate remains true if $s = -1$. We set

$$\bar{\varepsilon}_0 = \min\left\{\underline{\varepsilon}_0; \left(\frac{1}{\rho_1} 4\sigma_9\right)^{-2}\right\},$$

which guarantees the validity of the inequality

$$\|x(\tau, \overline{x}(0), \psi, \varepsilon) - \overline{x}(\tau)\| \le \frac{1}{4}\rho_1 \quad \forall \tau \in [0, L_1].$$

This inequality and the smoothness conditions for the right-hand side of system (4.1) allow one to extend the solution $(x(\tau, \overline{x}(0), \psi, \varepsilon); \varphi(\tau, \overline{x}(0), \psi, \varepsilon))$ to the entire segment $[0, L]$. In this case, inequality (4.19) does not change. Theorem 4.1 is proved.

As an example, we consider the Cauchy problem

$$\frac{dx}{d\tau} = \lambda[f_1(x, \widetilde{\varphi}, \tau) + f_2(x, \underset{\sim}{\varphi}, \tau) + \cos\varphi + 2.5], \quad x(0) = 0,$$

$$\frac{d\widetilde{\varphi}}{d\tau} = \frac{\lambda}{\varepsilon}, \quad \frac{d\underset{\sim}{\varphi}}{d\tau} = \frac{x}{\varepsilon}, \quad \frac{d\varphi}{d\tau} = \frac{\tau+2}{\varepsilon}, \quad \widetilde{\varphi}(0) = \underset{\sim}{\varphi}(0) = 0, \quad \varphi(0) = 0,$$

where x, $\widetilde{\varphi}$, and $\underset{\sim}{\varphi}$ are m-dimensional vectors, $m \ge 2$, $x \in \mathcal{D} = \{x\colon \|x\| \le 3\|\lambda\|\}$, φ is a scalar, $\tau \in [0, 1]$, $\lambda = (\lambda_1, \ldots, \lambda_m)$ is a constant nonzero vector, and f_1 and f_2 are scalar 2π-periodic (in $\widetilde{\varphi}$ and $\underset{\sim}{\varphi}$) functions satisfying conditions (4.4) for $\alpha = 0$ and $s = m + 1$. We also assume that the Fourier coefficients $f_{1,k}$ and $f_{2,k}$ of the functions f_1 and f_2 satisfy the relations

$$f_{1,0}(x, \tau) = f_{2,0}(x, \tau) \equiv 0, \quad \sum_{k\neq 0}[|f_{1,k}(x, \tau)| + |f_2(x, \tau)|] \le 2,$$

and the vector λ and every nonzero vector $k = (k_1, \ldots, k_m)$ with integer-valued coordinates satisfy the inequality

$$|(k, \lambda)| = \left|\sum_{\nu=1}^{m} k_\nu \lambda_\nu\right| \ge \frac{c}{\|k\|^{m+1}}, \quad c = \text{const} > 0.$$

It is known [Arn4] that, in the ball $\|\lambda\| \le 1$, the Lebesgue measure of the numbers $\lambda = (\lambda_1, \ldots, \lambda_m)$ for which the last inequality is not true tends to zero as $c \to 0$. In our case, the Cauchy problem for slow variables averaged with respect to all angular variables has the solution $\overline{x}(\tau) = 2.5\lambda\tau$, which lies in $\mathcal{D}$ together with its $\frac{1}{2}\|\lambda\|$-neighborhood for any $\tau \in [0, 1]$. It is easy to verify that, for the system under consideration, inequality (4.3) is satisfied for $s = m + 1$, $\alpha = 0$, and $\sigma_1 = \min\left\{\frac{c}{2}; 2\right\}$. Therefore, according to Theorem 4.1, for any $(\tau, \varepsilon) \in [0, 1] \times (0, \varepsilon_0]$ (ε_0 is sufficiently small), the following estimate is true:

$$\|x(\tau, \varepsilon) - \overline{x}(\tau)\| \le \overline{c}\sqrt{\varepsilon}, \quad \overline{c} = \text{const}.$$

Note that, in this special case, we cannot use the results of [Sam5, Kha1, Kha2] for the justification of the averaging method.

We now study the problem of the qualitative relationship between solutions of original and averaged equations on the infinite time interval $[0, \infty)$.

Theorem 4.2. *Suppose that the following conditions are satisfied:*

(i) conditions (4.3) and (4.4) are satisfied;

(ii) there exists an asymptotically stable solution $\overline{x} = \overline{x}(\tau)$, $\tau \in R_+$, *of Eqs. (4.2) that lies in* $\mathcal{D}$ *together with its* ρ*-neighborhood.*

Then the following assertions are true:

(a) there exist positive constants σ_{13}, $\overline{\sigma}_{13}$, *and* $\beta < \rho$ *such that, for all* $\tau \in R_+$, $\varepsilon \in (0, \overline{\sigma}_{13}]$, $\varphi^0 \in R^m$, *and* $x^0 \in \mathcal{D}_\beta(\overline{x}(0))$, *the slow variables* $x(\tau, x^0, \varphi^0, \varepsilon)$ *of every solution* $(x(\tau, x^0, \varphi^0, \varepsilon); \varphi(\tau, x^0, \varphi^0, \varepsilon))$ *of system (4.1) are uniformly bounded, i.e.,*

$$\|x(\tau, x^0, \varphi^0, \varepsilon)\| \le \sigma_{13}; \tag{4.22}$$

(b) for arbitrary $\eta \in (0, \beta)$, *there exists* $\varepsilon(\eta) > 0$ *such that*

$$\|x(\tau, \overline{x}(0), \varphi^0, \varepsilon) - \overline{x}(\tau)\| < \eta \tag{4.23}$$

for all $\tau \in R_+$, $\varphi^0 \in R^m$, *and* $\varepsilon \in (0, \varepsilon(\eta)]$.

Proof. According to the definition of uniform asymptotic stability [Fil], for the number $\frac{1}{2}\rho$ one can find $\beta > 0$ such that $\|\overline{x}(\tau, t, x^0) - \overline{x}(\tau)\| < \frac{1}{2}\rho$ for all

$\tau \geq t \in R_+$, provided that $\|x^0 - \overline{x}(t)\| < \beta$. Here, $\overline{x}(\tau, t, x^0)$ is a solution of system (4.2) that satisfies the condition $\overline{x}(t, t, x^0) = x^0$. Moreover, one can find a constant $L = L(\rho)$ such that $\|\overline{x}(\tau, t, x^0) - \overline{x}(\tau)\| \leq \frac{1}{2}\beta$ for $\tau \geq t + L$.

Using Theorem 4.1, for $\varepsilon \leq \min\left\{\left(\frac{\beta}{4\sigma_9}\right)^2; \overline{\varepsilon}_0\right\} = \overline{\sigma}_{13}$ we get

$$\begin{aligned}
&\|x_\tau(0, x^0, \varphi^0, \varepsilon) - \overline{x}(\tau)\| \\
&\qquad \leq \|x_\tau(0, x^0, \varphi^0, \varepsilon) - \overline{x}(\tau, 0, x^0)\| + \|\overline{x}(\tau, 0, x^0) - \overline{x}(\tau)\| \\
&\qquad < \sigma_9\sqrt{\varepsilon} + \frac{1}{2}\rho < \rho \qquad \forall \tau \in [0, L),
\end{aligned} \tag{4.24}$$

$$\|x_L(0, x^0, \varphi^0, \varepsilon) - \overline{x}(L)\| < \sigma_9\sqrt{\varepsilon} + \frac{1}{2}\beta < \beta,$$

i.e., the point $\widetilde{x}^0 = x_L(0, x^0, \varphi^0, \varepsilon)$ is located in the β-neighborhood of the point $\overline{x}(L)$. Here, $\overline{\varepsilon}_0$ and σ_9 are the constants defined in Theorem 4.1, and $(x_\tau(t, x^0, \varphi^0, \varepsilon); \varphi_\tau(t, x^0, \varphi^0, \varepsilon))$ is a solution of system (4.1) that passes through the point $(x^0; \varphi^0)$ at $\tau = t$.

We now consider the time interval $[L, 2L]$. By analogy with the above reasoning, we can establish the inequalities

$$\begin{aligned}
\|x_\tau(L, \widetilde{x}^0, \widetilde{\varphi}^0, \varepsilon) - \overline{x}(\tau)\| < \rho \quad \forall \tau \in [L, 2L), \\
\|x_{2L}(L, \widetilde{x}^0, \widetilde{\varphi}^0, \varepsilon) - \overline{x}(2L)\| < \beta,
\end{aligned} \tag{4.25}$$

where $\widetilde{\varphi}^0 = \varphi_L(0, x^0, \varphi^0, \varepsilon)$. Inequalities (4.24) and (4.25) yield

$$\begin{aligned}
\|x_\tau(0, x^0, \varphi^0, \varepsilon) - \overline{x}(\tau)\| < \rho \quad \forall \tau \in [0, 2L), \\
\|x_{2L}(0, x^0, \varphi^0, \varepsilon) - \overline{x}(2L)\| < \beta.
\end{aligned}$$

By induction, for an arbitrary natural $p \geq 3$ we get

$$\begin{aligned}
\|x_\tau(0, x^0, \varphi^0, \varepsilon) - \overline{x}(\tau)\| < \rho \quad \forall \tau \in [0, pL), \\
\|x_{pL}(0, x^0, \varphi^0, \varepsilon) - \overline{x}(pL)\| < \beta.
\end{aligned} \tag{4.26}$$

This yields

$$\|x_\tau(0, x^0, \varphi^0, \varepsilon)\| \leq \rho + \sup_{R_+} \|\overline{x}(\tau)\| \equiv \sigma_{13}$$

for all $\tau \in R_+$, $\varphi^0 \in R^m$, $x^0 \in \mathcal{D}_\beta(\overline{x}(0))$, and $\varepsilon \in (0, \overline{\sigma}_{13}]$. Thus, assertion (a) is proved.

We now fix an arbitrary $\eta \in (0, \beta)$. According to the definition of uniform asymptotic stability, for η there exist constants $\mu > 0$ and $L_1 = L_1(\eta) > 0$ such that the inequality $\|x^0 - \overline{x}(t)\| < \mu$ yields

$$\|\overline{x}(\tau, t, x^0) - \overline{x}(\tau)\| < \frac{1}{2}\eta \quad \forall \tau \geq t,$$

$$\|\overline{x}(\tau, t, x^0) - \overline{x}(\tau)\| < \frac{1}{2}\mu \quad \forall \tau \geq \tau + L_1.$$

The further proof of assertion (b) is based on inequalities (4.24)–(4.26) with $\beta,\ \rho,\ L$, and x^0 replaced by $\mu,\ \eta,\ L_1$, and $\overline{x}(0)$, respectively. In this case, $\varepsilon(\eta) = \min\left\{\overline{\varepsilon}_0; \left(\frac{\mu}{4\sigma_9}\right)^2\right\}$, and $\overline{\varepsilon}_0$ and σ_9 are defined in Theorem 4.1 for $L = L_1$.

Remark 5. If, in addition, we assume that $\overline{a}(x, \tau) \in C_x^2(\mathcal{D} \times R_+, \sigma_2)$ and the normal fundamental matrix $Q(\tau, t)$ of solutions of the variational equation $\dfrac{dz}{d\tau} = \dfrac{\partial}{\partial x}\overline{a}\left(\overline{x}(\tau), \tau\right) z$ satisfies the inequality

$$\|Q\left(\tau, t\right)\| \leq Ke^{-\gamma(\tau - t)} \quad \forall \tau \geq t \in R_+,$$

where K and γ are certain positive constants, then estimate (4.23) takes the form

$$\|x\left(\tau, \overline{x}(0), \psi, \varepsilon\right) - \overline{x}(\tau)\| < \sigma_{14}\sqrt{\varepsilon}, \quad \sigma_{14} = \text{const}.$$

The proof of this statement, in fact, repeats the proof of Theorem 2.4.

5. Averaging over All Fast Variables in Multifrequency Systems of Higher Approximation

Consider the case where system (4.1) can be represented in the form

$$\frac{dx}{d\tau} = \sum_{s=0}^{r} A_s(x, \tau)\varepsilon^s + \varepsilon^{r+1} a(x, \varphi, \tau, \varepsilon),$$

$$\frac{d\varphi}{d\tau} = \sum_{s=-1}^{r-1} B_s(x, \tau)\varepsilon^s + \varepsilon^r b\left(x, \varphi, \tau, \varepsilon\right), \tag{5.1}$$

where r is a nonnegative integer and $B_{-1}(x, \tau) \equiv \omega(x, \tau) \not\equiv 0,\ m \geq 2$. The principal difference between system (5.1) and (4.1) lies in the fact that the functions A_s and $B_{s-1},\ s = \overline{0, r}$, in (5.1) depend only on the slow variables x and

τ and do not depend on the angular variables φ. For $r = 0$, Grebenikov and Ryabov [GrR3] justified the method of averaging with respect to the time variable along a solution of the generating system under the assumption of isolated resonances. Since, in the case of the existence of resonances, the values obtained by averaging with respect to time and with respect to all fast variables do not coincide, the averaging scheme proposed in [GrR3] is, in fact, a scheme of averaging with respect to a part of fast variables. Below, we justify the averaging method for (5.1) with respect to all angular variables and establish the quantitative dependence of estimates on the value of the small parameter.

Assume that

$$[A_s(x,\tau); B_{s-1}(x,\tau)] \in C^l_{x,\tau}(\mathcal{D}\times[0,L],c_1), \quad s=\overline{0,r},$$

$$[a(x,\varphi,\tau,\varepsilon); b(x,\varphi,\tau,\varepsilon)] \in C^l_{x,\tau}(\overline{G},c_1), \quad l \geq m, \tag{5.2}$$

$$\sum_{k\neq 0}\|k\|^q\left[\sup_G\|c_k\| + \frac{1}{\|k\|}\left(\sup_G\left\|\frac{\partial c_k}{\partial \tau}\right\| + \sup_G\left\|\frac{\partial c_k}{\partial x}\right\|\right)\right] \leq c_1, \quad q \geq 0,$$

where c_1 is a constant independent of ε, $c_k = c_k(x,\tau,\varepsilon)$ are the Fourier coefficients of the function $c(x,\varphi,\tau,\varepsilon) = [a(x,\varphi,\tau,\varepsilon); b(x,\varphi,\tau,\varepsilon)]$ 2π-periodic in φ, $G = \mathcal{D}\times[0,L]\times(0,\varepsilon_0]$, and $\overline{G} = G\times R^m$.

Consider the system averaged with respect to all variables φ:

$$\frac{d\overline{x}}{d\tau} = \sum_{s=0}^{r} A_s(\overline{x},\tau)\varepsilon^s + \varepsilon^{r+1}\overline{a}(\overline{x},\tau,\varepsilon),$$

$$\frac{d\overline{\varphi}}{d\tau} = \sum_{s=-1}^{r-1} B_s(\overline{x},\tau)\varepsilon^s + \varepsilon^r\overline{b}(\overline{x},\tau,\varepsilon). \tag{5.3}$$

We denote by $(x(\tau,y,\psi,\varepsilon);\varphi(\tau,y,\psi,\varepsilon))$ and $(\overline{x}(\tau,y,\varepsilon);\overline{\varphi}(\tau,y,\psi,\varepsilon))$ the solutions of (5.1) and (5.3), respectively, that take a value $(y;\psi)\in\mathcal{D}_1\times R^m$ for $\tau=0$; here, $\mathcal{D}_1$ is a certain domain in $\mathcal{D}$.

Assume that, for all $\tau\in[0,L]$, $y\in\mathcal{D}_1$, and $\varepsilon\in(0,\varepsilon_0]$, the curve $\overline{x} = \overline{x}(\tau,y,\varepsilon)$ lies in $\mathcal{D}$ together with its ρ-neighborhood (ρ is a constant independent of ε and y).

Using the smoothness conditions (5.2) and the Gronwall–Bellman lemma, we deduce from (5.1) and (5.3) the *a priori* estimates

$$\|x(\tau,y,\psi,\varepsilon) - \overline{x}(\tau,y,\varepsilon)\| \leq 2Lc_1e^{Lnc_1(r+1)}\varepsilon^{r+1} \equiv \overline{c}\varepsilon^{r+1},$$

$$\|\varphi(\tau,y,\psi,\varepsilon) - \overline{\varphi}(\tau,y,\psi,\varepsilon)\| \leq [Lnc_1\overline{c}(1+r) + 2c_1]\varepsilon^r \equiv \underline{c}\varepsilon^r \tag{5.4}$$

for all $\tau \in [0, L]$, $y \in D_1$, $\psi \in R^m$, and $\varepsilon \in (0, \varepsilon_0]$. The condition $\varepsilon_0 \leq (\rho(2\overline{c})^{-1})^{\frac{1}{r+1}}$ guarantees that the solution $(x(\tau, y, \psi, \varepsilon); \varphi(\tau, y, \psi, \varepsilon))$ of system (5.1) is defined for all $\tau \in [0, L]$, and the curve $x = x(\tau, y, \psi, \varepsilon)$ lies in $\mathcal{D}$ together with its $\frac{1}{2}\rho$-neighborhood. The order of the second inequality in (5.4) with respect to ε is less by one than the order of the first inequality because, in system (5.1), ω depends on x, and the rate of the variation of angular variables is proportional to ε^{-1}.

The problem is to improve estimates (5.4) under certain additional restrictions by replacing r in these estimates by $r + \alpha$, $\alpha = \text{const} > 0$. Assume that, for all $(x, \tau) \in \mathcal{D} \times [0, L]$ and certain p, $m \leq p \leq l$, the following inequality is true:

$$\det(W_p^T(x, \tau) W_p(x, \tau)) \neq 0, \tag{5.5}$$

where

$$W_p(x, \tau) = \left(\frac{d^{j-1}}{d\tau^{j-1}} \omega_\nu (x, \tau) \right)_{j,\nu=1}^{p,m}$$

and the total derivatives of the functions $\omega_\nu(x, \tau)$ with respect to τ are calculated along the solutions of the equation $\dfrac{dx}{d\tau} = A_0(x, \tau)$.

Theorem 5.1. *If $\overline{x} = \overline{x}(\tau, y, \varepsilon)$ lies in $\mathcal{D}$ together with its ρ-neighborhood $\forall(\tau, y, \varepsilon) \in [0, L] \times \mathcal{D}_1 \times (0, \varepsilon_0]$ and conditions (5.2) for $q = 0$ and (5.5) are satisfied, then there exists a constant c_2 such that*

$$\|U(\tau, y, \psi, \varepsilon)\| \leq c_2 \varepsilon^{r+1+\frac{1}{p}} \quad \forall(\tau, y, \psi, \varepsilon) \in [0, L] \times \mathcal{D}_1 \times R^m \times (0, \varepsilon_0], \tag{5.6}$$

where ε_0 is sufficiently small and $U = (x(\tau, y, \psi, \varepsilon) - \overline{x}(\tau, y, \varepsilon); \varepsilon\varphi(\tau, y, \psi, \varepsilon) - \varepsilon\overline{\varphi}(\tau, y, \psi, \varepsilon))$.

Proof. Denote by $\overline{\mathcal{D}}_{\frac{1}{2}\rho}$ the closure of the set of points that lie in $\mathcal{D}$ together with their $\frac{1}{2}\rho$-neighborhoods. Under the conditions of the theorem, we have $\overline{\mathcal{D}}_{\frac{1}{2}\rho} \neq \varnothing$. By virtue of the continuity of the functions

$$\frac{d^{j-1}}{d\tau^{j-1}} \omega_\nu(x, \tau), \quad \nu = \overline{1, m}, \quad j = \overline{1, p},$$

on the set $\overline{\mathcal{D}}_{\frac{1}{2}\rho} \times [0, L]$ and inequality (5.5), there exists a constant $c_3 > 0$ such that

$$\det(W_p^T(x, \tau) W_p(x, \tau)) \geq c_3 \quad \forall(x, \tau) \in \overline{\mathcal{D}}_{\frac{1}{2}\rho} \times [0, L]. \tag{5.7}$$

Further, we consider the matrix

$$W_p(\overline{x},\tau,\varepsilon) = \Big(\frac{d^{j-1}}{d\tau^{j-1}}\,\omega_\nu\,(\overline{x},\tau)\Big)_{j,\nu=1}^{p,m},$$

where the total derivatives of $\omega_\nu(\overline{x},\tau)$ with respect to τ are calculated along the solutions of the averaged equations (5.3). It is clear that

$$\det(W_p^T(\overline{x},\tau,\varepsilon)W_p(\overline{x},\tau,\varepsilon)) = \det(W_p^T(\overline{x},\tau)W_p(\overline{x},\tau)) + \varepsilon\Delta(\overline{x},\tau,\varepsilon), \quad (5.8)$$

where $\Delta(\overline{x},\tau,\varepsilon)$ is expressed in terms of the functions $\omega_\nu(\overline{x},\tau)$, $\nu = \overline{1,m}$, $A_s(\overline{x},\tau)$, $s = \overline{0,r}$, and $\overline{a}(\overline{x},\tau,\varepsilon)$ and their derivatives with respect to τ and x up to the $(p-1)$th order. Therefore, according to conditions (5.2), we have $|\Delta(\overline{x},\tau,\varepsilon)| \le \overline{c}_3 = \text{const}$ $\forall(\overline{x},\tau,\varepsilon) \in G$. It follows from (5.7) and (5.8) for $\varepsilon_0 \le c_3(2\overline{c}_3)^{-1}$ that

$$\det(W_p^T(\overline{x},\tau,\varepsilon)W_p(\overline{x},\tau,\varepsilon)) \ge \frac{1}{2}c_3$$

$$\forall(\overline{x},\tau,\varepsilon) \in \overline{\mathcal{D}}_{\frac{1}{2}\rho} \times [0,L] \times (0,\varepsilon_0].$$

This inequality, together with (5.2), yields

$$\|(W_p^T(\overline{x},\tau,\varepsilon)W_p(\overline{x},\tau,\varepsilon))^{-1}W_p^T(\overline{x},\tau,\varepsilon)\| \le c_4 \quad (5.9)$$

for all $\overline{x} \in \overline{\mathcal{D}}_{\frac{1}{2}\rho}$, $\tau \in [0,L]$, and $\varepsilon \in (0,\varepsilon_0]$; here, c_4 is a constant independent of ε.

Subtracting Eqs. (5.3) from Eqs. (5.1) and multiplying the equality for the angular variables by ε, we get

$$\|U(\tau,y,\psi,\varepsilon)\| \le 2nc_1(1+r)\int_0^\tau \|U(t,y,\psi,\varepsilon)\|\,dt$$

$$+\,\varepsilon^{r+1}\Big\|\int_0^\tau \sum_{k\ne 0} c_k(\overline{x},t,\varepsilon)\exp\{i(k,\overline{\varphi})\}\,dt\Big\|,$$

whence

$$\|U(\tau,y,\psi,\varepsilon)\| \le e^{2nc_1(1+r)L}\sum_{k\ne 0}\sup_{\tau\in[0,L]}\Big\|\int_0^\tau c_k(\overline{x},t,\varepsilon)$$

$$\times\exp\{i(k,\overline{\theta})\}\exp\Big\{\frac{i}{\varepsilon}\int_0^t (k,\omega(\overline{x},t))\,dt\Big\}\,dt\Big\|\,\varepsilon^{r+1}, \quad (5.10)$$

where

$$\overline{\theta} = \overline{\varphi} - \frac{1}{\varepsilon} \int_0^t \omega(\overline{x}, t) dt, \quad \overline{x} = \overline{x}(t, y, \varepsilon), \quad \overline{\varphi} = \overline{\varphi}(t, y, \psi, \varepsilon).$$

Since the curve $\overline{x} = \overline{x}(\tau, y, \varepsilon)$ lies in $\overline{\mathcal{D}}_{\frac{1}{2}\rho}$, condition (5.9) is satisfied for every solution $\overline{x} = \overline{x}(\tau, y, \varepsilon)$ of the first equation of system (5.3) for $\tau \in [0, L]$, $y \in \mathcal{D}_1$, and $\varepsilon \in (0, \varepsilon_0]$. Moreover, according to conditions (5.2), the total derivatives of the functions $\omega_\nu(\overline{x}(\tau, y, \varepsilon), \tau)$, $\nu = \overline{1, m}$, with respect to τ up to the order $l \geq p$ inclusive are uniformly bounded from above by a constant independent of ε and y; therefore, the functions $\omega_\nu(\overline{x}(\tau, y, \varepsilon), \tau)$ and their derivatives with respect to τ up to the order $(p - 1)$ are uniformly continuous in τ for all $(y, \varepsilon) \in \mathcal{D}_1 \times (0, \varepsilon_0]$. These arguments enable one to apply Theorem 1.2 for $\omega = \omega(\overline{x}(\tau, y, \varepsilon), \tau)$ and $f = c_k(\overline{x}(\tau, y, \varepsilon), \tau, \varepsilon) \exp\{i(k, \overline{\theta})\}$ to the estimation of the oscillation integral on the right-hand side of inequality (5.10). Thus, according to conditions (5.2) and (5.9), we get

$$\|U(\tau, y, \psi, \varepsilon)\| \leq e^{2nc_1(1+r)L} c_1 \sigma_2 [2 + (1 + r)c_1] \varepsilon^{r+1+\frac{1}{p}}$$

$$\forall (\tau, y, \psi, \varepsilon) \in [0, L] \times \mathcal{D}_1 \times R^m \times (0, \varepsilon_0],$$

which yields estimate (5.6). Here, σ_2 is the constant determined by inequality (1.12). Theorem 5.1 is proved.

Theorem 5.2. *Suppose that the conditions of Theorem 5.1 and conditions (5.2) for* $q = 1$ *are satisfied. Then one can find constants* $c_5 > 0$ *and* $\overline{\varepsilon}_0 > 0$ *such that*

$$\left\| \frac{\partial}{\partial \psi} U(\tau, y, \psi, \varepsilon) \right\| + \varepsilon \left\| \frac{\partial}{\partial y} U(\tau, y, \psi, \varepsilon) \right\| \leq c_5 \varepsilon^{r+1+\frac{1}{p}} \tag{5.11}$$

for all $(\tau, y, \psi, \varepsilon) \in [0, L] \times \mathcal{D}_1 \times R^m \times (0, \overline{\varepsilon}_0]$.

The proof of estimate (5.11), in fact, coincides with the proof of Theorem 2.2. The only difference is that the order of the estimate for $\left\| \frac{\partial}{\partial y} U \right\|$ with respect to ε is less by one than the order of the estimate for $\left\| \frac{\partial}{\partial \psi} U \right\|$ because the frequencies ω_ν depend on x and, therefore, $\left\| \frac{\partial}{\partial y} \varphi(\tau, y, \psi, \varepsilon) \right\| \sim \varepsilon^{-1}$.

Remark 6. The main assumption in Theorems 5.1 and 5.2 is inequality (5.5) [or the equivalent inequality (5.9)], which is a restriction imposed on the averaged system. Condition (5.5) guarantees the fast passage of the averaged system [and system (5.1) with regard for the *a priori* estimates (5.4)] through a small neighborhood of the resonance surface $(k, \omega(x, \tau)) = 0, \;\; k \neq 0$. Note that, generally speaking, this is not the case for multifrequency systems of the general form [Arn4, GrR1, GrR3, Sam5].

Let us formulate a theorem on the justification of the averaging method on a semiaxis for the system of $n + m$ equations

$$\frac{dx}{d\tau} = a(x, \tau\varepsilon) + \varepsilon A(x, \varphi, \tau, \varepsilon),$$

$$\frac{d\varphi}{d\tau} = \frac{\omega(x, \tau, \varepsilon)}{\varepsilon} + B(x, \varphi, \tau, \varepsilon), \tag{5.12}$$

where $a, \;\; A, \;\; \omega$, and B are $p \geq m$ times continuously differentiable with respect to $x, \;\; \varphi$, and τ for every fixed ε, and all partial derivatives of these functions are uniformly bounded in $\overline{G}$ by a constant c_6 independent of ε. Assume that A and B belong to the class of functions almost periodic in $\varphi_j, \;\; j = \overline{1, m}$,

$$[A(x, \varphi, \tau, \varepsilon); B(x, \varphi, \tau, \varepsilon)] = \sum_{\nu=0}^{\infty} [A_\nu(x, \tau, \varepsilon); B_\nu(x, \tau, \varepsilon)] \exp\{i(\lambda_\nu, \varphi)\},$$

$\lambda_0 = 0, \;\; \lambda_\nu \neq 0$ for $\nu \geq 1$, and

$$\sum_{\nu=1}^{\infty} \left[\left(1 + \frac{1}{\|\lambda_\nu\|} \right) \sup_G \|C_\nu\| + \frac{1}{\|\lambda_\nu\|} \left(\sup_G \left\| \frac{\partial C_\nu}{\partial \tau} \right\| + \sup_G \left\| \frac{\partial C_\nu}{\partial x} \right\| \right) \right] \leq c_6. \tag{5.13}$$

Here, $C_\nu = [A_\nu(x, \tau, \varepsilon); B_\nu(x, \tau, \varepsilon)]$ and (λ_ν, φ) is the scalar product of the vectors $(\lambda_\nu^{(1)}, \ldots, \lambda_\nu^{(m)})$ and $(\varphi_1, \ldots, \varphi_m)$.

The system averaged with respect to φ has the form

$$\frac{d\overline{x}}{d\tau} = a(\overline{x}, \tau, \varepsilon) + \varepsilon A_0(\overline{x}, \tau, \varepsilon), \quad \frac{d\overline{\varphi}}{d\tau} = \frac{\omega(\overline{x}, \tau, \varepsilon)}{\varepsilon} + B_0(\overline{x}, \tau, \varepsilon), \tag{5.14}$$

where

$$[A_0; B_0] = \lim_{t \to \infty} t^{-m} \int_0^t \ldots \int_0^t [A(\overline{x}, \varphi, \tau, \varepsilon); B(\overline{x}, \varphi, \tau, \varepsilon)] d\varphi_1 \ldots d\varphi_m.$$

As for frequencies, we assume that

$$\|(W_p^T(\overline{x},\tau,\varepsilon)W_p(\overline{x},\tau,\varepsilon))^{-1}W_p^T(\overline{x},\tau,\varepsilon)\| \le c_7 \quad \forall(\overline{x},\tau,\varepsilon)\in G, \tag{5.15}$$

where

$$W_p^T(\overline{x},\tau,\varepsilon) = \Big(\frac{d^{j-1}}{d\tau^{j-1}}\,\omega_\nu(\overline{x},\tau,\varepsilon)\Big)_{j,\nu=1}^{m,p}$$

and the total derivatives of the functions $\omega_\nu(\overline{x},\tau,\varepsilon)$ with respect to τ are calculated with regard for the equation $\dfrac{d\overline{x}}{d\tau} = a(\overline{x},\tau,\varepsilon)$.

We also assume that there exists a solution $\overline{x} = \overline{x}(\tau,\varepsilon)$ of the averaged equations (5.14) for slow variables that is defined and lies in $\mathcal{D}$ together with its ρ-neighborhood $\forall(\tau,\varepsilon)\in R_+\times(0,\varepsilon_0]$, and the normal fundamental matrix $Q(\tau,t,\varepsilon)$ of solutions of the variational system

$$\frac{dz}{d\tau} = H(\tau,\varepsilon)z, \quad H(\tau,\varepsilon) = \frac{\partial}{\partial x}[a(\overline{x}(\tau,\varepsilon),\tau,\varepsilon) + \varepsilon A(\overline{x}(\tau,\varepsilon),\tau,\varepsilon)],$$

satisfies the estimate

$$\|Q(\tau,t,\varepsilon)\| \le K\varepsilon^{-l_1}e^{-\gamma\varepsilon^{l_2}(\tau-t)} \quad \forall\tau\ge t\ge 0, \quad \varepsilon\in(0,\varepsilon_0], \tag{5.16}$$

where $K>0$, $\gamma>0$, $r_1\ge 0$, and $r_2\ge 0$ are certain constants independent of ε.

Theorem 5.3 [PeP]. *If conditions (5.13), (5.15), and (5.16) for* $l = l_1 + l_2 < \dfrac{1}{2} + \dfrac{1}{2p}$ *are satisfied, then the solution* $(x(\tau,\overline{x}(0,\varepsilon),\psi,\varepsilon);\varphi(\tau,\overline{x}(0,\varepsilon),\psi,\varepsilon))$ *of system (5.12) is defined for all* $\tau\in R_+$, $\psi\in R^m$, *and* $\varepsilon\in(0,\varepsilon_0]$ (ε_0 *is sufficiently small) and satisfies the following inequalities:*

$$\|x(\tau,\overline{x}(0,\varepsilon),\psi,\varepsilon) - \overline{x}(\tau,\varepsilon)\| \le c_8\varepsilon^{1+\frac{1}{p}-l}, \quad c_8 = \text{const},$$

$$\|\varphi(\tau,\overline{x}(0,\varepsilon),\psi,\varepsilon) - \overline{\varphi}(\tau,\overline{x}(0,\varepsilon),\psi,\varepsilon)\| \le c_8(1+\tau)\varepsilon^{\frac{1}{p}-l}.$$

2. AVERAGING METHOD IN MULTIPOINT PROBLEMS

6. Boundary-Value Problems for Oscillation Systems with Frequencies Dependent on Time Variable

Consider the multifrequency system

$$\frac{dx}{d\tau} = a(x, \varphi, \tau, \varepsilon), \quad \frac{d\varphi}{d\tau} = \frac{\omega(\tau)}{\varepsilon} + b(x, \varphi, \tau, \varepsilon) \tag{6.1}$$

whose right-hand side is defined in $\overline{G} = \mathcal{D} \times R^m \times [0, L] \times (0, \varepsilon_0]$. For this system, we introduce the boundary conditions

$$x|_{\tau=0} = y \in \mathcal{D}_1, \quad \varphi|_{\tau=L} = f(x|_{\tau=0}, x|_{\tau=L}, \varepsilon), \tag{6.2}$$

where $f(y, z, \varepsilon)$ is a known m-dimensional vector function of the variables $(y, z, \varepsilon) \in \mathcal{D}_1 \times \mathcal{D} \times (0, \varepsilon_0] \equiv A$ and $\mathcal{D}_1$ is a certain domain $(\mathcal{D}_1 \subset \mathcal{D})$.

Parallel with (6.1), (6.2), we consider the following boundary-value problem averaged over all angular variables φ:

$$\frac{d\overline{x}}{d\tau} = \overline{a}(\overline{x}, \tau, \varepsilon), \quad \overline{x}|_{\tau=0} = y, \tag{6.3}$$

$$\frac{d\overline{\varphi}}{d\tau} = \frac{\omega(\tau)}{\varepsilon} + \overline{b}(\overline{x}, \tau, \varepsilon), \quad \overline{\varphi}|_{\tau=L} = f(\overline{x}|_{\tau=0}, \overline{x}|_{\tau=L}, \varepsilon). \tag{6.4}$$

It is obvious that the solution of problem (6.3), (6.4) is much simpler than the solution of problem (6.1), (6.2) because problem (6.3), (6.4) decomposes into two Cauchy problems. If problem (6.3) has a solution $\overline{x} = \overline{x}(\tau, y, \varepsilon)$ defined and lying in $\mathcal{D}$ $\forall (\tau, y, \varepsilon) \in [0, L] \times \mathcal{D}_1 \times (0, \varepsilon_0]$, then a solution $\overline{\varphi} = \overline{\varphi}(\tau, y, \psi^0, \varepsilon)$

of problem (6.4) is given by the formulas

$$\overline{\varphi}(\tau, y, \psi^0, \varepsilon) = \psi^0 + \frac{1}{\varepsilon}\int\limits_0^{\tau} [\omega(t) + \varepsilon\overline{b}(\overline{x}(t, y, \varepsilon), t, \varepsilon)]\, dt,$$

$$\psi^0 = -\frac{1}{\varepsilon}\int\limits_0^{L} [\omega(t) + \varepsilon\overline{b}(\overline{x}(t, y, \varepsilon), t, \varepsilon)]dt + f(y, \overline{x}(L, y, \varepsilon), \varepsilon).$$

In the next section, we use the following theorem for the justification of the averaging method on the entire axis:

Theorem 6.1. *Suppose that system (6.1) satisfies all conditions of Theorem 2.2 and the function* $f(y, z, \varepsilon)$ *is continuously differentiable with respect to* $z \in \mathcal{D}$ *for every fixed* $y \in \mathcal{D}_1$ *and* $\varepsilon \in (0, \varepsilon_0]$ *and such that*

$$\sup_{z\in\mathcal{D}} \|f(y, z, \varepsilon)\| < \infty, \qquad \sup_{(y,z,\varepsilon)\in A} \left\|\frac{\partial}{\partial z} f(y, z, \varepsilon)\right\| < \infty. \tag{6.5}$$

Then there exists a unique solution $(x(\tau, y, \psi, \varepsilon); \varphi(\tau, y, \psi, \varepsilon))$ *of the boundary-value problem (6.1), (6.2), and, furthermore, this solution satisfies the inequality*

$$\|x(\tau, y, \psi, \varepsilon) - \overline{x}(\tau, y, \varepsilon)\| + \|\varphi(\tau, y, \psi, \varepsilon) - \overline{\varphi}(\tau, y, \psi^0, \varepsilon)\| \le c_1 \varepsilon^{\frac{1}{p}} \tag{6.6}$$

$$\forall (\tau, y, \varepsilon) \in [0, L] \times \mathcal{D}_1 \times (0, \varepsilon_0],$$

where ε_0 *is positive and sufficiently small.*

Proof. According to Theorem 2.1, the solution $(x(\tau, y, \psi, \varepsilon); \varphi(\tau, y, \psi, \varepsilon))$ of system (6.1) is defined for all $\tau \in [0, L]$, $y \in \mathcal{D}_1$, $\psi \in R^m$, and $\varepsilon \in (0, \varepsilon_0]$. Therefore, we seek a solution of the boundary-value problem (6.1), (6.2) in the form $x = x(\tau, y, \psi, \varepsilon)$, $\varphi = \varphi(\tau, y, \psi, \varepsilon)$, where $\psi = \psi(y, \varepsilon)$ is unknown. To determine ψ, we substitute this solution into the boundary conditions (6.2). As a result, we get

$$\varphi\,(L, y, \psi, \varepsilon) = f\,(y, x\,(L, y, \psi, \varepsilon), \varepsilon),$$

or

$$\psi = f(y, x(L, y, \psi, \varepsilon), \varepsilon) - \frac{1}{\varepsilon}\int\limits_0^{L} [\omega(t) + \varepsilon\overline{b}(\overline{x}(t, y, \varepsilon), t, \varepsilon)]dt - \Delta\varphi(L, y, \psi, \varepsilon)$$

$$\equiv \Phi\,(y, \psi, \varepsilon), \tag{6.7}$$

where

$$\Delta\varphi(\tau, y, \psi, \varepsilon) = \varphi(\tau, y, \psi, \varepsilon) - \overline{\varphi}(\tau, y, \psi, \varepsilon).$$

It follows from the first inequality in (6.5) that there exists a constant $c_2 = c_2(y, \varepsilon)$ such that

$$\|f(y, z, \varepsilon)\| \leq c_2(y, \varepsilon) \quad \forall z \in \mathcal{D}$$

for every fixed $y \in \mathcal{D}_1$ and $\varepsilon \in (0, \varepsilon_0]$. Furthermore, Theorem 2.1 yields

$$\|\Delta\varphi(\tau, y, \psi, \varepsilon)\| \leq c_3\varepsilon^{\frac{1}{p}} \quad \forall(\tau, y, \psi, \varepsilon) \in [0, L] \times \mathcal{D}_1 \times R^m \times (0, \varepsilon_0],$$

where $c_3 = \max\{\sigma_2; \sigma_3\}$, and σ_2 and σ_3 are the constants defined by Theorems 2.1 and 2.2. Thus,

$$\begin{aligned}\|\Phi(y, \psi, \varepsilon)\| &\leq c_2(y, \varepsilon) + \frac{1}{\varepsilon}L\Big[\max_{[0,L]}\|\omega(\tau)\| + \varepsilon\sup\|\overline{b}(x, \tau, \varepsilon)\|\Big] + c_3\varepsilon^{\frac{1}{p}} \\ &\equiv c_4(y, \varepsilon).\end{aligned}$$

Therefore, for every fixed $y \in \mathcal{D}_1$ and $\varepsilon \in (0, \varepsilon_0]$, the function $\Phi(y, \psi, \varepsilon)$ maps the set $\psi \in R^m$ into the set $T = \{\psi\colon \psi \in R^m, \|\psi\| \leq c_4(y, \varepsilon)\}$. Moreover, according to (6.5) and Theorem 2.2, we get

$$\begin{aligned}\left\|\frac{\partial}{\partial\psi}\Phi(y, \psi, \varepsilon)\right\| &\leq \sup_A\left\|\frac{\partial}{\partial z}f(y, z, \varepsilon)\right\|\left\|\frac{\partial}{\partial\psi}\left(x(L, y, \psi, \varepsilon) - \overline{x}(L, y, \varepsilon)\right)\right\| \\ &\quad + \left\|\frac{\partial}{\partial\psi}\Delta\varphi(L, y, \psi, \varepsilon)\right\| \\ &\leq c_3\varepsilon^{\frac{1}{p}}\left(1 + \sup_A\left\|\frac{\partial}{\partial z}f(y, z, \varepsilon)\right\|\right) \leq \frac{1}{2}\end{aligned}$$

for

$$\varepsilon_0 \leq \left[2c_3\left(1 + \sup_A\left\|\frac{\partial}{\partial z}f(y, z, \varepsilon)\right\|\right)\right]^{-p}. \tag{6.8}$$

Thus, the equation $\psi = \Phi(y, \psi, \varepsilon)$ has the unique solution $\psi = \psi(y, \varepsilon) \in R^m$, and the boundary-value problem (6.1), (6.2) has the unique solution $(x(\tau, y, \psi(y, \varepsilon), \varepsilon); \varphi(\tau, y, \psi(y, \varepsilon), \varepsilon))$.

It remains to prove estimate (6.6). Using Theorem 2.1 and Eq. (6.7), we obtain the inequality

$$\begin{aligned}\|\psi(y,\varepsilon)-\psi^0\| &\leq \|f(y,x(L,y,\psi(y,\varepsilon),\varepsilon),\varepsilon)-f(y,\overline{x}(L,y,\varepsilon),\varepsilon)\| \\ &\quad + \|\Delta\varphi(L,y,\psi(y,\varepsilon),\varepsilon)\| \\ &\leq \Big(1+\sup_A\Big\|\frac{\partial}{\partial z}f(y,z,\varepsilon)\Big\|\Big)c_3\varepsilon^{\frac{1}{p}} \equiv c_5\varepsilon^{\frac{1}{p}},\end{aligned}$$

which yields

$$\begin{aligned}&\|\varphi(\tau,y,\psi(y,\varepsilon),\varepsilon)-\overline{\varphi}(\tau,y,\psi^0,\varepsilon)\| \\ &\quad\leq \|\Delta\varphi(\tau,y,\psi(y,\varepsilon),\varepsilon)\| + \|\overline{\varphi}(\tau,y,\psi(y,\varepsilon),\varepsilon)-\overline{\varphi}(\tau,y,\psi^0,\varepsilon)\| \\ &\quad\leq c_3\varepsilon^{\frac{1}{p}} + \|\psi(y,\varepsilon)-\psi^0\| \leq (c_3+c_5)\varepsilon^{\frac{1}{p}} \\ &\qquad\qquad \forall(\tau,y,\varepsilon)\in[0,L]\times\mathcal{D}_1\times(0,\varepsilon_0]\,.\end{aligned}$$

Combining the last inequality and the inequality

$$\|x(\tau,y,\psi(y,\varepsilon),\varepsilon)-\overline{x}(\tau,y,\varepsilon)\| \leq c_3\varepsilon^{\frac{1}{p}},$$

we obtain estimate (6.6) for $c_1 = 2c_3 + c_5$. The restrictions for ε_0 are specified by Theorems 2.1 and 2.2 and condition (6.8). Theorem 6.1 is proved.

Remark 1. Assume that, the function $f(y,z,\varepsilon)$ in the boundary conditions (6.2) is independent of y, i.e., $f(y,z,\varepsilon)\equiv\widetilde{f}(z,\varepsilon)$. Then, differentiating (6.7) with respect to y, we get

$$\begin{aligned}\frac{\partial\psi}{\partial y} &= \Big[E_m - \frac{\partial}{\partial\psi}\Delta\varphi(L,y,\psi,\varepsilon) - \frac{\partial}{\partial x}\widetilde{f}\left(x(L,y,\psi,\varepsilon),\varepsilon\right)\frac{\partial}{\partial\psi}x(L,y,\psi,\varepsilon)\Big]^{-1} \\ &\quad\times\Big[\frac{\partial}{\partial x}\widetilde{f}\left(x(L,y,\psi,\varepsilon),\varepsilon\right)\frac{\partial}{\partial y}x\left(L,y,\psi,\varepsilon\right) \\ &\quad + \frac{\partial}{\partial y}\Delta\varphi(L,y,\psi,\varepsilon) - \int_0^L \frac{\partial}{\partial x}\overline{b}(\overline{x}(t,y,\varepsilon),t,\varepsilon)\frac{\partial}{\partial y}\overline{x}\left(t,y,\varepsilon\right)dt\Big],\end{aligned}$$

which yields

$$\left\|\frac{\partial\psi(y,\varepsilon)}{\partial y}\right\| \leq 2m\left[\left\|\frac{\partial}{\partial x}\widetilde{f}(x(L,y,\psi,\varepsilon),\varepsilon)\right\| \left\|\frac{\partial}{\partial y}\overline{x}\,(L,y,\varepsilon)\right\| + \overline{c}_3\varepsilon^{\frac{1}{p}} + L\sup\left\|\frac{\partial}{\partial x}\overline{b}(x,\tau,\varepsilon)\right\| \sup\left\|\frac{\partial}{\partial y}\overline{x}\,(\tau,y,\varepsilon)\right\|\right] \tag{6.9}$$

for

$$\varepsilon_0 \leq \left[2c_3\left(1+\sup\left\|\frac{\partial\widetilde{f}(x,\varepsilon)}{\partial x}\right\|\right)\right]^{-p}, \quad \overline{c}_3 = c_3\left(1+\sup\left\|\frac{\partial\widetilde{f}(x,\varepsilon)}{\partial x}\right\|\right).$$

We now consider more general [as compared with (6.2)] boundary conditions of the form [VaB]

$$F(x|_{\tau=0},\varphi|_{\tau=0},x|_{\tau=L},\varphi|_{\tau=L},\varepsilon) = 0, \tag{6.10}$$

where $F(y,\psi,z,\theta,\varepsilon)$ is an $(n+m)$-dimensional vector function. Problem (6.2), (6.10) is a two-point boundary-value problem that contains slow and fast variables and possesses resonance properties. Note that there is a fairly complete theory of singularly perturbed boundary-value problems (see [VaD]), which is based on the method of boundary-layer functions developed in [VaB].

Assume that the following conditions are satisfied:

(a) for every $\varepsilon \in (0,\varepsilon_0]$, the averaged boundary-value problem

$$\frac{d\overline{x}}{d\tau} = \overline{a}(\overline{x},\tau,\varepsilon), \quad \frac{d\overline{\varphi}}{d\tau} = \frac{\omega(\tau)}{\varepsilon} + \overline{b}(\overline{x},\tau,\varepsilon),$$
$$F(\overline{x}|_{\tau=0},\overline{\varphi}|_{\tau=0},\overline{x}|_{\tau=L},\overline{\varphi}|_{\tau=L},\varepsilon) = 0 \tag{6.11}$$

has a unique solution

$$(\overline{x}(\tau,x^0(\varepsilon),\varepsilon);\overline{\varphi}(\tau,x^0(\varepsilon),\varphi^0(\varepsilon),\varepsilon) \equiv (\overline{x}(\tau,\varepsilon);\overline{\varphi}(\tau,\varepsilon)),$$

which lies in $\mathcal{D}\times R^m$ together with its ρ_1-neighborhood;

(b) there exist constants $c_6 > 0$ and $c_7 > 0$ independent of ε and such that

$$F(y,\psi,z,\theta,\varepsilon) \in C^2_{y,\psi,z,\theta}(B,c_6), \quad B = \overline{B}\times(0,\varepsilon_0],$$

where $\overline{B}$ denotes the c_7-neighborhood of the point $(x^0(\varepsilon),\varphi^0(\varepsilon), \overline{x}(L,\varepsilon),\overline{\varphi}(L,\varepsilon)) \in R^{2(n+m)}$;

(c) $\|S^{-1}(x^0(\varepsilon), \varphi^0(\varepsilon), \varepsilon)\| \le c_8 = \text{const} \;\; \forall \varepsilon \in (0, \varepsilon_0]$, where S is the quadratic $(n+m)$-dimensional matrix defined by the equality

$$S(x^0(\varepsilon), \varphi^0(\varepsilon), \varepsilon)$$

$$= \left(\frac{\partial F^0}{\partial y} + \frac{\partial F^0}{\partial z} \frac{\partial \overline{x}(L, x^0(\varepsilon), \varepsilon)}{\partial x^0} \right.$$

$$+ \frac{\partial F^0}{\partial \theta} \int\limits_0^L \frac{\partial}{\partial x} \overline{b}(\overline{x}(t, x^0(\varepsilon), \varepsilon), t, \varepsilon) \frac{\partial \overline{x}(t, x^0(\varepsilon), \varepsilon)}{\partial x^0} dt,$$

$$\left. \frac{\partial F^0}{\partial \psi} + \frac{\partial F^0}{\partial \theta} \right).$$

In this case, the values of the derivatives of the function $F(y, \psi, z, \theta, \varepsilon)$ are taken for $y = x^0(\varepsilon)$, $\varphi = \varphi^0(\varepsilon)$, $z = \overline{x}(L, x^0(\varepsilon), \varepsilon)$, and $\theta = \overline{\varphi}(L, x^0(\varepsilon), \varphi^0(\varepsilon), \varepsilon)$.

Theorem 6.2. *Suppose that the following conditions are satisfied:*

(i) $\omega(\tau) \in C^{p-1}_{[0,L]}$, $p \ge m$, *and* $\det(W_p^T(\tau) W_p(\tau)) \ne 0 \;\; \forall \tau \in [0, L]$;

(ii) conditions (a)–(c) and inequality (2.6) are satisfied.

Then, for every $\varepsilon \in (0, \varepsilon_0]$ *(*ε_0 *is sufficiently small), the boundary-value problem (6.1), (6.10) has a unique solution* $(x(\tau, \varepsilon); \varphi(\tau, \varepsilon))$, *which lies in a* $c_9 \varepsilon^{\frac{1}{p}}$*-neighborhood of the solution* $(\overline{x}(\tau, \varepsilon); \overline{\varphi}(\tau, \varepsilon))$ *of problem (6.11).*

Proof. According to condition (a), the curve $\overline{x} = \overline{x}(\tau, x^0, \varepsilon)$ lies in $\mathcal{D}$ together with its ρ_1-neighborhood $\forall (\tau, \varepsilon) \in [0, L] \times (0, \varepsilon_0]$. We now determine from which set one must choose $\widetilde{x}$ in order that the curve $\overline{x} = \overline{x}(\tau, x^0 + \widetilde{x}, \varepsilon)$ belong to $\mathcal{D}$ together with its $\dfrac{1}{2}\rho_1$-neighborhood. Using the averaged equations for slow variables, we get

$$\|x(\tau, x^0 + \widetilde{x}, \varepsilon) - \overline{x}(\tau, x^0, \varepsilon)\|$$

$$\le \|\widetilde{x}\| + \int\limits_0^\tau \|\overline{x}(t, x^0 + \widetilde{x}, \varepsilon) - \overline{x}(t, x^0, \varepsilon)\| \sup_{\overline{G}} \left\| \frac{\partial}{\partial x} \overline{a}(x, \tau, \varepsilon) \right\| dt$$

or

$$\|x(\tau, x^0 + \widetilde{x}, \varepsilon) - \overline{x}(\tau, x^0, \varepsilon)\| \leq \|\widetilde{x}\| e^{L\sigma_1} \quad \forall(\tau, \varepsilon) \in [0, L] \times (0, \varepsilon_0].$$

Here, σ_1 is the constant defined by inequality (2.6). The condition $\|\widetilde{x}\| \leq c_{10} = \frac{1}{2}\rho_1 e^{-L\sigma_1}$ guarantees that $\overline{x} = \overline{x}(\tau, x^0 + \widetilde{x}, \varepsilon)$ lies in $\mathcal{D}$ together with its $\frac{1}{2}\rho_1$-neighborhood. For every solution $(\overline{x}(\tau, x^0 + \widetilde{x}, \varepsilon); \overline{\varphi}(\tau, x^0 + \widetilde{x}, \varphi^0 + \widetilde{\psi}, \varepsilon))$, $\|\widetilde{x}\| \leq c_{10}$, $\widetilde{\psi} \in R^m$, of the averaged equations, we write the representation

$$\overline{x}(\tau, x^0 + \widetilde{x}, \varepsilon) = \overline{x}(\tau, x^0, \varepsilon) + \frac{\partial \overline{x}(\tau, x^0, \varepsilon)}{\partial x^0}\widetilde{x} + X(\tau, \widetilde{x}, \varepsilon),$$

$$\begin{aligned} &\overline{\varphi}(\tau, x^0 + \widetilde{x}, \varphi^0 + \widetilde{\psi}, \varepsilon) \\ &\qquad = \overline{\varphi}(\tau, x^0, \varphi^0, \varepsilon) + \int\limits_0^\tau \frac{\partial}{\partial x}\overline{b}(\overline{x}(t, x^0, \varepsilon), t, \varepsilon)\frac{\partial \overline{x}(\tau, x^0, \varepsilon)}{\partial x^0}\, dt\, \widetilde{x} \\ &\qquad\qquad + \widetilde{\psi} + Y(\tau, \widetilde{x}, \varepsilon), \end{aligned}$$

where [according to conditions (2.6)] the functions X and Y satisfy the following inequality for all $\tau \in [0, L]$, $\|\widetilde{x}\| \leq c_{10}$, and $\varepsilon \in (0, \varepsilon_0]$:

$$\|X(\tau, \widetilde{x}, \varepsilon)\| + \|Y(\tau, \widetilde{x}, \varepsilon)\| \leq c_{11}\|\widetilde{x}\|^2,$$

where the constant c_{11} is independent of τ, $\widetilde{x}$, and ε.

We seek a solution of problem (6.1), (6.10) in the form

$$x(\tau, \varepsilon) = x(\tau, x^0 + \widetilde{x}, \varphi^0 + \widetilde{\psi}, \varepsilon), \quad \varphi(\tau, \varepsilon) = \varphi(\tau, x^0 + \widetilde{x}, \varphi^0 + \widetilde{\psi}, \varepsilon),$$

where the unknown parameters $\widetilde{x}$, $\|\widetilde{x}\| \leq c_{10}$, and $\widetilde{\psi} \in R^m$ can be determined from the boundary conditions (6.10). After the substitution of the solution thus chosen in (6.10), we obtain

$$\begin{aligned} F\Big(&x^0 + \widetilde{x}, \varphi^0 + \widetilde{\psi}, \overline{x}(L, x^0, \varepsilon) + \frac{\partial \overline{x}(L, x^0, \varepsilon)}{\partial x^0}\,\widetilde{x} \\ &+ X(L, x^0, \varepsilon) + \Delta x(L, x^0 + \widetilde{x}, \varphi^0 + \widetilde{\psi}, \varepsilon), \overline{\varphi}(L, x^0, \varphi^0, \varepsilon) \\ &+ \widetilde{\psi} + Y(L, x^0, \varepsilon) + \Delta\varphi(L, x^0 + \widetilde{x}, \varphi^0 + \widetilde{\psi}, \varepsilon) \\ &+ \int\limits_0^L \frac{\partial}{\partial x}\overline{b}(\overline{x}(t, x^0, \varepsilon), t, \varepsilon)\frac{\partial}{\partial x^0}\overline{x}\,(t, x^0, \varepsilon)\, dt\, \widetilde{x}, \varepsilon\Big) = 0, \end{aligned} \tag{6.12}$$

where

$$\Delta x\,(\tau, y, \psi, \varepsilon) = x\,(\tau, y, \psi, \varepsilon) - \overline{x}\,(\tau, y, \varepsilon),$$

$$\Delta \varphi\,(\tau, y, \psi, \varepsilon) = \varphi\,(\tau, y, \psi, \varepsilon) - \overline{\varphi}\,(\tau, y, \psi, \varepsilon).$$

Taking into account condition (b), we specify admissible values of $\widetilde{x}$ and $\widetilde{\psi}$ for staying within the domain of definition of the function F. Since Δx and $\Delta\varphi$ satisfy estimate (2.5), it suffices to impose the restrictions

$$\|\widetilde{x}\|\Big(1 + \sup\Big\|\frac{\partial}{\partial x^0}\overline{x}(\tau, x^0, \varepsilon)\Big\|\Big(1 + L\sup\Big\|\frac{\partial}{\partial x}\overline{b}(x, \tau, \varepsilon)\Big\|\Big)\Big)$$
$$+ 2\|\widetilde{\psi}\| + \|\Delta x\| + \|\Delta\varphi\| + \|X\| + \|Y\| < c_7$$

or

$$\|\widetilde{x}\| + \|\widetilde{\psi}\| \le c_{12} = \min\Big\{\frac{1}{c_{11}}; c_{10}; \frac{1}{2}c_7[3 + (1 + L\sigma_1)ne^{\sigma_1 L}]^{-1}\Big\}$$

for $\varepsilon_0 \le c_7^p(2c_3)^{-p}$.

For such $(\widetilde{x}; \widetilde{\psi}) = \xi$, we expand the function F on the left-hand side of (6.12) into a Taylor series, taking into account the smoothness condition (b). After obvious transformations, we get

$$\xi = S^{-1}(x^0(\varepsilon), \varphi^0(\varepsilon), \varepsilon)\widetilde{F}\,(\xi, \varepsilon), \tag{6.13}$$

where $\widetilde{F}(\xi, \varepsilon)$ is defined for any $\varepsilon \in (0, \varepsilon_0]$ and all ξ satisfying the inequality $\|\xi\| \le c_{12}$ and

$$\|\widetilde{F}(\xi, \varepsilon)\| \le c_{13}(\|\xi\|^2 + \varepsilon^{\frac{1}{p}}), \quad \Big\|\frac{\partial}{\partial \xi}\widetilde{F}(\xi, \varepsilon)\Big\| \le c_{13}(\|\xi\| + \varepsilon^{\frac{1}{p}}). \tag{6.14}$$

Here, c_{13} is a constant independent of ξ and ε. The presence of the term $\varepsilon^{\frac{1}{p}}$ on the right-hand sides of the inequalities is a consequence of estimates (2.5) and (2.7) for Δx and $\Delta\varphi$ and their derivatives with respect to $\widetilde{x}$ and $\widetilde{\psi}$.

Condition (c) imposed on the matrix S and the first inequality in (6.14) guarantee that $S^{-1}(x^0(\varepsilon), \varphi^0(\varepsilon), \varepsilon)\widetilde{F}(\xi, \varepsilon)$ maps the set $M_\varepsilon = \{\xi \colon \xi \in R^{n+m}, \|\xi\| \le 2c_8c_{13}\varepsilon^{\frac{1}{p}}\}$ into itself for $\varepsilon \in (0, \varepsilon_0]$,

$$\varepsilon_0 \le \Big[\min\Big\{\frac{1}{4c_8^2c_{13}^2}; \frac{c_{12}}{2c_8c_{13}}\Big\}\Big]^p.$$

Moreover, the second inequality in (6.14) yields

$$\left\| \frac{\partial}{\partial \xi}(S^{-1}(x^0(\varepsilon), \varphi^0(\varepsilon), \varepsilon)\widetilde{F}(\xi, \varepsilon)) \right\| \leq c_8 c_{13}(\|\xi\| + \varepsilon^{\frac{1}{p}}) \leq \frac{1}{2}$$

for all $\xi \in M_\varepsilon$ and $\varepsilon \in (0, \varepsilon_0]$, $\varepsilon_0 \leq [2c_8c_{13}(1 + 2c_8c_{13})]^{-p}$. By virtue of the fixed-point theorem, for every $\varepsilon \in (0, \varepsilon_0]$ Eq. (6.13) has a unique solution $\xi = \xi(\varepsilon) = (\widetilde{x}(\varepsilon); \widetilde{\psi}(\varepsilon)) \in M_\varepsilon$, and the boundary-value problem (6.1), (6.10) has the unique solution

$$x(\tau, \varepsilon) = x(\tau, x^0(\varepsilon) + \widetilde{x}(\varepsilon), \varphi^0(\varepsilon) + \widetilde{\psi}(\varepsilon), \varepsilon), \varphi(\tau, \varepsilon)$$
$$= \varphi(\tau, x^0(\varepsilon) + \widetilde{x}(\varepsilon), \varphi^0(\varepsilon) + \widetilde{\psi}(\varepsilon), \varepsilon),$$

whose initial data $x^0(\varepsilon) + \widetilde{x}(\varepsilon), \varphi^0(\varepsilon) + \widetilde{\psi}(\varepsilon)$ lie in the $2c_8c_{13}\varepsilon^{\frac{1}{p}}$-neighborhood of the initial data $(x^0(\varepsilon); \varphi^0(\varepsilon))$ of the solution of the averaged boundary-value problem (6.11). Also note that, according to Theorem 2.1 and conditions (2.6), the following inequalities are true:

$$\|x(\tau, \varepsilon) - \overline{x}(\tau, \varepsilon)\| + \|\varphi(\tau, \varepsilon) - \overline{\varphi}(\tau, \varepsilon)\|$$
$$\leq \|x(\tau, x^0 + \widetilde{x}, \varphi^0 + \widetilde{\psi}, \varepsilon) - \overline{x}(\tau, x^0 + \widetilde{x}, \varepsilon)\|$$
$$+ \|\varphi(\tau, x^0 + \widetilde{x}, \varphi^0 + \widetilde{\psi}, \varepsilon) - \overline{\varphi}(\tau, x^0 + \widetilde{x}, \varphi^0 + \widetilde{\psi}, \varepsilon)\|$$
$$+ \|\overline{x}(\tau, x^0 + \widetilde{x}, \varepsilon) - \overline{x}(\tau, x^0, \varepsilon)\|$$
$$+ \|\overline{\varphi}(\tau, x^0 + \widetilde{x}, \varphi^0 + \widetilde{\psi}, \varepsilon) - \overline{\varphi}(\tau, x^0, \varphi^0, \varepsilon)\|$$
$$\leq \{c_3 + [(1 + L\sigma_1)ne^{\sigma_1 L} + 1]2c_8c_{13}\}\varepsilon^{\frac{1}{p}} \equiv c_9\varepsilon^{\frac{1}{p}}$$
$$\forall (\tau, \varepsilon) \in [0, L] \times (0, \varepsilon_0].$$

Theorem 6.2 is proved.

Example. Consider the boundary-value problem

$$\frac{dx}{d\tau} = x(1 + x^2\cos(\varphi_1 - \varphi_2) + \cos\varphi_2),$$
$$\frac{d\varphi_1}{d\tau} = \frac{\tau}{\varepsilon} + x^2\cos\varphi_2, \quad \frac{d\varphi_2}{d\tau} = \frac{\tau^2}{\varepsilon} + x^3\sin\varphi_1, \tag{6.15}$$
$$x|_{\tau=0} + x|_{\tau=1} = 1, \quad \varphi_1|_{\tau=0} + x|_{\tau=1} = 0, \quad \varphi_2|_{\tau=1} + x|_{\tau=0} = 0$$

and the corresponding problem averaged with respect to φ_1 and φ_2:

$$\frac{d\overline{x}}{d\tau} = \overline{x}, \quad \frac{d\overline{\varphi}_1}{d\tau} = \frac{\tau}{\varepsilon}, \quad \frac{d\overline{\varphi}_2}{d\tau} = \frac{\tau^2}{\varepsilon},$$

$$\overline{x}|_{\tau=0} + \overline{x}|_{\tau=1} = 1, \quad \overline{\varphi}_1|_{\tau=0} + \overline{x}|_{\tau=1} = 0, \quad \overline{\varphi}_2|_{\tau=1} + \overline{x}|_{\tau=0} = 0.$$

It can easily be verified that the last problem has the unique solution

$$\overline{x}(\tau,\varepsilon) = \frac{e^\tau}{1+e},$$

$$\overline{\varphi}_1(\tau,\varepsilon) = \frac{\tau^2}{2\varepsilon} - \frac{e}{e+1}, \quad \overline{\varphi}_2(\tau,\varepsilon) = \frac{\tau^3-1}{3\varepsilon} - \frac{1}{e+1},$$

and, for this solution, we have

$$S(x^0(\varepsilon),\varphi^0(\varepsilon),\varepsilon) = \begin{pmatrix} 1+e & 0 & 0 \\ e & 1 & 0 \\ 1 & 0 & 1 \end{pmatrix},$$

$$\|S^{-1}(x^0(\varepsilon),\varphi^0(\varepsilon),\varepsilon)\| = 3 + \frac{1}{e+1}.$$

Since $\det(W_3^T(\tau)W_3(\tau)) = (\tau^2+2)^2 \neq 0 \quad \forall\tau \in [0,1]$, all conditions of Theorem 6.2 are satisfied. Thus, for every sufficiently small $\varepsilon > 0$, there exists a unique solution $(x(\tau,\varepsilon); \varphi_1(\tau,\varepsilon); \varphi_2(\tau,\varepsilon))$ of problem (6.15) that satisfies the inequality

$$\left|x(\tau,\varepsilon) - \frac{e^\tau}{e+1}\right| + \left|\varphi_1(\tau,\varepsilon) - \frac{\tau^2}{2\varepsilon} + \frac{e}{e+1}\right| + \left|\varphi_2(\tau,\varepsilon) - \frac{\tau^3-1}{3\varepsilon} + \frac{1}{e+1}\right| \leq c_9\varepsilon^{\frac{1}{3}}$$

for all $\tau \in [0,1]$.

Finally, note that Theorem 6.1 guarantees the global uniqueness of a solution of the boundary-value problem (6.1), (6.2), whereas Theorem 6.2 establishes the uniqueness of a solution of problem (6.1), (6.10) only in a certain small neighborhood of the solution of the averaged problem (6.11).

7. Theorem on Justification of Averaging Method on Entire Axis

In this section, we establish the existence of a solution (defined on the entire axis) of an oscillation system using the combination of the averaging method on a segment and the solution of certain boundary-value problems. Note that, in this case, we do not use the method of integral manifolds, which requires additional restrictions on the equations of the system.

Consider the system of $n + m$ equations

$$\frac{dx}{d\tau} = a(x, \varphi, \tau, \varepsilon), \quad \frac{d\varphi}{d\tau} = \frac{\omega(\tau)}{\varepsilon} + b(x, \varphi, \tau, \varepsilon), \tag{7.1}$$

where the functions a, b, and ω are defined on the set $(x, \varphi, \tau, \varepsilon) \in \mathcal{D} \times R^m \times R \times [0, \varepsilon_0] \equiv \overline{G}$ $(R^n \supset \mathcal{D}$ is a bounded domain) and 2π-periodic in φ_ν, $\nu = \overline{1, m}$. For this system, we write the corresponding system of equations of the first approximation for slow variables averaged with respect to all angular variables φ, namely

$$\frac{d\overline{x}}{d\tau} = \overline{a}(\overline{x}, \tau, 0), \tag{7.2}$$

$$\overline{a}(\overline{x}, \tau, \varepsilon) = (2\pi)^{-m} \int_0^{2\pi} \ldots \int_0^{2\pi} a(\overline{x}, \varphi, \tau, \varepsilon) d\varphi_1 \ldots d\varphi_m.$$

In this section, we denote by $(x_\tau(t, y, \psi, \varepsilon); \varphi_\tau(t, y, \psi, \varepsilon))$ and $(\overline{x}_\tau(t, y, \varepsilon);$ $\overline{\varphi}_\tau(t, y, \psi, \varepsilon))$, respectively, the solutions of system (7.1) and the averaged system

$$\frac{d\overline{x}}{d\tau} = \overline{a}(\overline{x}, \tau, \varepsilon), \quad \frac{d\overline{\varphi}}{d\tau} = \frac{\omega(\tau)}{\varepsilon} + \overline{b}(\overline{x}, \tau, \varepsilon) \tag{7.3}$$

that take the value $(y; \psi)$ for $\tau = t$. Assume that $\overline{a}(x, \tau, \varepsilon)$ satisfies the inequality

$$\|\overline{a}(x, \tau, \varepsilon) - \overline{a}(x, \tau, 0)\| + \left\| \frac{\partial \overline{a}(x, \tau, \varepsilon)}{\partial x} - \frac{\partial \overline{a}(x, \tau, 0)}{\partial x} \right\| \leq \sigma_1 \varepsilon \tag{7.4}$$

$$\forall (x, \tau, \varepsilon) \in \mathcal{D} \times R \times [0, \varepsilon_0] = G.$$

Theorem 7.1. *Suppose that the following conditions are satisfied:*

(i) the function $c(x, \varphi, \tau, \varepsilon) = [a(x, \varphi, \tau, \varepsilon); b(x, \varphi, \tau, \varepsilon)]$ *is twice continuously differentiable with respect to* x, φ, *and* τ *for every fixed* ε, *and its Fourier coefficients* $c_k(x, \tau, \varepsilon)$ *satisfy inequalities (2.6) and (7.4);*

(ii) $\|(W_p^T(\tau) W_p(\tau))^{-1} W_p^T(\tau)\|$ *is uniformly bounded, and*

$$\omega_\nu^{(j-1)}(\tau), \quad \nu = \overline{1, m}, \quad j = \overline{1, p}, \quad p \geq m,$$

are uniformly continuous $\forall \tau \in R$;

(iii) there exists a solution $\overline{x} = \xi(\tau)$ *of the averaged equations of the first approximation (7.2) that is defined* $\forall\tau \in R$ *and lies in* $\mathcal{D}$ *together with its* ρ*-neighborhood;*

(iv) the normal fundamental matrix $Q(\tau,t)$ *of solutions of the variational equation* $\frac{dz}{d\tau} = \frac{\partial}{\partial x}\overline{a}(\xi(\tau),\tau,0)z$ *satisfies the estimate*

$$\|Q\,(\tau,t)\| \le Ke^{-\gamma(\tau-t)} \tag{7.5}$$

$$\forall\tau \ge t \in R, \quad K = \text{const} \ge 1, \quad \gamma = \text{const} > 0.$$

Then, for sufficiently small $\varepsilon_0 > 0$ *and every* $(\psi,\varepsilon) \in R^m \times (0,\varepsilon_0]$, *there exists a point* $x^0(\psi,\varepsilon) \in \mathcal{D}$ *such that the solution*

$$(x_\tau(0,x^0(\psi,\varepsilon),\psi,\varepsilon);\varphi_\tau(0,x^0(\psi,\varepsilon),\psi,\varepsilon))$$

of system (7.1) is defined $\forall\tau \in R$ *and satisfies the inequality*

$$\|x_\tau(0,x^0(\psi,\varepsilon),\psi,\varepsilon) - \xi(\tau)\| \le \sigma_2\varepsilon^{\frac{1}{p}} \quad \forall(\psi,\tau,\varepsilon) \in R^m \times R \times (0,\varepsilon_0], \tag{7.6}$$

where the constant σ_2 *is independent of* ψ *and* ε.

Remark 2. Inequality (7.6) can be interpreted as an estimate of the error of the averaging method $\forall\tau \in R$ under the condition that the slow variables take the value $x^0(\psi,\varepsilon)$ at the initial moment of time.

We now establish several facts necessary for the proof of Theorem 7.1.

Lemma 7.1. *If the conditions of Theorem 7.1 are satisfied and*

$$\varepsilon_0 < \frac{\gamma\sigma_3}{4\sigma_1 K}, \quad \sigma_3 = \min\left\{\frac{1}{2}\rho; \frac{1}{\gamma}(2Kn^2\sigma_1 + \gamma)^{-1}\right\},$$

then

$$\|\overline{x}_\tau(t,y+\xi(t),\varepsilon) - \xi(\tau)\| \le K\Big(\|y\|e^{-\frac{\gamma}{2}(\tau-t)} + \frac{2}{\gamma}\sigma_1\varepsilon\Big), \tag{7.7}$$

$$\left\|\frac{\partial}{\partial y}\overline{x}_\tau(t,y+\xi(t),\varepsilon)\right\| \le Ke^{-\frac{\gamma}{2}(\tau-t)} \tag{7.8}$$

for all $\tau \ge t$, $\varepsilon \in (0,\varepsilon_0]$, *and* $\|y\| \le \sigma_3(4K)^{-1}$.

Proof. It follows from the averaged equations (7.3) and inequality (7.5) for the function $z_\tau(t, y + \xi(t), \varepsilon) = \overline{x}_\tau(t, y + \xi(t), \varepsilon) - \xi(\tau)$ that

$$\|z_\tau(t, y + \xi(t), \varepsilon)\|$$
$$\leq K\|y\|e^{-\gamma(\tau - t)} + \int_t^\tau Ke^{-\gamma(\tau - l)}\Big[\varepsilon\sigma_1 + n^2\sigma_1\|z_l(t, y + \xi(t), \varepsilon)\|\Big]dl. \quad (7.9)$$

Assume that the inequality $\|z_\tau(t, y+\xi(t), \varepsilon)\| < \sigma_3$ holds on the maximum half-interval $[t, T)$. Then relation (7.9) yields the following estimate for the function $v_\tau(t, y, \varepsilon) = \|z_\tau(t, y + \xi(t), \varepsilon)\|e^{\gamma(\tau - t)}$:

$$v_\tau(t, y, \varepsilon) \leq K\|y\| + \frac{1}{\gamma}K\sigma_1 e^{\gamma(\tau - t)}\varepsilon + \frac{\gamma}{2}\int_t^\tau v_l(t, y, \varepsilon)dl \quad \forall\tau \in [t, T).$$

In the last inequality, we replace the sign $\leq$ by $=$. The function $\overline{v}_\tau(t, y, \varepsilon)$ that is a solution of the equation constructed is determined by the formula

$$\overline{v}_\tau(t, y, \varepsilon) = K\Big(\|y\| - \frac{1}{\gamma}\sigma_1\varepsilon\Big)e^{\frac{\gamma}{2}(\tau - t)} + \frac{2}{\gamma}K\sigma_1 e^{\gamma(\tau - t)}.$$

This yields

$$v_\tau(t, y, \varepsilon) \leq \overline{v}_\tau(t, y, \varepsilon) < K\|y\|e^{\frac{\gamma}{2}(\tau - t)} + \frac{2\varepsilon}{\gamma}K\sigma_1 e^{\gamma(\tau - t)},$$

or

$$\|z_\tau(t, y + \xi(t), \varepsilon)\| \leq K\Big(\|y\|e^{-\frac{\gamma}{2}(\tau - t)} + \frac{2}{\gamma}\sigma_1\varepsilon\Big) \quad \forall\tau \in [t, T). \quad (7.10)$$

Since

$$K\Big(\|y\| + \frac{2}{\gamma}\sigma_1\varepsilon\Big) \leq \frac{3}{4}\sigma_3$$

for $\|y\| \leq \sigma_3(4K)^{-1}$ and $\varepsilon \leq \varepsilon_0 \leq \gamma\sigma_3(4K\sigma_1)^{-1}$, we can set $T = \infty$ in (7.10). Hence, inequality (7.7) is proved.

We differentiate the averaged equations for slow variables over y. Taking into account that $\dfrac{\partial}{\partial y}\overline{x}_t(t, y + \xi(t), \varepsilon) = E_n$ (E_n is the n-dimensional identity matrix), we get

$$\frac{\partial}{\partial y}\overline{x}_\tau(t, y+\xi(t), \varepsilon)$$

$$= Q(\tau,t) + \int_t^\tau Q(\tau,l)\Big[\Big(\frac{\partial}{\partial x}\overline{a}(\overline{x}_l(t,y+\xi(t),\varepsilon),l,\varepsilon\Big)$$

$$- \frac{\partial}{\partial x}\overline{a}(\overline{x}_l(t,y+\xi(t),\varepsilon),l,0)) + \Big(\frac{\partial}{\partial x}\overline{a}(\overline{x}_l(t,y+\xi(t),\varepsilon),l,0\Big)$$

$$- \frac{\partial}{\partial x}\overline{a}(\xi(l),l,0)\Big]\frac{\partial}{\partial y}\overline{x}_l(t,y+\xi(t),\varepsilon)\,dl,$$

whence

$$\Big\|\frac{\partial}{\partial y}\overline{x}_\tau(t,y+\xi(t),\varepsilon)\Big\|$$

$$\leq Ke^{-\gamma(\tau-t)} + K\Big[\varepsilon + n^2\Big(\|y\| + \frac{2}{\gamma}\sigma_1\varepsilon\Big)K\Big]\sigma_1$$

$$\times \int_t^\tau e^{-\gamma(\tau-l)}\Big\|\frac{\partial}{\partial y}\overline{x}_l(t,y+\xi(t),\varepsilon)\Big\|\,dl.$$

Solving this inequality, we obtain

$$\Big\|\frac{\partial}{\partial y}\overline{x}_\tau(t,y+\xi(t),\varepsilon)\Big\|$$

$$\leq K\exp\Big\{-(\tau-t)\Big[\gamma - K\sigma_1\Big(1+Kn^2\sigma_1\frac{2}{\gamma}\Big)\varepsilon - K^2\sigma_1 n^2\|y\|\Big]\Big\}$$

$$\leq Ke^{-\frac{\gamma}{2}(\tau-t)}$$

for $\tau \geq t$, $\|y\| \leq \sigma_3(4K)^{-1}$, and $\varepsilon \leq \varepsilon_0 \leq \gamma\sigma_3(4K\sigma_1)^{-1}$. Lemma 7.1 is proved.

It follows from estimate (7.7) and the restrictions imposed on ε_0 and y that the slow variables $\overline{x}_\tau(t,y+\xi(t),\varepsilon)$ of every solution of the averaged equations (7.3) lie in $\mathcal{D}$ together with their $\frac{1}{2}\rho$-neighborhoods $\forall\tau \geq t$. Then, using Theorems 2.1 and 2.2, we can write the following inequality for the function $U =$

$(x_\tau(t, y+\xi(t), \varepsilon)-\overline{x}_\tau(t, y+\xi(t), \varepsilon); \varphi_\tau(t, y+\xi(t), \psi, \varepsilon)-\overline{\varphi}_\tau(t, y++\xi(t), \psi, \varepsilon))$:

$$\|U\| + \left\|\frac{\partial}{\partial y}U\right\| + \left\|\frac{\partial}{\partial \psi}U\right\| \leq \sigma_4 \varepsilon^{\frac{1}{p}} \tag{7.11}$$

$$\forall \tau \in [t, t+L], \quad \|y\| \leq \sigma_3 (4K)^{-1}, \quad \varepsilon \in (0, \varepsilon_0], \quad \psi \in R^m,$$

where the constant σ_4 depends on L and does not depend on t, y, ψ, and ε.

Proof of Theorem 7.1. Let

$$L = \frac{2}{\gamma}\ln(8mK) \quad \text{and} \quad \|y\| \leq 2\sigma_4 \varepsilon^{\frac{1}{p}}.$$

For $\varepsilon_0 \leq \left(\frac{1}{\sigma_3} 8\sigma_4 K\right)^{-p}$, inequalities (7.7) and (7.8) yield

$$\begin{gathered} \|\overline{x}_\tau(t, y+\xi(t), \varepsilon) - \xi(\tau)\| \leq K\left(2\sigma_4 \varepsilon^{\frac{1}{p}} e^{-\frac{\gamma}{2}(\tau - t)} + \frac{2}{\gamma}\sigma_1 \varepsilon\right) \quad \forall \tau \geq t, \\ \left\|\frac{\partial}{\partial y}\overline{x}_\tau(t, y+\xi(t), \varepsilon)\right\| \leq \frac{1}{8m} \quad \forall \tau \geq t + L. \end{gathered} \tag{7.12}$$

We fix an arbitrary $\psi \in R^m$ and consider the boundary conditions

$$x|_{\tau=-L} = y + \xi(-L), \quad \varphi|_{\tau=0} = \psi. \tag{7.13}$$

According to Theorem 7.1, there exists a unique solution

$$\begin{gathered} (x_\tau(-L, y+\xi(-L), \psi^{(1)}, \varepsilon); \varphi_\tau(-L, y+\xi(-L), \psi^{(1)}, \varepsilon)), \\ \psi^{(1)} = \psi^{(1)}(y + \xi(-L), \psi, \varepsilon), \end{gathered}$$

of the boundary-value problem (7.1), (7.13), whose slow variables, with regard for (7.11) and (7.12), satisfy the following conditions for $\varepsilon_0 \leq \left(\frac{\gamma\sigma_4}{2\sigma_1 K}\right)^{\frac{p}{p-1}}$:

$$\begin{aligned} &\|x_\tau(-L, y+\xi(-L), \psi^{(1)}, \varepsilon) - \xi(\tau)\| \\ &\quad \leq \|x_\tau(-L, y+\xi(-L), \psi^{(1)}, \varepsilon) - \overline{x}_\tau(-L, y+\xi(-L), \varepsilon)\| \\ &\qquad + \|\overline{x}_\tau(-L, y+\xi(-L), \varepsilon) - \xi(\tau)\| \\ &\quad \leq \sigma_4 \varepsilon^{\frac{1}{p}} + K\left(\frac{2}{\gamma}\sigma_1 \varepsilon + 2\sigma_4 \varepsilon^{\frac{1}{p}}\right) \leq 2(K+1)\sigma_4 \varepsilon^{\frac{1}{p}} \end{aligned} \tag{7.14}$$

for all $\tau \in [-L, 0)$ and

$$\|x_0(-L, y + \xi(-L), \psi^{(1)}, \varepsilon) - \xi(0)\| \\ \leq \sigma_4 \varepsilon^{\frac{1}{p}} + K\Big(\frac{2}{\gamma}\sigma_1 \varepsilon + 2\sigma_4 \varepsilon^{\frac{1}{p}} e^{-\frac{\gamma}{2}L}\Big) < 2\sigma_4 \varepsilon^{\frac{1}{p}}. \quad (7.15)$$

Note that, for $\varepsilon \leq \varepsilon_0 \leq \min\{(8m\sigma_4)^{-p}; (2\sigma_4(1+\sigma_5))^{-p}\}$, the function $\psi^{(1)} = \psi^{(1)}(y + \xi(-L), \psi, \varepsilon)$ satisfies inequality (6.9), namely

$$\Big\|\frac{\partial}{\partial y}\psi^{(1)}\Big\| \leq 2m[L\sigma_1 K + \sigma_4 \varepsilon^{\frac{1}{p}}] < 2m\Big[L\sigma_1 K + \sigma_5 \frac{1}{8m} + \sigma_4 \varepsilon^{\frac{1}{p}}(1 + \sigma_5)\Big] \\ \leq \sigma_5 = 4mLK\sigma_1 + \frac{1}{4}. \quad (7.16)$$

We now consider the boundary conditions

$$x|_{\tau=-2L} = y + \xi(-2L), \quad \varphi|_{\tau=-L} = \psi^{(1)}(x|_{\tau=-L}, \psi, \varepsilon). \quad (7.17)$$

By analogy with the above reasoning, we find the unique solution

$$(x_\tau(-2L, y + \xi(-2L), \psi^{(2)}, \varepsilon); \varphi_\tau(-2L, y + \xi(-2L), \psi^{(2)}, \varepsilon)), \\ \psi^{(2)} = \psi^{(2)}(y + \xi(-2L), \psi, \varepsilon),$$

of the boundary-value problem (7.1), (7.17), for which the following estimates are true:

$$\|x_\tau(-2L, y + \xi(-2L), \psi^{(2)}, \varepsilon) - \xi(\tau)\| \leq 2(K+1)\sigma_4 \varepsilon^{\frac{1}{p}} \quad \forall \tau \in [-2L, -L), \\ \|x_{-L}(-2L, y + \xi(-2L), \psi^{(2)}, \varepsilon) - \xi(-L)\| < 2\sigma_4 \varepsilon^{\frac{1}{p}}. \quad (7.18)$$

Further, we estimate $\dfrac{\partial \psi^{(2)}}{\partial y}$. Taking into account inequalities (6.9), (7.11), (7.12), and (7.16), we get

$$\Big\|\frac{\partial}{\partial y}\psi^{(2)}\Big\| \leq 2m\Big[\sigma_5 \Big\|\frac{\partial}{\partial y}\overline{x}_{-L}(-2L, y + \xi(-2L), \varepsilon)\Big\| \\ + L\sigma_1 \max_{[-2L,-L]} \Big\|\frac{\partial}{\partial y}\overline{x}_\tau(-2L, y + \xi(-2L), \varepsilon)\Big\| + \sigma_4 \varepsilon^{\frac{1}{p}} + \sigma_4 \sigma_5 \varepsilon^{\frac{1}{p}}\Big] \\ \leq 2m\left[\frac{\sigma_5}{8m} + L\sigma_1 K + \sigma_4(1 + \sigma_5)\Big]\varepsilon^{\frac{1}{p}}\right] \leq \sigma_5$$

for $\varepsilon \le \varepsilon_0 \le \min\{(8m\sigma_4)^{-p}; (2\sigma_4(1+\sigma_5))^{-p}\}$. Note that the restriction $\varepsilon_0 \le (2\sigma_4(1+\sigma_5))^{-p}$ is determined by conditions for the validity of inequality (6.9). Combining (7.14), (7.15), and (7.18), we establish that

$$(x_\tau(-2L, y+\xi(-2L), \psi^{(2)}, \varepsilon); \varphi_\tau(-2L, y+\xi(-2L), \psi^{(2)}, \varepsilon))$$

is a solution of system (7.1) for $\tau \in [-2L, 0]$ and satisfies the boundary conditions

$$x_{-2L}(-2L, y+\xi(-2L), \psi^{(2)}, \varepsilon) = y + \xi(-2L),$$
$$\varphi_0(-2L, y+\xi(-2L), \psi^{(2)}, \varepsilon) = \psi$$

and the inequalities

$$\|x_\tau(-2L, y+\xi(-2L), \psi^{(2)}, \varepsilon) - \xi(\tau)\| \le 2(K+1)\sigma_4\varepsilon^{\frac{1}{p}} \quad \forall \tau \in [-2L, 0),$$
$$\|x_0(-2L, y+\xi(-2L), \psi^{(2)}, \varepsilon) - \xi(0)\| < 2\sigma_4\varepsilon^{\frac{1}{p}}.$$

By induction, for an arbitrary integer $r > 2$ and $\tau \in [-rL, -(r-1)L]$ we obtain the solution

$$(x_\tau(-rL, y+\xi(-rL), \psi^{(r)}, \varepsilon); \varphi_\tau(-rL, y+\xi(-rL), \psi^{(r)}, \varepsilon))$$

of Eqs. (7.1) that satisfies the boundary conditions

$$x_{|\tau=-rL} = y + \xi(-rL), \varphi_{|\tau=-(r-1)L} = \psi^{(r-1)}(x_{|\tau=-(r-1)L}, \psi, \varepsilon)$$

and the inequalities

$$\|x_\tau(-rL, y+\xi(-rL), \psi^{(r)}, \varepsilon) - \xi(\tau)\| \le 2(K+1)\sigma_4\varepsilon^{\frac{1}{p}}$$
$$\forall \tau \in [-rL, -(r-1)L),$$
$$\|x_{-(r-1)}(-rL, y+\xi(-rL), \psi^{(r)}, \varepsilon) - \xi(-(r-1)L)\| < 2\sigma_4\varepsilon^{\frac{1}{p}},$$
$$\left\|\frac{\partial}{\partial y}\psi^{(r)}(y+\xi(-rL), \psi, \varepsilon)\right\| \le \sigma_5.$$

Thus,

$$(x_\tau(-rL, y+\xi(-rL), \psi^{(r)}, \varepsilon); \varphi_\tau(-rL, y+\xi(-rL), \psi^{(r)}, \varepsilon))$$

is a solution of system (7.1) for all $\tau \in [-rL, 0]$, and

$$\|x_\tau(-rL, y + \xi(-rL), \psi^{(r)}, \varepsilon) - \xi(\tau)\| \le 2(K+1)\sigma_4 \varepsilon^{\frac{1}{p}} \quad \forall \tau \in [-rL, 0),$$

$$\|x_0(-rL, y + \xi(-rL), \psi^{(r)}, \varepsilon) - \xi(0)\| < 2\sigma_4 \varepsilon^{\frac{1}{p}}, \tag{7.19}$$

$$\varphi_0(-rL, y + \xi(-rL), \psi^{(r)}, \varepsilon) = \psi.$$

We now fix an arbitrary $y \in R^n$, $\|y\| \le 2\sigma_4 \varepsilon^{\frac{1}{p}}$, and consider the sequence

$$\{x_0(-rL, y + \xi(-rL), \psi^{(r)}(y + \xi(-rL)), \psi, \varepsilon), \varepsilon)\}_{r=1}^{\infty} \equiv \{x^{(r)}(\psi, \varepsilon)\}_{r=1}^{\infty}.$$

By virtue of the uniform boundedness of the norm of every element of this sequence by the number $\|\xi(0)\| + 2\sigma_4 \varepsilon^{\frac{1}{p}}$, we can select a convergent subsequence of this sequence, namely

$$\{x^{(r_j)}(\psi, \varepsilon)\}_{j=1}^{\infty}, \quad r_j = r_j(\psi, \varepsilon), \quad \lim_{j\to\infty} x^{(r_j)}(\psi, \varepsilon) = x^0(\psi, \varepsilon),$$

$$\|x^0(\psi, \varepsilon) - \xi(0)\| \le 2\sigma_4 \varepsilon^{\frac{1}{p}}.$$

Let us prove that a solution $(x_\tau(0, x^0(\psi, \varepsilon), \psi, \varepsilon); \varphi_\tau(0, x^0(\psi, \varepsilon), \psi, \varepsilon))$ of system (7.1) is defined $\forall \tau \in (-\infty; 0]$ and

$$\|x_\tau(0, x^0(\psi, \varepsilon), \psi, \varepsilon) - \xi(\tau)\| \le 2(K+1)\sigma_4 \varepsilon^{\frac{1}{p}}.$$

Assume the contrary, i.e., let

$$\|x_{\tau_0}(0, x^0(\psi, \varepsilon), \psi, \varepsilon) - \xi(\tau_0)\| > 2(K+1)\sigma_4 \varepsilon^{\frac{1}{p}} \tag{7.20}$$

for certain $\tau_0 < 0$. Taking into account that

$$x_\tau(-rL, y + \xi(-rL), \psi^{(r)}(y + \xi(-rL), \psi, \varepsilon), \varepsilon) = x_\tau(0, x^{(r)}(\psi, \varepsilon), \psi, \varepsilon)$$

for all $\tau \in [-rL, 0]$, we derive from (7.19) for $r_j L > -\tau_0$ that

$$\|x_{\tau_0}(0, x^{(r_j)}(\psi, \varepsilon), \psi, \varepsilon) - \xi(\tau_0)\| \le 2(K+1)\sigma_4 \varepsilon^{\frac{1}{p}}. \tag{7.21}$$

Using the continuous dependence of a solution on the initial data and passing to the limit as $j \to \infty$ in (7.21), we arrive at a contradiction with (7.20).

For $\tau \in [0, \infty)$, estimate (7.5) follows from Theorem 2.4 and inequality (7.7). The restriction $\sigma_2 \varepsilon_0^{\frac{1}{p}} \le \frac{1}{2}\rho$, $\sigma_2 = 2(K+1)\sigma_4$, which guarantees that the curve $x = x_\tau(0, x^0(\psi, \varepsilon), \psi, \varepsilon)$ lies in $\mathcal{D}$ $\forall \tau \in R$, completes the proof of Theorem 7.1.

8. Multipoint Problem for Resonance Multifrequency Systems

Consider a nonlinear system of the form

$$\begin{aligned} \frac{dx}{d\tau} &= a(x,\tau,\varepsilon) + \varepsilon A(x,\varphi,\tau,\varepsilon), \\ \frac{d\varphi}{d\tau} &= \frac{\omega(x,\tau,\varepsilon)}{\varepsilon} + B(x,\varphi,\tau,\varepsilon), \end{aligned} \tag{8.1}$$

where $a,\ A,\ \omega,$ and B are defined for $(x,\varphi,\tau,\varepsilon) \in \mathcal{D}\times R^m\times[0,L]\times(0,\varepsilon_0] = \overline{G}\ \ (m \geq 2)$, 2π-periodic in each component $\varphi_\nu,\ \nu = \overline{1,m},$ of the vector φ, and $l \geq m$ times continuously differentiable with respect to $x,\ \varphi,$ and τ for every fixed $\varepsilon \in (0,\varepsilon_0]$, and all their partial derivatives are uniformly bounded in $\overline{G}$ by a constant c_1 independent of ε. In addition, we assume that

$$\sum_k \left[\|k\| \sup_G \|c_k\| + \sup_G \left\|\frac{\partial c_k}{\partial \tau}\right\| + \sup_G \left\|\frac{\partial c_k}{\partial x}\right\|\right] \leq c_1. \tag{8.2}$$

Here, $G = \mathcal{D}\times[0,L]\times(0,\varepsilon_0]$ and $c_k = c_k(x,\tau,\varepsilon)$ are the Fourier coefficients of the function $[A(x,\varphi,\tau,\varepsilon); B(x,\varphi,\tau,\varepsilon)]$.

For Eqs. (8.1), we introduce the multipoint conditions

$$\Phi(x|_{\tau=\tau_1},\ldots,x|_{\tau=\tau_r},\varepsilon\varphi|_{\tau=\tau_1},\ldots,\varepsilon\varphi|_{\tau=\tau_r},\varepsilon) = 0, \tag{8.3}$$

where $0 \leq \tau_1 < \tau_2 < \ldots < \tau_r \leq L,\ r \geq 2,\ \Phi = (\Phi_1,\ldots,\Phi_{n+m})$, and $\Phi_j(x|_{\tau=\tau_1},\ldots,\varepsilon\varphi|_{\tau=\tau_r},\varepsilon),\ j = \overline{1,m+n},$ are certain functionals.

Problem (8.1), (8.3) is a multipoint problem that possesses resonance properties. In the case of a one-frequency system with nonzero frequency, there are no resonance modes, but if the number of frequencies is $m \geq 2$, then the resonance phenomenon is typical of the problems under consideration. Note that, for $r = 2$, $\tau_1 = 0$, and $\tau_2 = L$, problem (8.1), (8.3) is a boundary-value problem

Assume that $\Phi = \Phi(p_1,\ldots,p_r,q_1,\ldots,q_r,\varepsilon)$ is an $(n+m)$-dimensional vector function of $p_j \in \mathcal{D},\ q_j \in R^m,\ j = \overline{1,r},$ and $\varepsilon \in (0,\varepsilon_0]$, that is twice continuously differentiable for every fixed ε and such that

$$\sum_{s=1}^{2} \|D^s\Phi\| \leq c_2 = \text{const} \tag{8.4}$$

for all $p_j = (p_j^{(1)}, \dots, p_j^{(n)}) \in \mathcal{D}$, $q_j = (q_j^{(1)}, \dots, q_j^{(m)}) \in R^m$, and $\varepsilon \in (0, \varepsilon]$. Here, D^s is an arbitrary partial derivative with respect to $p_j^{(\nu)}$ and $q_j^{(\mu)}$ $(j = \overline{1,r},\ \nu = \overline{1,n},\ \mu = \overline{1,m})$ of order s.

Parallel with (8.1), (8.3), we consider the following problem averaged with respect to all angular variables φ:

$$\frac{d\overline{x}}{d\tau} = a(\overline{x}, \tau, \varepsilon) + \varepsilon\overline{A}(\overline{x}, \tau, \varepsilon), \quad \frac{d\overline{\theta}}{d\tau} = \omega(\overline{x}, \tau, \varepsilon) + \varepsilon\overline{B}(\overline{x}, \tau, \varepsilon), \tag{8.5}$$

$$\Phi(\overline{x}|_{\tau=\tau_1}, \dots, \overline{x}|_{\tau=\tau_r}, \overline{\theta}|_{\tau=\tau_1}, \dots, \overline{\theta}|_{\tau=\tau_r}, \varepsilon) = 0, \tag{8.6}$$

where $\overline{\theta} = \varepsilon\overline{\varphi}$ and

$$[\overline{A}; \overline{B}] = (2\pi)^{-m} \int_0^{2\pi} \dots \int_0^{2\pi} [A(\overline{x}, \varphi, \tau, \varepsilon); B(\overline{x}, \varphi, \tau, \varepsilon)] d\varphi_1 \dots d\varphi_m.$$

In order that the averaging operator be efficient for the investigation of oscillation processes, it is necessary to impose certain restrictions on the components $\omega_\nu(\overline{x}, \tau, \varepsilon)$, $\nu = \overline{1,m}$, of the frequency vector ω. In what follows, we assume that

$$\|(W_l^T(\overline{x}, \tau, \varepsilon) W_l(\overline{x}, \tau, \varepsilon))^{-1} W_l^T(\overline{x}, \tau, \varepsilon)\| \le c_3 \quad \forall (\overline{x}, \tau, \varepsilon) \in G, \tag{8.7}$$

where W_l and W_l^T denote the matrix

$$W_l(\overline{x}, \tau, \varepsilon) = \left(\frac{d^{j-1}}{d\tau^{j-1}} \omega_\nu(\overline{x}, \tau, \varepsilon) \right)_{j,\nu=1}^{l,m}$$

and its transpose, respectively; here, the total derivatives with respect to τ of the functions $\omega_\nu(\overline{x}, \tau, \varepsilon)$ are calculated with regard for the averaged system (8.5). Conditions (8.2) and (8.7) guarantee (Theorems 5.1 and 5.2) that

$$\|\widetilde{U}\| + \varepsilon \left\| \frac{\partial}{\partial y} \widetilde{U} \right\| + \varepsilon \left\| \frac{\partial}{\partial \psi} \widetilde{U} \right\| \le c_4 \varepsilon^{1+\alpha}, \quad \alpha = \frac{1}{l}, \tag{8.8}$$

for all $\tau \in [0, L]$, $y \in \mathcal{D}_1$, $\psi \in R^m$, and $\varepsilon \in (0, \varepsilon_0]$ for sufficiently small $\varepsilon_0 > 0$. Here,

$$\widetilde{U} = (x(\tau, y, \psi, \varepsilon) - \overline{x}(\tau, y, \varepsilon); \theta(\tau, y, \psi, \varepsilon) - \overline{\theta}(\tau, y, \psi, \varepsilon)), \quad \theta = \varepsilon\varphi,$$

$(x; \theta)$ and $(\overline{x}; \overline{\theta})$ are solutions of systems (8.1) and (8.5), respectively, that take the values $(y; \psi)$ for $\tau = 0$, and $\mathcal{D}_1$ is the set of points $y \in \mathcal{D}$ for which the curve $\overline{x} = \overline{x}(\tau, y, \varepsilon)$ lies in $\mathcal{D}$ together with a certain ρ_1-neighborhood $\forall(\tau, \varepsilon) \in [0, L] \times (0, \varepsilon_0]$.

Denote by $P(y^0, \psi^0, \varepsilon)$ the $(m+n)$-dimensional square matrix

$$\sum_{j=1}^{r}\left(\frac{\partial\Phi^0}{\partial p_j}\frac{\partial\overline{x}(\tau_j, y^0, \varepsilon)}{\partial y^0} + \frac{\partial\Phi^0}{\partial q_j}\int_0^{\tau_j}\frac{\partial\omega(\overline{x}(\tau, y^0, \varepsilon), \tau, \varepsilon)}{\partial\overline{x}}\frac{\partial\overline{x}(\tau, y^0, \varepsilon)}{\partial y^0}d\tau, \frac{\partial\Phi^0}{\partial q_j}\right).$$

Here, the values of the derivatives $\dfrac{\partial\Phi^0}{\partial p_j}$ and $\dfrac{\partial\Phi^0}{\partial q_j}$ of $\Phi(p_1, \ldots, q_r, \varepsilon)$ are taken for $p_\nu = \overline{x}(\tau_\nu, y^0, \varepsilon)$ and $q_\nu = \overline{\theta}(\tau_\nu, y^0, \psi^0, \varepsilon)$, $\nu = \overline{1, r}$.

Theorem 8.1. *Suppose that the following conditions are satisfied:*

(i) conditions (8.2), (8.4), and (8.7) are satisfied;

(ii) for every $\varepsilon \in (0, \varepsilon_0]$, *the averaged problem (8.5), (8.6) has a unique solution* $(\overline{x}(\tau, y^0, \varepsilon); \overline{\theta}(\tau, y^0, \psi^0, \varepsilon))$ *that lies in* $\mathcal{D} \times R^m$ *together with its* ρ*-neighborhood* $\forall(\tau, \varepsilon) \in [0, L] \times (0, \varepsilon_0]$;

(iii) for a given solution, the matrix $P(y^0, \psi^0, \varepsilon)$ *is nondegenerate and*

$$\|P^{-1}(y^0, \psi^0, \varepsilon)\| \leq c_5 = \text{const} \quad \forall\varepsilon \in (0, \varepsilon_0]. \tag{8.9}$$

Then one can find constants $c_6 > 0$ *and* $\varepsilon_1 > 0$ *such that, for all* $\varepsilon \in (0, \varepsilon_0]$, *where* $\varepsilon_0 \leq \varepsilon_1$, *the multipoint problem (8.1), (8.3) has a unique solution* $(x(\tau, \varepsilon); \theta(\tau, \varepsilon))$ *that satisfies the inequality*

$$\|x(\tau, \varepsilon) - \overline{x}(\tau, y^0, \varepsilon)\| + \|\theta(\tau, \varepsilon) - \overline{\theta}(\tau, y^0, \psi^0, \varepsilon)\| \leq c_6\varepsilon^{1+\alpha}. \tag{8.10}$$

Proof. It follows from the smoothness conditions for the right-hand side of the averaged system (8.5) that, for $\|y\| < \frac{1}{2}\rho e^{-2c_1 L}$, the curve $\overline{x} = \overline{x}(\tau, y + y^0, \varepsilon)$ lies in $\mathcal{D}$ together with its $\rho_1 = \frac{1}{2}\rho$-neighborhood $\forall(\tau, \varepsilon) \in [0, L] \times (0, \varepsilon_0]$. Therefore, the domain $\mathcal{D}_1 \neq \varnothing$ is not empty, and we can use inequality (8.8).

We seek a solution of problem (8.1), (8.3) in the form $(x(\tau, y^0 + y, \psi^0 + \psi, \varepsilon); \theta(\tau, y^0 + y, \psi^0 + \psi, \varepsilon))$, where the unknown vector $z = (y, \psi)$ is determined from conditions (8.3), namely,

$$
\begin{aligned}
z = -P^{-1}(y^0, \psi^0, \varepsilon)\Big\{ \Big[&\Phi(\overline{x}(\tau_1, y^0 + y, \varepsilon), \dots, \\
&\overline{\theta}(\tau_r, y^0 + y, \psi^0 + \psi, \varepsilon), \varepsilon) - P(y^0, \psi^0, \varepsilon) z \Big] \\
&+ \Big[\Phi(x(\tau_1, y^0 + y, \psi^0 + \psi, \varepsilon), \dots, \theta(\tau_r, y^0 + y, \psi^0 + \psi, \varepsilon), \varepsilon) \\
&- \Phi(\overline{x}(\tau_1, y^0 + y, \varepsilon), \dots, \overline{\theta}(\tau_r, y^0 + y, \psi^0 + \psi, \varepsilon), \varepsilon) \Big] \Big\} \\
\equiv M(z, \varepsilon). &\qquad (8.11)
\end{aligned}
$$

Taking into account conditions (8.4) and estimate (8.8), we get

$$
\begin{aligned}
\|\Phi(x(\tau_1, y^0 + y, \psi^0 + \psi, \varepsilon), \dots, \varepsilon) - \Phi(\overline{x}(\tau_1, y^0 + y, \varepsilon), \dots, \varepsilon)\| \\
\le c_2 c_4 \varepsilon^{1+\alpha}. \qquad (8.12)
\end{aligned}
$$

Using the smoothness conditions for the right-hand side of system (8.5), we get

$$
\begin{aligned}
\overline{x}(\tau, y^0 + y, \varepsilon) &= \overline{x}(\tau, y^0, \varepsilon) + \frac{\partial \overline{x}(\tau, y^0, \varepsilon)}{\partial y^0} y + X(\tau, y, \varepsilon), \\
\overline{\theta}(\tau, y^0 + y, &\psi^0 + \psi, \varepsilon) \\
&= \int_0^\tau \frac{\partial \omega(\overline{x}(t, y^0, \varepsilon), t, \varepsilon)}{\partial \overline{x}} \frac{\partial \overline{x}(t, y^0, \varepsilon)}{\partial y^0} dt y \\
&\qquad + \overline{\theta}(\tau, y^0, \psi^0, \varepsilon) + Y(\tau, y, \varepsilon), \qquad (8.13)
\end{aligned}
$$

where

$$
\begin{aligned}
\|X(\tau, y, \varepsilon)\| &\le c_7 \|y\|^2, \\
\|Y(\tau, y, \varepsilon)\| &\le c_7 (\|y\|^2 + \varepsilon \|y\|),
\end{aligned}
$$

and c_7 is a constant independent of ε.

We expand the function

$$
\Phi(\overline{x}(\tau_1, y^0 + y, \varepsilon), \dots, \overline{\theta}(\tau, y^0 + y, \psi^0 + \psi, \varepsilon), \varepsilon)
$$

according to the Taylor formula by using equalities (8.13) and inequality (8.4). After obvious transformations, we get

$$\Phi(\overline{x}(\tau_1, y^0 + y, \varepsilon), \ldots, \overline{\theta}(\tau_r, y^0 + y, \psi^0 + \psi, \varepsilon), \varepsilon)$$
$$= P(y^0, \psi^0, \varepsilon) z + R(z, \varepsilon), \quad (8.14)$$

where $\|R(z, \varepsilon)\| \le c_8(\|z\|^2 + \varepsilon\|z\|)$ and c_8 is a constant. Combining (8.12)–(8.14), we obtain

$$\|M(z, \varepsilon)\| \le c_5[c_2 c_4 \varepsilon^{1+\alpha} + c_8(\|z\|^2 + \varepsilon\|z\|)],$$

which implies that M maps the set

$$V = \{z\colon z \in R^{n+m}, \|z\| \le 2c_2 c_4 c_5 \varepsilon^{1+\alpha}\}$$

into itself for

$$\varepsilon \le \varepsilon_0 = \min\{(2c_2 c_4 c_5)^{-\frac{1}{\alpha}}; (4c_5 c_8)^{-1}\}.$$

Let us prove that the mapping $M\colon V \to V$ is contracting. For this purpose, we represent $\dfrac{\partial M}{\partial z}$ in the form

$$\frac{\partial M(z, \varepsilon)}{\partial z} = -P^{-1}(y^0, \psi^0, \varepsilon)\Big\{\sum_{j=1}^{r}\Big[\frac{\partial \Phi}{\partial p_j}\frac{\partial}{\partial y}(x(\tau_j, y^0 + y, \psi^0 + \psi, \varepsilon)$$
$$- \overline{x}(\tau_j, y^0 + y, \varepsilon)) + \frac{\partial \Phi}{\partial q_j}\frac{\partial}{\partial y}(\theta(\tau_j, y^0 + y, \psi^0 + \psi, \varepsilon)$$
$$- \overline{\theta}(\tau_j, y^0 + y, \psi^0 + \psi, \varepsilon), \frac{\partial \Phi}{\partial p_j}\frac{\partial}{\partial \psi}(\overline{x}(\tau_j, y^0 + y, \psi^0 + \psi, \varepsilon))$$
$$+ \frac{\partial \Phi}{\partial q_j}\frac{\partial}{\partial \psi}(\theta(\tau_j, y^0 + y, \psi^0 + \psi, \varepsilon) - \overline{\theta}(\tau_j, y^0 + y, \psi^0 + \psi, \varepsilon))\Big]\Big\}$$
$$- P^{-1}(y^0, \psi^0, \varepsilon)\Big\{\sum_{j=1}^{r}\Big[\frac{\partial \Phi}{\partial p_j}\frac{\partial}{\partial y}\overline{x}(\tau_j, y^0 + y, \varepsilon)$$
$$+ \frac{\partial \Phi}{\partial q_j}\frac{\partial}{\partial y}\overline{\theta}(\tau_j, y^0 + y, \psi^0 + \psi, \varepsilon), \frac{\partial \Phi}{\partial q_j}\Big] - P(y^0, \psi^0, \varepsilon)\Big\}. \quad (8.15)$$

Using inequalities (8.4) and (8.8), one can estimate the norm of the matrix in the first braces on the right-hand side of the last equality from above by the value $(n + m)^2 c_2 c_4 \varepsilon^{\alpha}$. The smoothness conditions for the right-hand side of system (8.5) yield the following representation:

$$\frac{\partial}{\partial y}\overline{x}(\tau, y^0 + y, \varepsilon) = \frac{\partial}{\partial y^0}\overline{x}(\tau, y^0, \varepsilon) + \widetilde{X}(\tau, y, \varepsilon),$$

$$\frac{\partial}{\partial \psi}\overline{\theta}(\tau, y^0 + y, \psi^0 + \psi, \varepsilon) = E_m, \tag{8.16}$$

$$\frac{\partial}{\partial y}\overline{\theta}(\tau, y^0 + y, \psi^0 + \psi, \varepsilon)$$

$$= \int\limits_0^\tau \frac{\partial}{\partial x}\omega(\overline{x}(t, y^0, \varepsilon), t, \varepsilon)\frac{\partial}{\partial y^0}\overline{x}(t, y^0, \varepsilon)dt + \widetilde{Y}(\tau, y, \varepsilon),$$

where E_m is the m-dimensional identity matrix,

$$\|\widetilde{X}(\tau, y, \varepsilon)\| \leq c_9\|y\|, \quad \|\widetilde{Y}(\tau, y, \varepsilon)\| \leq c_9(\|y\| + \varepsilon), \quad c_9 = \text{const},$$

The smoothness conditions for the function Φ and inequalities (8.4) and (8.8) yield

$$\frac{\partial \Phi}{\partial p_j} = \frac{\partial \Phi^0}{\partial p_j} + \widetilde{\Phi}_j(z, \varepsilon), \qquad \frac{\partial \Phi}{\partial q_j} = \frac{\partial \Phi^0}{\partial q_j} + \underset{\sim j}{\Phi}(z, \varepsilon),$$

where

$$\|\widetilde{\Phi}_j(z, \varepsilon)\| + \|\underset{\sim j}{\Phi}(z, \varepsilon)\| \leq c_{10}(\|z\| + \varepsilon^{1+\alpha}), \quad c_{10} = \text{const}.$$

Therefore, the norm of the matrix in the second braces on the right-hand side of (8.15) can be estimated from above by the value

$$(n+m)^2\Big[2c_2c_9\|y\| + c_2c_9\varepsilon + c_{10}(\|y\| + \|\psi\| + \varepsilon^{1+\alpha})$$

$$\times\left(2c_9\|y\| + \sup_G\left\|\frac{\partial\overline{x}(\tau, y, \varepsilon)}{\partial y}\right\|\left(1 + L\sup_G\left\|\frac{\partial\omega(x, \tau, \varepsilon)}{\partial x}\right\|\right)\right)\Big] \leq c_{11}\varepsilon,$$

$$c_{11} = \text{const},$$

for $z \in V$. Thus,

$$\left\|\frac{\partial M(z, \varepsilon)}{\partial z}\right\| \leq c_5[c_2c_4(n+m)^2\varepsilon^\alpha + c_{11}\varepsilon] \leq \frac{1}{2}$$

for

$$\varepsilon \leq \varepsilon_0 \leq [2c_5(c_2c_4(n+m)^2 + c_{11})]^{-\frac{1}{\alpha}},$$

i.e., the mapping $M\colon V \to V$ is contracting. Thus, there exists a unique solution $z = z(\varepsilon) = (y(\varepsilon), \psi(\varepsilon))$ of Eq. (8.11) that satisfies the condition $\|z(\varepsilon)\| \le 2c_2c_4c_5\varepsilon^{1+\alpha}$ and, therefore, there exists a unique solution

$$(x(\tau,\varepsilon);\theta(\tau,\varepsilon)) = (x(\tau,y^0+y(\varepsilon),\psi^0+\psi(\varepsilon),\varepsilon);\theta(\tau,y^0+y(\varepsilon),\psi^0+\psi(\varepsilon),\varepsilon))$$

of the multipoint problem (8.1), (8.3) whose initial data lie in a small neighborhood of the point (y^0,ψ^0). Estimate (8.10) follows from the inequalities

$$\begin{aligned}
&\|x(\tau,\varepsilon)-\overline{x}(\tau,y^0,\varepsilon)\| + \|\theta(\tau,\varepsilon)-\overline{\theta}(\tau,y^0,\psi^0,\varepsilon)\| \\
&\qquad \le \|x(\tau,\varepsilon)-\overline{x}(\tau,y^0+y(\varepsilon),\varepsilon)\| \\
&\qquad\qquad + \|\theta(\tau,\varepsilon)-\overline{\theta}(\tau,y^0+y(\varepsilon),\psi^0+\psi(\varepsilon),\varepsilon)\| \\
&\qquad\qquad + \|\overline{x}(\tau,y^0+y(\varepsilon),\varepsilon)-\overline{x}(\tau,y^0,\varepsilon)\| \\
&\qquad\qquad + \|\overline{\theta}(\tau,y^0+y(\varepsilon),\psi^0+\psi(\varepsilon),\varepsilon)-\overline{\theta}(\tau,y^0,\psi^0,\varepsilon)\| \\
&\qquad \le c_6\varepsilon^{1+\alpha},
\end{aligned}$$

where $c_6 = 2c_4 + 2c_2c_4c_5(2mc_1L+1)ne^{2c_1L}$.

It remains to impose the condition $c_6\varepsilon_0^{1+\alpha} \le \dfrac{1}{2}\rho$ in order that the solution $(x(\tau,\varepsilon),\theta(\tau,\varepsilon))$ of problem (8.1), (8.3) do not leave the domain $D\times R^m$. Theorem 8.1 is proved.

As an example, we consider the three-point problem

$$\begin{aligned}
\frac{dx}{d\tau} &= -x + \varepsilon x^2(\cos\varphi_2 + \cos(5\varphi_2-\varphi_1)), \\
\frac{d\varphi_1}{d\tau} &= \frac{2x^2+2}{\varepsilon} + \sin\varphi_2, \\
\frac{d\varphi_2}{d\tau} &= \frac{\tau}{\varepsilon} + x\sin(5\varphi_2-\varphi_1), \\
x_{|\tau=0} + \varepsilon\varphi_{2|\tau=0} &= -1.5; \quad \varepsilon\varphi_{1|\tau=\frac{1}{2}} = 0, \\
x_{|\tau=1} + \varepsilon\varphi_{1|\tau=1} &+ \varepsilon\varphi_{2|\tau=1} = 0,
\end{aligned} \tag{8.17}$$

where $x \in \left(\frac{1}{2},4\right)$, $\varphi_1 \in R$, $\varphi_2 \in R$, $\tau \in [0,1]$, and ε is a small positive parameter. The corresponding problem averaged with respect to all angular variables

$$\frac{d\overline{x}}{d\tau} = -\overline{x}, \quad \frac{d\overline{\theta}_1}{d\tau} = 2\overline{x}^2 + 2, \quad \frac{d\overline{\theta}_2}{d\tau} = \tau, \quad \overline{\theta}_1 = \varepsilon\overline{\varphi}_1, \quad \overline{\theta}_2 = \varepsilon\overline{\varphi}_2,$$

$$\overline{x}|_{\tau=0} + \overline{\theta}_2|_{\tau=0} = -1.5, \quad \overline{\theta}_1|_{\tau=\frac{1}{2}} = 0, \quad \overline{x}|_{\tau=1} + \overline{\theta}_1|_{\tau=1} + \overline{\theta}_2|_{\tau=1} = 0$$

has the unique solution

$$\overline{x}(\tau) = e^{-\tau+1}, \quad \overline{\theta}_1(\tau) = 2\tau - e^{2(1-\tau)} + e - 1, \quad \overline{\theta}_2(\tau) = \frac{\tau^2}{2} - 1.5 - e,$$

which belongs to the set $\left(\frac{1}{2}, 4\right) \times R^2$ together with its $\frac{1}{2}$-neighborhood. For this solution, we have

$$P(y^0, \psi^0, \varepsilon) = \begin{pmatrix} 1 & 0 & 1 \\ 2(e-1) & 1 & 0 \\ 2e - \frac{1}{e} & 1 & 1 \end{pmatrix}, \quad \det P = -1 + \frac{1}{e} \neq 0.$$

Moreover,

$$\det W_2(\overline{x}, \tau, \varepsilon) = \begin{vmatrix} 2\overline{x}^2 + 2 & \tau \\ -4\overline{x}^2 & 1 \end{vmatrix} \geq 2.5 \quad \forall (\tau, \overline{x}) \in [0, 1] \times \left(\frac{1}{2}, 4\right).$$

Therefore, according to Theorem 8.1, for every $\varepsilon \in (0, \varepsilon_0]$ (ε_0 is sufficiently small) there is a unique solution $(x(\tau, \varepsilon), \varphi_1(\tau, \varepsilon), \varphi_2(\tau, \varepsilon))$ of problem (8.17) that satisfies the inequality

$$\begin{aligned} |x(\tau, \varepsilon) - e^{1-\tau}| &+ \left|\varphi_1(\tau, \varepsilon) - \frac{2\tau - e^{2(1-\tau)} + e - 1}{\varepsilon}\right|\varepsilon \\ &+ \left|\varphi_2(\tau, \varepsilon) - \frac{\tau^2 - 3 - 2e}{2\varepsilon}\right|\varepsilon \leq c_6 \varepsilon^{1+\frac{1}{2}} \end{aligned}$$

for all $(\tau, \varepsilon) \in [0, 1] \times (0, \varepsilon_0]$.

Remark 3. The verification of the restrictions imposed by conditions (ii) and (iii) of Theorem 8.1 and related to the value of the small parameter ε can be a fairly difficult problem. Assume, in addition, that the functions a, ω, and Φ are smooth with respect to $\varepsilon \in [0, \varepsilon_0]$, and consider the problem

$$\frac{d\xi}{d\tau} = a(\xi, \tau, 0), \quad \frac{d\eta}{d\tau} = \omega(\xi, \tau, 0), \quad \Phi(\xi|_{\tau=\tau_1}, \ldots, \eta|_{\tau=\tau_r}, 0) = 0.$$

Assume that this problem has the unique solution

$$(\xi(\tau,\xi^0);\eta(\tau,\xi^0,\eta^0)),\xi(0,\xi^0)=\xi^0,\quad \eta(0,\xi^0,\eta^0)=\eta^0,$$

that lies in $\mathcal{D}\times R^m\ \ \forall\tau\in[0,L]$ and satisfies the condition

$$\det\sum_{j=1}^{r}\Big(\frac{\partial\Phi_0^0}{\partial p_j}\frac{\partial\xi(\tau_j,\xi^0)}{\partial\xi^0}+\frac{\partial\Phi_0^0}{\partial q_j}\int_0^{\tau_j}\frac{\partial\omega(\xi(t,\xi^0),t,0)}{\partial\xi}\frac{\partial\xi(t,\xi^0)}{\partial\xi^0}\,dt,\frac{\partial\Phi_0^0}{\partial q_j}\Big)\neq 0.$$

Here, the values of the derivatives $\dfrac{\partial\Phi_0^0}{\partial p_j}$ and $\dfrac{\partial\Phi_0^0}{\partial q_j}$ of $\Phi(p_1,\ldots,q_r,0)$ are taken for $p_\nu=\xi(\tau_\nu,\xi^0)$ and $q_\nu=\eta(\tau_\nu,\xi^0,\eta^0)$, $\nu=\overline{1,r}$. It is easy to verify that these assumptions are sufficient for the existence of a solution $(\overline{x}(\tau,y^0,\varepsilon);\overline{\theta}(\tau,y^0,\psi^0,\varepsilon))$ of problem (8.5), (8.6) that satisfies condition (8.9) and the inequality

$$\|\overline{x}(\tau,y^0,\varepsilon)-\xi(\tau,\xi^0)\|+\|\overline{\theta}(\tau,y^0,\psi^0,\varepsilon)-\eta(\tau,\xi^0,\eta^0)\|\leq\widetilde{c}_6\varepsilon.$$

Remark 4. It follows from estimates (8.10) that it suffices to impose restrictions (8.2), (8.4), and (8.7) on the functions c, Φ, and ω not in the entire domain of their definition, but only in a certain $\mu(\varepsilon)$-neighborhood ($\mu(\varepsilon)\to 0$ as $\varepsilon\to 0$) of the solution of the averaged problem.

An analog of Theorem 8.1 is also true for the multipoint problem

$$\frac{dx}{d\tau}=a(x,\varphi,\tau,\varepsilon),\quad \frac{d\varphi}{d\tau}=\frac{\omega(\tau)}{\varepsilon}+b(x,\varphi,\tau,\varepsilon),$$
$$\Phi(x|_{\tau=\tau_1},\ldots,x|_{\tau=\tau_r},\varphi|_{\tau=\tau_1},\ldots,\varphi|_{\tau=\tau_r},\varepsilon)=0, \tag{8.18}$$

where the frequencies ω depend only on the time variable, and Φ is an $(n+m)$-dimensional functional. Consider the averaged problem

$$\frac{d\overline{x}}{d\tau}=\overline{a}(\overline{x},\varphi,\tau,\varepsilon),\quad \frac{d\overline{\varphi}}{d\tau}=\frac{\omega(\tau)}{\varepsilon}+\overline{b}(\overline{x},\tau,\varepsilon),$$
$$\Phi(\overline{x}|_{\tau=\tau_1},\ldots,\overline{x}|_{\tau=\tau_r},\overline{\varphi}|_{\tau=\tau_1},\ldots,\overline{\varphi}|_{\tau=\tau_r},\varepsilon)=0 \tag{8.19}$$

and denote by $S(y^0, \psi^0, \varepsilon)$ the $(n+m)$-dimensional square matrix

$$S(y^0, \psi^0, \varepsilon) = \sum_{j=1}^{r} \Big(\frac{\partial \Phi^0}{\partial p_j} \frac{\partial \overline{x}(\tau_j, y^0, \varepsilon)}{\partial y^0}$$

$$+ \frac{\partial \Phi^0}{\partial q_j} \int_0^{\tau_j} \frac{\partial \overline{b}(\overline{x}(t, y^0, \varepsilon), t, \varepsilon)}{\partial \overline{x}} \frac{\partial \overline{x}(t, y^0, \varepsilon)}{\partial y^0} dt, \frac{\partial \Phi^0}{\partial q_j} \Big). \quad (8.20)$$

Here, the values of the derivatives of the function $\Phi(p_1, \ldots, q_r, \varepsilon)$ with respect to p_j and q_j are taken for $p_\nu = \overline{x}(\tau_\nu, y^0, \varepsilon)$ and $q_\nu = \overline{\varphi}(\tau_\nu, y^0, \psi^0, \varepsilon)$, $\nu = \overline{1, r}$, and $(\overline{x}(\tau, y^0, \varepsilon); \overline{\varphi}(\tau, y^0, \psi^0, \varepsilon))$ is a solution of the averaged system for which $\overline{x}(0, y^0, \varepsilon) = y^0$ and $\overline{\varphi}(0, y^0, \psi^0, \varepsilon) = \psi^0$.

Theorem 8.2. *Suppose that the following conditions are satisfied:*

(i) there exists a unique solution $(\overline{x}(\tau, y^0, \varepsilon); \overline{\varphi}(\tau, y^0, \psi^0, \varepsilon))$ *of the averaged problem (8.19) whose slow variables belong to* $\mathcal{D}$ *together with their* ρ*-neighborhoods;*

(ii) for this solution, the matrix $S(y^0, \psi^0, \varepsilon)$ *is nondegenerate and, furthermore,* $\|S^{-1}(y^0, \psi^0, \varepsilon)\| \leq c = \text{const}$ $\forall \varepsilon \in (0, \varepsilon_0]$;

(iii) $\det(W_p^T(\tau) W_p(\tau)) \neq 0$ *for any* $\tau \in [0, L]$ *and certain* $p \geq m$;

(iv) conditions (2.6) and (8.4) are satisfied.

Then there exist constants $\overline{c} > 0$ *and* $\overline{\varepsilon} > 0$ *such that, for any* $\varepsilon \in (0, \overline{\varepsilon}]$, *problem (8.18) has a unique solution* $(x(\tau, \varepsilon); \varphi(\tau, \varepsilon))$ *that satisfies the inequality*

$$\|x(\tau, \varepsilon) - \overline{x}(\tau, y^0, \varepsilon)\| + \|\varphi(\tau, \varepsilon) - \overline{\varphi}(\tau, y^0, \psi^0, \varepsilon)\| \leq \overline{c} \varepsilon^{1/p}. \quad (8.21)$$

9. Estimates of the Error of Averaging Method for Multipoint Problems in Critical Case

In this section, we consider a multipoint problem of the form

$$\frac{dx}{d\tau} = a(x, \tau, \varepsilon) + \varepsilon A(x, \varphi, \tau, \varepsilon), \quad \frac{d\varphi}{d\tau} = \frac{\omega(x, \tau, \varepsilon)}{\varepsilon} + B(x, \varphi, \tau, \varepsilon), \tag{9.1}$$

$$F(x|_{\tau=\tau_1}, \ldots, x|_{\tau=\tau_r}, \varepsilon) = 0, \tag{9.2}$$

$$\Phi(x|_{\tau=\tau_1}, \ldots, x|_{\tau=\tau_r}, \varphi|_{\tau=\tau_1}, \ldots, \varphi|_{\tau=\tau_r}, \varepsilon) = 0, \tag{9.3}$$

where $F(p_1, \ldots, p_r, \varepsilon)$ and $\Phi(p_1, \ldots, p_r, q_1, \ldots, q_r, \varepsilon)$ are, respectively, n-dimensional and m-dimensional vector functions of the variables $p_j \in \mathcal{D}$, $q_j \in R^m$, $j = \overline{1, r}$, and $\varepsilon \in (0, \varepsilon_0]$, and $0 \leq \tau_1 < \tau_2 < \ldots < \tau_r \leq L$, $r \geq 2$. The main difference between this problem and problem (8.1), (8.3) lies in the fact that, first, in problem (9.1)–(9.3) the group of boundary conditions dependent only on slow variables is selected, and, second, in conditions (9.3) the function Φ depends only on the arguments $\varphi|_{\tau=\tau_j}$, whereas in conditions (8.3) it depends on $\varepsilon\varphi|_{\tau=\tau_j}$.

We also consider the corresponding averaged problem

$$\frac{d\overline{x}}{d\tau} = a(\overline{x}, \tau, \varepsilon) + \varepsilon \overline{A}(\overline{x}, \tau, \varepsilon), \quad F(\overline{x}|_{\tau=\tau_1}, \ldots, \overline{x}|_{\tau=\tau_r}, \varepsilon) = 0, \tag{9.4}$$

$$\frac{d\overline{\varphi}}{d\tau} = \frac{\omega(\overline{x}, \tau, \varepsilon)}{\varepsilon} + \overline{B}(\overline{x}, \tau, \varepsilon),$$

$$\Phi(\overline{x}|_{\tau=\tau_1}, \ldots, \overline{x}|_{\tau=\tau_r}, \overline{\varphi}|_{\tau=\tau_1}, \ldots, \overline{\varphi}|_{\tau=\tau_r}, \varepsilon) = 0 \tag{9.5}$$

and assume that there exists a solution $\overline{x} = \overline{x}(\tau, y^0, \varepsilon)$ of problem (9.4) for which the matrix

$$P_1(y^0, \varepsilon) = \sum_{j=1}^{r} \frac{\partial F^0}{\partial p_j} \frac{\partial \overline{x}(\tau_j, y^0, \varepsilon)}{\partial y^0}$$

satisfies the inequality

$$\|P_1^{-1}(y^0, \varepsilon)\| \leq c_{12} \varepsilon^{-\alpha_1} \quad \forall \varepsilon (0, \varepsilon_0] \tag{9.6}$$

where $c_{12} > 0$ and $\alpha_1 \geq 0$ are certain constants. Here, $\dfrac{\partial F^0}{\partial p_j}$ denotes the matrix of the partial derivatives of the function $F(p_1, \ldots, p_r, \varepsilon)$ with respect to $p_j = (p_j^{(1)}, \ldots, p_j^{(n)})$ for $p_\nu = \overline{x}(\tau_\nu, y^0, \varepsilon)$, $\nu = \overline{1, r}$.

To solve problem (9.5), it suffices to solve the equation

$$\Phi(\overline{x}(\tau_1, y^0, \varepsilon), \ldots, \overline{x}(\tau_r, y^0, \varepsilon), \psi + \frac{1}{\varepsilon}\int_0^{\tau_1}[\omega(\overline{x}(t, y^0, \varepsilon), t, \varepsilon) + \varepsilon\overline{B}(\overline{x}(t, y^0, \varepsilon), t, \varepsilon)]dt, \ldots, \psi + \frac{1}{\varepsilon}\int_0^{\tau_r}[\omega(\overline{x}(t, y^0, \varepsilon), t, \varepsilon) + \varepsilon\overline{B}(\overline{x}(t, y^0, \varepsilon), t, \varepsilon)]dt, \varepsilon) = 0$$

with respect to ψ. Assume that there exists a unique solution $\psi = \psi^0(\varepsilon)$ of this equation, i.e., there exists a unique solution

$$\overline{\varphi}(\tau, y^0, \psi^0, \varepsilon) = \psi^0 + \int_0^{\tau}\left[\omega(\overline{x}(t, y^0, \varepsilon), t, \varepsilon) + \varepsilon\overline{B}(\overline{x}(t, y^0, \varepsilon), t, \varepsilon)\right] dt$$

of problem (9.5). We also assume that the matrix

$$P_2(y^0, \psi^0, \varepsilon) = \sum_{j=1}^{r} \frac{\partial\Phi^0}{\partial q_j}$$

satisfies the inequality

$$\|P_2^{-1}(y^0, \psi^0, \varepsilon)\| \leq c_{13}\varepsilon^{-\alpha_2}, \quad c_{13} > 0, \quad \alpha_2 \geq 0, \tag{9.7}$$

where $\dfrac{\partial\Phi^0}{\partial q_j}$ denotes the matrix of the partial derivatives of $\Phi(p_1, \ldots, q_r, \varepsilon)$ with respect to $q_j = (q_j^{(1)}, \ldots, q_j^{(m)})$ for $p_\nu = \overline{x}(\tau_\nu, y^0, \varepsilon)$ and $q_\nu = \overline{\varphi}(\tau_\nu, y^0, \psi^0, \varepsilon)$, $\nu = \overline{1, r}$.

If the numbers α_1 and α_2 in inequalities (9.6) and (9.7) are positive, then the norms of the matrices P_1^{-1} and P_2^{-1} may tend to infinity as $\varepsilon \to 0$. It is natural to call this case critical. Below, we study the question of the solvability of problem (9.1)–(9.3) in the critical case and establish estimates for the deviation of solutions of the original and averaged problems.

In what follows, we assume that, for every fixed $\varepsilon \in (0, \varepsilon_0]$, the functions $F(p_1, \ldots, p_r, \varepsilon)$ and $\Phi(p_1, \ldots, q_r, \varepsilon)$ are twice continuously differentiable with respect to $p_j \in \mathcal{D}$ and $q_j \in R^m$, $j = \overline{1, r}$, and

$$\sum_{j=1}^{r}\Big(\Big\|\frac{\partial F}{\partial p_j}\Big\| + \sum_{\nu=1}^{r}\sum_{k=1}^{n}\Big\|\frac{\partial^2 F}{\partial p_j \partial p_\nu^{(k)}}\Big\|\Big) \le c_{14},$$

$$\sum_{j=1}^{r}\Big[\varepsilon\Big\|\frac{\partial \Phi}{\partial p_j}\Big\| + \Big\|\frac{\partial \Phi}{\partial q_j}\Big\| + \sum_{\nu=1}^{r}\sum_{s=1}^{m}\Big(\varepsilon\Big\|\frac{\partial^2 \Phi}{\partial p_j \partial q_\nu^{(s)}}\Big\| + \Big\|\frac{\partial^2 \Phi}{\partial q_j \partial q_\nu^{(s)}}\Big\|\Big)\Big] \le c_{14} \quad (9.8)$$

for all $p_j \in \mathcal{D},\ \ q_j \in R^m,\ \ j = \overline{1, r},$ and $\varepsilon \in (0, \varepsilon_0]$.

Theorem 9.1. *Suppose that the following conditions are satisfied:*

(i) for every $\varepsilon \in (0, \varepsilon_0]$, the averaged problem (9.4), (9.5) has a unique solution $(\overline{x}(\tau, y^0, \varepsilon),\ \overline{\varphi}(\tau, y^0, \psi^0, \varepsilon))$ whose slow variables $\overline{x}(\tau, y^0, \varepsilon)$ belong to $\mathcal{D}$ together with their ρ-neighborhoods;

(ii) conditions (8.2), (8.7), and (9.6)–(9.8) for $\alpha > \alpha_1 + 2\alpha_2$ are satisfied.

Then, for sufficiently small $\varepsilon_0 > 0$ and every $\varepsilon \in (0, \varepsilon_0]$, problem (9.1)–(9.3) has a unique solution $(x(\tau, \varepsilon); \varphi(\tau, \varepsilon))$ that satisfies the following inequalities for any $\tau \in [0, L]$:

$$\begin{aligned} \|x(\tau, \varepsilon) - \overline{x}(\tau, y^0, \varepsilon)\| &\le c_{15}\varepsilon^{1+\alpha-\alpha_1}, \\ \|\varphi(\tau, \varepsilon) - \overline{\varphi}(\tau, y^0, \psi^0, \varepsilon)\| &\le c_{15}\varepsilon^{\alpha-\alpha_1-\alpha_2}, \end{aligned} \quad (9.9)$$

where the constant c_{15} does not depend on ε.

Proof. We determine the unknown parameters $(y; \psi)$ of the solution $(x(\tau, y^0 + y, \psi^0 + \psi, \varepsilon); \varphi(\tau, y^0 + y, \psi^0 + \psi, \varepsilon))$ of system (9.1) from conditions (9.2) and (9.3). We rewrite (9.2) in the form

$$\begin{aligned} y = -P_1^{-1}(y^0, \varepsilon)\Big\{&\Big[F(\overline{x}(\tau_1, y^0 + y, \varepsilon), \ldots, \overline{x}(\tau_r, y^0 + y, \varepsilon), \varepsilon) - P_1(y^0, \varepsilon)y\Big] \\ &+ \Big[F(x(\tau_1, y^0 + y, \psi^0 + \psi, \varepsilon), \ldots, x(\tau_r, y^0 + y, \psi^0 + \psi, \varepsilon), \varepsilon) \\ &- F(\overline{x}(\tau_1, y^0 + y, \varepsilon), \ldots, \overline{x}(\tau_r, y^0 + y, \varepsilon), \varepsilon)\Big]\Big\} \\ \equiv T(y, \psi, \varepsilon). \end{aligned} \quad (9.10)$$

Using the estimate of the error of the averaging method (8.8) and inequalities (9.8), we obtain

$$\|F(x(\tau_1, y^0+y, \psi^0+\psi, \varepsilon), \dots, \varepsilon) - F(\overline{x}(\tau_1, y^0+y, \varepsilon), \dots, \varepsilon)\| \\ \leq c_4 c_{14} r \varepsilon^{1+\alpha}. \tag{9.11}$$

Further, using representations (8.13), inequalities (9.8), and condition (9.4), we get

$$F(\overline{x}(\tau_1, y^0+y, \varepsilon), \dots, \varepsilon) = P_1(y^0, \varepsilon) y + \widetilde{F}(y, \varepsilon),$$

where $\|\widetilde{F}(y,\varepsilon)\| \leq c_{16}\|y\|^2$ and $c_{16} = \text{const}$. Using this equality and inequalities (9.6) and (9.9), we obtain the following estimate for $T(y, \psi, \varepsilon)$:

$$\|T(y,\psi,\varepsilon)\| \leq c_{12}(c_4 c_{14} r + c_{16}\|y\|^2 \varepsilon^{-1-\alpha})\varepsilon^{1+\alpha-\alpha_1}.$$

This estimate implies that $T(y,\psi,\varepsilon)$ maps the set

$$\|y\| \leq c_{17}\varepsilon^{1+\alpha-\alpha_1}, \quad c_{17} = 2 r c_4 c_{12} c_{14},$$

into itself for

$$\varepsilon \leq \varepsilon_0 \leq (4 c_4 c_{12}^2 c_{14} c_{16} r)^{\frac{1}{2\alpha_1 - 1 - \alpha}}, \quad \psi \in R^m.$$

Let us calculate $\dfrac{\partial T}{\partial y}$. We have

$$\frac{\partial T}{\partial y} = -P_1^{-1}(y^0,\varepsilon)\Bigg\{\Bigg[\sum_{j=1}^{r} \frac{\partial}{\partial p_j} F(\overline{x}(\tau_1, y^0+y, \varepsilon), \dots, \varepsilon) \\ \times \frac{\partial}{\partial y}\overline{x}(\tau_j, y^0+y, \varepsilon) - P_1(y^0,\varepsilon)\Bigg] \\ + \Bigg[\sum_{j=1}^{r}\Bigg(\frac{\partial}{\partial p_j} F(x(\tau_1, y^0+y, \psi^0+\psi, \varepsilon), \dots, \varepsilon)\frac{\partial}{\partial y} x(\tau_j, y^0+y, \psi^0+\psi, \varepsilon) \\ - \frac{\partial}{\partial p_j} F(\overline{x}(\tau_1, y^0+y, \varepsilon), \dots, \varepsilon)\frac{\partial}{\partial y}\overline{x}(\tau_j, y^0+y, \varepsilon)\Bigg)\Bigg]\Bigg\}. \tag{9.12}$$

In view of inequalities (8.8) and (9.8), the norm of the matrix in the second square brackets on the right-hand side of (9.12) can be estimated from above by the value $c_{18}\varepsilon^{\alpha}$, $c_{18} = \text{const}$. Using (8.16) and writing the equality

$$\sum_{j=1}^{r} \frac{\partial}{\partial p_j} F(\overline{x}(\tau_1, y^0 + y, \varepsilon), \ldots, \varepsilon) \frac{\partial}{\partial y} \overline{x}(\tau_j, y^0 + y, \varepsilon)$$

$$= \sum_{j=1}^{r} \Big(\frac{\partial F^0}{\partial p_j} + F_j(y, \varepsilon) \Big) \Big(\frac{\partial}{\partial y^0} \overline{x}(\tau_j, y^0, \varepsilon) + \widetilde{X}(\tau_j, y, \varepsilon) \Big)$$

$$\equiv P_1(y^0, \varepsilon) + K(y, \varepsilon),$$

where $\|K(y, \varepsilon)\| \leq c_{19}\|y\|$, $c_{19} = \text{const}$, we can estimate the norm of the matrix in the first square brackets on the right-hand side of equality (9.12) by the value $c_{20}\|y\|$, $c_{20} = \text{const}$. Thus,

$$\Big\| \frac{\partial}{\partial y} T(y, \psi, \varepsilon) \Big\| \leq c_{12}\varepsilon^{-\alpha_1}(c_{18}\varepsilon^{\alpha} + c_{20}\|y\|) \leq \frac{1}{2}$$

for $\alpha > \alpha_1$, $\|y\| \leq c_{17}\varepsilon^{1+\alpha-\alpha_1}$, $\psi \in R^m$, and $\varepsilon \leq \varepsilon_0 \leq [2c_{12}(c_{18} + c_{17}c_{20})]^{\frac{1}{\alpha_1 - \alpha}}$. Consequently, Eq. (9.10) has a unique solution $y = y(\psi, \varepsilon)$, $\|y(\psi, \varepsilon)\| \leq c_{17}\varepsilon^{1+\alpha-\alpha_1}$, which can be determined by the method of successive approximations:

$$y_{k+1}(\psi, \varepsilon) = T(y_k(\psi, \varepsilon), \psi, \varepsilon), \quad k \geq 1,$$

$$y_0(\psi, \varepsilon) \equiv 0, \quad y(\psi, \varepsilon) = \lim_{k \to \infty} y_k(\psi, \varepsilon).$$

Using equality (9.10), we get

$$\frac{\partial y_k(\psi, \varepsilon)}{\partial \psi} = \frac{\partial}{\partial \psi} T(y_{k-1}(\psi, \varepsilon), \psi, \varepsilon)$$

$$= -P_1^{-1}(y^0, \varepsilon) \Big\{ \Big[\sum_{j=1}^{r} \frac{\partial}{\partial p_j} F(\overline{x}(\tau_1, y^0 + y_{k-1}, \varepsilon), \ldots, \varepsilon)$$

$$\times \frac{\partial}{\partial y} \overline{x}(\tau_j, y^0 + y_{k-1}, \varepsilon) \frac{\partial y_{k-1}}{\partial \psi} - P_1(y^0, \varepsilon) \frac{\partial y_{k-1}}{\partial \psi} \Big]$$

$$+ \Big[\sum_{j=1}^{r} \Big(\frac{\partial}{\partial p_j} F(x(\tau_1, y^0 + y_{k-1}, \psi^0 + \psi, \varepsilon), \ldots, \varepsilon)$$

$$\times \Big(\frac{\partial}{\partial y}x(\tau_j, y^0 + y_{k-1}, \psi^0 + \psi, \varepsilon)\frac{\partial y_{k-1}}{\partial \psi} + \frac{\partial}{\partial \psi}x(\tau_j, y^0 + y_{k-1}, \psi, \varepsilon)\Big)$$

$$- \frac{\partial}{\partial p_j}F(\overline{x}(\tau_1, y^0 + y_{k-1}, \varepsilon), \dots, \varepsilon)\frac{\partial}{\partial y}\overline{x}(\tau_j, y^0 + y_{k-1}, \varepsilon)\frac{\partial y_{k-1}}{\partial \psi}\Big)\Big]\Big\}.$$

Further, using the methods proposed in the course of the investigation of the properties of $T(y, \psi, \varepsilon)$, we obtain

$$\Big\|\frac{\partial y_k(\psi, \varepsilon)}{\partial \psi}\Big\| \le c_{21}\varepsilon^{1+\alpha-\alpha_1} + c_{22}\varepsilon^{\alpha-\alpha_1}\Big\|\frac{\partial y_{k-1}(\psi, \varepsilon)}{\partial \psi}\Big\|,$$

where c_{21} and c_{22} are certain constants independent of ε. This yields

$$\Big\|\frac{\partial y_k(\psi, \varepsilon)}{\partial \psi}\Big\| \le c_{23}\varepsilon^{1+\alpha-\alpha_1} \quad \forall k \ge 0, \quad c_{23} = 2c_{21},$$

provided that $\varepsilon_0 \le (2c_{22})^{\frac{1}{\alpha_1-\alpha}}$. Hence, the sequence $\Big\{\frac{\partial}{\partial \psi}y_k(\psi, \varepsilon)\Big\}$ is uniformly bounded by the constant $c_{23}\varepsilon^{1+\alpha-\alpha_1}$. This is sufficient to guarantee that the function $y(\psi, \varepsilon)$ satisfies the Lipschitz condition

$$\|y(\psi^1, \varepsilon) - y(\psi^2, \varepsilon)\| \le c_{23}\varepsilon^{1+\alpha-\alpha_1}\|\psi^1 - \psi^2\| \quad \forall \psi^1, \psi^2 \in R^m. \tag{9.13}$$

We rewrite equality (8.3) in the form

$$\psi = -P_2^{-1}(y^0, \psi^0, \varepsilon)\{[\Phi - \overline{\Phi}] + [\overline{\Phi} - P_2(y^0, \psi^0, \varepsilon)\psi]\} \equiv \widetilde{T}(\psi, \varepsilon), \tag{9.14}$$

where $\Phi = \Phi(x(\tau_1, y^0 + y(\psi, \varepsilon), \psi^0 + \psi, \varepsilon), \dots, \varphi(\tau_r, y^0 + y(\psi, \varepsilon), \psi^0 + \psi, \varepsilon), \varepsilon)$ and $\overline{\Phi} = \Phi(\overline{x}(\tau_1, y^0 + y(\psi, \varepsilon), \varepsilon), \dots, \overline{\varphi}(\tau_r, y^0 + y(\psi, \varepsilon), \psi^0 + \psi, \varepsilon), \varepsilon)$, and estimate each term on the right-hand side of equality (9.14). Using inequalities (8.8) and (9.8), we get

$$\|\Phi - \overline{\Phi}\| \le \sum_{j=1}^{r}\Big(\frac{1}{\varepsilon}c_4c_{14}\varepsilon^{1+\alpha} + c_4c_{14}\varepsilon^{\alpha}\Big) = 2rc_4c_{14}\varepsilon^{\alpha}, \tag{9.15}$$

$$\|\overline{\Phi} - P_2(y^0, \psi^0, \varepsilon)\psi\|$$

$$\le rc_{14}(ne^{2c_1L} + c_{24})\frac{\|y(\psi, \varepsilon)\|}{\varepsilon} + rc_{14}\Big(c_{24}\frac{1}{\varepsilon}\|y(\psi, \varepsilon)\| + \|\psi\|\Big)^2,$$

where $c_{24} = 2c_1Lne^{2c_1L}$. Taking into account that $\|y(\psi, \varepsilon)\| \le c_{17}\varepsilon^{1+\alpha-\alpha_1}$, we finally obtain

$$\|\overline{\Phi} - P_2(y^0, \psi^0, \varepsilon)\psi\| \le c_{25}(\varepsilon^{\alpha-\alpha_1} + \|\psi\|^2), \quad c_{25} = \text{const}. \tag{9.16}$$

Inequalities (9.15) and (9.16) yield

$$\|\widetilde{T}(\psi,\varepsilon)\| \leq c_{26}(\varepsilon^{\alpha-\alpha_1-\alpha_2} + \varepsilon^{-\alpha_2}\|\psi\|^2), \quad c_{26} = c_{13}(2rc_4c_{14} + c_{25}). \quad (9.17)$$

This implies that $\widetilde{T}(\psi,\varepsilon)$ maps the set

$$U = \{\psi \colon \psi \in R^m, \ \ \|\psi\| \leq 2c_{26}\varepsilon^{\alpha-\alpha_1-\alpha_2}\}$$

into itself for $\alpha > \alpha_1 + 2\alpha_2$ and $4c_{26}\varepsilon_0^{\alpha-\alpha_1-2\alpha_2} \leq 1$.

Let $\psi^{(1)}$ and $\psi^{(2)}$ be arbitrary points of the set U. Using the error estimates of the averaging method (8.8) and inequalities (9.7), (9.8), and (9.13), we get

$$\|\widetilde{T}(\psi^{(1)},\varepsilon) - \widetilde{T}(\psi^{(2)},\varepsilon)\|$$

$$\leq c_{27}\left(\varepsilon^{\alpha-\alpha_1-\alpha_2} + \sum_{j,\nu=1}^{r}\sum_{s=1}^{m}\sup\left\|\frac{\partial^2\Phi}{\partial q_j \partial q_\nu^{(s)}}\right\|\varepsilon^{\alpha-\alpha_1-2\alpha_2}\right)\|\psi^{(1)} - \psi^{(2)}\|, \tag{9.18}$$

$$c_{27} = \text{const.}$$

Since $\alpha > \alpha_1 + 2\alpha_2$, the last inequality implies that the mapping $\widetilde{T}\colon U \to U$ is contracting. Thus, there exists a unique solution $\psi = \psi_0(\varepsilon) \in U$ of Eq. (9.14), and, hence, there exists the solution

$$(x(\tau,\varepsilon); \varphi(\tau,\varepsilon)) = (x(\tau, y^0 + y(\psi_0(\varepsilon),\varepsilon), \psi^0 + \psi_0(\varepsilon),\varepsilon);$$
$$\varphi(\tau, y^0 + y(\psi_0(\varepsilon),\varepsilon), \psi^0 + \psi_0(\varepsilon),\varepsilon))$$

of problem (9.1)–(9.3). Estimates (9.9) follow from estimates (8.8) and the inequalities

$$\|y(\psi_0(\varepsilon),\varepsilon)\| \leq c_{17}\varepsilon^{1+\alpha-\alpha_1}, \quad \|\psi_0(\varepsilon)\| \leq 2c_{26}\varepsilon^{\alpha-\alpha_1-\alpha_2},$$

and the condition $c_{15}\varepsilon_0^{1+\alpha-\alpha_1} \leq \frac{1}{2}\rho$ guarantees that $x = x(\tau,\varepsilon)$ lies in $\mathcal{D}$ $\forall(\tau,\varepsilon) \in [0,L] \times (0,\varepsilon_0]$. Theorem 9.1 is proved.

Remark 5. If the vector function $\Phi(p_1,\ldots,q_r,\varepsilon)$ in condition (9.3) linearly depends on q_j, $j = \overline{1,r}$, i.e.,

$$\Phi = \sum_{j=1}^{r} A_j(x|_{\tau=\tau_1},\ldots,x|_{\tau=\tau_r},\varepsilon)\varphi|_{\tau=\tau_j} + A_0(x|_{\tau=\tau_1},\ldots,x|_{\tau=\tau_r},\varepsilon),$$

then the inequality $\alpha > \alpha_1 + 2\alpha_2$ in Theorem 9.1 can be weakened to the inequality $\alpha > \alpha_1 + \alpha_2$. Indeed, in this case, the analysis of inequalities (9.17) and (9.18) shows that all conditions of the principle of contracting mappings are satisfied for $\alpha > \alpha_1 + \alpha_2$.

We apply the results obtained to the solution of the multipoint problem

$$\frac{dx}{d\tau} = P(\tau)x + \varepsilon A(x, \varphi, \tau, \varepsilon), \quad \frac{d\varphi}{d\tau} = \frac{\omega(x, \tau, \varepsilon)}{\varepsilon} + B(x, \varphi, \tau, \varepsilon), \tag{9.19}$$

$$\sum_{j=1}^{r} A_j(\varepsilon)x|_{\tau=\tau_j} = x^0(\varepsilon),$$
$$\sum_{j=1}^{r} B_j(\varepsilon)\varphi|_{\tau=\tau_j} = f(x|_{\tau=\tau_1}, \ldots, x|_{\tau=\tau_r}, \varepsilon), \tag{9.20}$$

where the right-hand sides of Eqs. (9.19) satisfy the same conditions as the right-hand sides of Eqs. (8.1), $A_j(\varepsilon)$ and $B_j(\varepsilon)$, $j = \overline{1, r}$, are uniformly bounded (by a constant c_{28}) n-dimensional and m-dimensional, respectively, square matrices, and $f(p_1, \ldots, p_r, \varepsilon)$ is a function continuously differentiable with respect to $p_j \in \mathcal{D}$, $j = \overline{1, r}$, for every fixed ε and such that

$$\|f\| + \sum_{j=1}^{r} \left\| \frac{\partial f}{\partial p_j} \right\| \leq \frac{1}{\varepsilon} c_{28} \qquad \forall p_j \in \mathcal{D}, \quad j = \overline{1, r}, \quad \varepsilon \in (0, \varepsilon_0]. \tag{9.21}$$

We write the problem averaged with respect to φ:

$$\frac{d\overline{x}}{d\tau} = P(\tau)\overline{x} + \varepsilon \overline{A}(\overline{x}, \tau, \varepsilon),$$
$$\sum_{j=1}^{r} A_j(\varepsilon)\overline{x}|_{\tau=\tau_j} = x^0(\varepsilon), \tag{9.22}$$

$$\frac{d\overline{\varphi}}{d\tau} = \frac{\omega(\overline{x}, \tau, \varepsilon)}{\varepsilon} + \overline{B}(\overline{x}, \tau, \varepsilon),$$
$$\sum_{j=1}^{r} B_j(\varepsilon)\overline{\varphi}|_{\tau=\tau_j} = f(\overline{x}|_{\tau=\tau_1}, \ldots, \overline{x}|_{\tau=\tau_r}, \varepsilon). \tag{9.23}$$

Let $Q(\tau, t)$ denote the normal fundamental matrix of the linear system $\dfrac{dx}{d\tau} = P(\tau)x$.

Lemma 9.1. *Suppose that the following conditions are satisfied:*

(i) for all $\varepsilon \in (0, \varepsilon_0]$,

$$\Big\|\Big(\sum_{j=1}^{r} A_j(\varepsilon)Q(\tau_j, 0)\Big)^{-1}\Big\| \le c_{29} = \text{const}, \quad \det \sum_{j=1}^{r} B_j(\varepsilon) \ne 0;$$

(ii) for all $(\tau, \varepsilon) \in [0, L] \times (0, \varepsilon_0]$, *the curve*

$$y(\tau, \varepsilon) = Q(\tau, 0)\Big(\sum_{j=1}^{r} A_j(\varepsilon)Q(\tau_j, 0)\Big)^{-1} x^0(\varepsilon)$$

lies in $\mathcal{D}$ *together with its* ρ*-neighborhood.*

Then, for sufficiently small $\varepsilon_0 > 0$ *and every* $\varepsilon \in (0, \varepsilon_0]$, *there exists a unique solution* $\overline{x} = \overline{x}(\tau, \varepsilon)$, $\overline{\varphi} = \overline{\varphi}(\tau, \varepsilon)$ *of the averaged problem (9.22), (9.23) for which the curve* $\overline{x} = \overline{x}(\tau, \varepsilon)$ *lies in a certain small neighborhood of the curve* $y = y(\tau, \varepsilon)$.

Proof. Since the right-hand sides of Eqs. (9.22) are smooth, there exists a solution $\overline{x} = \overline{x}(\tau, z + \widetilde{x}, \varepsilon)$ of the Cauchy problem

$$\frac{d\overline{x}}{d\tau} = P(\tau)\overline{x} + \varepsilon\overline{A}(\overline{x}, \tau, \varepsilon),$$

$$\overline{x}|_{\tau=0} = z + \widetilde{x},$$

$$\widetilde{x} = \Big(\sum_{j=1}^{r} A_j(\varepsilon)Q(\tau_j, 0)\Big)^{-1} x^0(\varepsilon),$$

which satisfies the equality

$$\overline{x}(\tau, z + \widetilde{x}, \varepsilon) = y(\tau, \varepsilon) + Q(\tau, 0)z + \int_0^\tau Q(\tau, t)\overline{A}(\overline{x}(t, z + \widetilde{x}, \varepsilon), t, \varepsilon)dt.$$

The last relation yields

$$\|x(\tau, z + \widetilde{x}, \varepsilon) - y(\tau, \varepsilon)\| \le K\|z\| + \varepsilon KLc_1 < \frac{1}{2}\rho$$

for $\|z\| \le \rho(4K)^{-1}$ and $\varepsilon \le \varepsilon_0 \le \rho(4KLc_1)^{-1}$. Here, K is a constant that bounds the norm of the matrix $Q(\tau, t)$ $\forall(\tau, t) \in [0, L] \times [0, L]$. This implies that, for indicated z and $\varepsilon \in (0, \varepsilon_0]$, the solution $\overline{x} = \overline{x}(\tau, z + \widetilde{x}, \varepsilon)$ of the Cauchy problem can be extended for any $\tau \in [0, L]$ and lies in $\mathcal{D}$ together with its $\frac{1}{2}\rho$-neighborhood.

Let us prove that z can be chosen so that the function $\overline{x} = \overline{x}(\tau, z + \widetilde{x}, \varepsilon)$ is a solution of problem (9.22). Indeed, if the boundary conditions are satisfied, then we get

$$z = -\varepsilon\Big(\sum_{j=1}^{r} A_j(\varepsilon)Q(\tau_j, 0)\Big)^{-1}\sum_{j=1}^{r} A_j(\varepsilon)\int_0^{\tau_j} Q(\tau_j, t)\overline{A}(\overline{x}(t, z + \widetilde{x}, \varepsilon), t, \varepsilon)\,dt$$
$$\equiv T(z, \varepsilon). \tag{9.24}$$

Taking into account the first inequality in condition (i) of the lemma and restrictions imposed on the matrices $A_j(\varepsilon)$, one can easily establish that $T(z, \varepsilon)$ maps the set of vectors $z \in R^n$ such that $\|z\| \le c\varepsilon$, $c = c_1 c_{28} c_{29} r K L$, into itself, provided that $c\varepsilon_0 \le \rho(4K)^{-1}$. Moreover, using the inequality

$$\Big\|\frac{\partial}{\partial z}\overline{x}(\tau, z + \widetilde{x}, \varepsilon)\Big\| \le ne^{2c_1 L},$$

we obtain

$$\Big\|\frac{\partial}{\partial z}T(z, \varepsilon)\Big\| \le c_{30}\varepsilon \le \frac{1}{2} \qquad \forall\varepsilon \le \varepsilon_0 \le \frac{1}{2c_{30}}, \qquad c_{30} = cne^{2c_1 L}.$$

Therefore, Eq. (9.24) has a unique solution $z = z(\varepsilon)$ that satisfies the inequality $\|z(\varepsilon)\| \le c\varepsilon$, and the boundary-value problem (9.22) has a unique solution $\overline{x}(\tau, \varepsilon) = \overline{x}(\tau, z(\varepsilon) + \widetilde{x}, \varepsilon)$ for which

$$\|\overline{x}(\tau, \varepsilon) - y(\tau, \varepsilon)\| < K(c + c_1 L)\varepsilon \quad \forall(\tau, \varepsilon) \in [0, L] \times (0, \varepsilon_0].$$

To obtain $\varphi(\tau, \varepsilon)$, we substitute the value of $\overline{x} = \overline{x}(\tau, \varepsilon)$ into (9.23). As a result, we get

$$\overline{\varphi}(\tau, \varepsilon) = \frac{1}{\varepsilon}\int_0^{\tau}[\omega(\overline{x}(t, \varepsilon), t, \varepsilon) + \varepsilon\overline{B}(\overline{x}(t, \varepsilon), t, \varepsilon)]\,dt$$
$$+ \Big(\sum_{j=1}^{r} B_j(\varepsilon)\Big)^{-1}\Big[f(\overline{x}(\tau_1, \varepsilon), \ldots, \overline{x}(\tau_r, \varepsilon), \varepsilon)$$
$$- \frac{1}{\varepsilon}\sum_{j=1}^{r} B_j(\varepsilon)\int_0^{\tau_j}(\omega(\overline{x}(t, \varepsilon), t, \varepsilon) + {+\varepsilon}\overline{B}(\overline{x}(t, \varepsilon), t, \varepsilon))dt\Big].$$

Lemma 9.1 is proved.

The proof of the theorem below, in fact, repeats the proof of Theorem 9.1, and, therefore, we present only its formulation.

Theorem 9.2. *Suppose that conditions (8.2), (8.7), and (9.21) and the conditions of Lemma 9.1 are satisfied and, furthermore,*

$$\Big\|\Big(\sum_{j=1}^{r} B_j(\varepsilon)\Big)^{-1}\Big\| \le c_{31}\varepsilon^{-\beta} \quad \forall \varepsilon \in (0, \varepsilon_0], \quad \beta < \alpha.$$

Then there exist constants c_{32} *and* $\overline{\varepsilon}_0 \le \varepsilon_0$ *such that, for every* $\varepsilon \in (0, \overline{\varepsilon}_0]$, *problem (9.19), (9.20) has a unique solution* $(x(\tau,\varepsilon); \varphi(\tau,\varepsilon))$ *that satisfies the inequalities*

$$\|x(\tau,\varepsilon) - \overline{x}(\tau,\varepsilon)\| \le c_{32}\varepsilon^{1+\alpha},$$

$$\|\varphi(\tau,\varepsilon) - \overline{\varphi}(\tau,\varepsilon)\| \le c_{32}\varepsilon^{\alpha-\beta}$$

$$\forall (\tau,\varepsilon) \in [0, L] \times (0, \overline{\varepsilon}_0].$$

10. Theorems on Existence of Solutions of Boundary-Value Problems

In Sections 6–9, using the principle of contracting mappings, we have proved the existence and uniqueness of solutions of certain boundary-value problems for multifrequency systems. This has been done on the basis of the fact that, for the oscillation system (6.1) with $\omega = \omega(\tau)$ or system (8.1) with $\omega = \omega(x,\tau,\varepsilon)$, we have efficient estimates for the difference of solutions of the original and averaged equations and their partial derivatives with respect to the initial data [inequalities (2.5), (2.7), and (8.8)]. For multifrequency systems of the general form (4.1) in which $a(x,\varphi,\tau)$ depends on angular variables and the frequencies depend on the variables x, the justification of the averaging method can be reduced to the proof of the estimate $\|x - \overline{x}\| \le c(\varepsilon)$, where $c(\varepsilon) \to 0$ as $\varepsilon \to 0$. In this case, for time $\tau \in [0, L]$, the difference of the angular variables $\varphi - \overline{\varphi}$ can reach an arbitrarily large value as $\varepsilon \to 0$ [Arn4, Bak1, GrR3, Kha2]; the same is true for the behavior of the partial derivatives of the functions $x - \overline{x}$ and $\varphi - \overline{\varphi}$ with respect to the initial data. Therefore, the combination of the principle of contracting mappings and the averaging method in the solution of boundary-value problems for systems of the form (4.1) loses its sense.

In the present section, we prove only the existence of solutions of boundary-value problems by using the Schauder fixed-point theorem [Har, Sch]. According to this theorem, for the existence of a (not necessarily unique) solution of the equation $Ty = y$ it is sufficient that the mapping T of the ball $K \subset R^n$ into itself be continuous.

Consider a nonlinear system

$$\frac{dx}{d\tau} = a(x, \varphi, \tau, \varepsilon), \quad \frac{d\varphi}{d\tau} = \frac{\omega(x,\tau)}{\varepsilon} + b(x, \varphi, \tau, \varepsilon) \tag{10.1}$$

whose right-hand side is defined for $(x, \varphi, \tau, \varepsilon) \in \mathcal{D} \times R^m \times [0, L] \times (0, \varepsilon_0] \equiv \overline{G}$ $\big(\mathcal{D}$ is a bounded domain of the real Euclidean space $R^n\big)$ and continuously differentiable with respect to x, φ, and τ for every fixed ε and belongs to the class of almost periodic (with respect to φ_ν, $\nu = \overline{1, m}$) functions

$$a(x, \varphi, \tau, \varepsilon) = \sum_{\nu=0}^{\infty} a_\nu(x, \tau, \varepsilon) e^{i(\lambda_\nu, \varphi)},$$

$$b(x, \varphi, \tau, \varepsilon) = \sum_{\nu=0}^{\infty} b_\nu(x, \tau, \varepsilon) e^{i(\lambda_\nu, \varphi)},$$

$$\lambda_0 = 0, \quad \lambda_\nu = (\lambda_\nu^{(1)}, \ldots, \lambda_\nu^{(m)}) \neq 0 \quad \forall \nu \geq 1,$$

$$i^2 = -1, \quad (\lambda_\nu, \varphi) = \sum_{j=1}^{m} \lambda_\nu^{(j)} \varphi_j,$$

for which

$$\sum_{\nu=1}^{\infty} \left[\left(1 + \frac{1}{\|\lambda_\nu\|}\right) \sup_G \|a_\nu\| + \frac{1}{\|\lambda_\nu\|}\left(\sup_G \left\|\frac{\partial a_\nu}{\partial \tau}\right\| + \sup_G \left\|\frac{\partial a_\nu}{\partial x}\right\|\right)\right] \leq c_1. \tag{10.2}$$

Here, c_1 is a constant independent of ε and $G = \mathcal{D} \times [0, L] \times (0, \varepsilon_0]$. We also assume that the first-order partial derivatives of the functions a, b, and ω with respect to x, φ, and τ are uniformly bounded in $\overline{G}$ by the constant c_1.

For Eqs. (10.1), we introduce boundary conditions of the form

$$F(x|_{\tau=\tau_1}, \ldots, x|_{\tau=\tau_r}, \varepsilon) = 0, \quad \varphi|_{\tau=\tau_{\nu_0}} = \varphi^0, \tag{10.3}$$

where $0 \leq \tau_1 < \tau_2 < \ldots < \tau_r \leq L$, $r \geq 2$, ν_0 is fixed $(1 \leq \nu_0 \leq r)$, $\varphi^0 \in R^m$ is a constant vector, and $F(p_1, \ldots, p_r, \varepsilon)$ is an n-dimensional vector function of the variables $p_j \in \mathcal{D}$, $j = \overline{1, r}$, and $\varepsilon \in (0, \varepsilon_0]$ that has continuous and bounded (by the constant c_1) first-order partial derivatives with respect to all variables p_j, $j = \overline{1, r}$, for every ε.

To investigate the solvability of the multipoint problem (10.1), (10.3), we use the method of averaging with respect to all fast variables φ. Parallel with (10.1), (10.3), we consider the averaged problem

$$\frac{d\overline{x}}{d\tau} = a_0(\overline{x}, \tau, \varepsilon), \tag{10.4^{I}}$$

$$F(\overline{x}|_{\tau=\tau_1}, \ldots, \overline{x}|_{\tau=\tau_r}, \varepsilon) = 0, \tag{10.4^{II}}$$

$$\frac{d\overline{\varphi}}{d\tau} = \frac{\omega(\overline{x}, \tau)}{\varepsilon} + b_0(\overline{x}, \tau, \varepsilon), \tag{10.4^{III}}$$

$$\overline{\varphi}|_{\tau=\tau_{\nu_0}} = \varphi^0, \tag{10.4^{IV}}$$

where

$$[a_0; b_0] = \lim_{T\to\infty} T^{-m} \int\limits_0^T \ldots \int\limits_0^T [a(\overline{x}, \varphi, \tau, \varepsilon); b(\overline{x}, \varphi, \tau, \varepsilon)] d\varphi_1 \ldots d\varphi_m.$$

In order that the averaging method correctly describe the evolution of the slow variables x on the time interval $[0, L]$, it is necessary to impose certain restrictions on the frequency vector $\omega(x, \tau) = (\omega_1(x, \tau), \ldots, \omega_m(x, \tau))$. Assume that, for any $(x, \varphi, \tau, \varepsilon) \in \overline{G}$ and $\nu \geq 1$ and certain $\alpha \in \left[0, \frac{1}{2}\right)$, the following inequality holds:

$$|(\lambda_\nu, \omega(x, \tau))| + |(\lambda_\nu, \Omega(x, \varphi, \tau, \varepsilon))| \geq c_2 \|\lambda_\nu\|, \quad c_2 = \mathrm{const} > 0, \tag{10.5}$$

where

$$\Omega = \frac{\partial \omega(x, \tau)}{\partial \tau} + \frac{\partial \omega(x, \tau)}{\partial x}\Big(a_0(x, \tau, \varepsilon) + \sum_{j=1}^{\infty} a_j(x, \tau, \varepsilon) h_{\varepsilon^\alpha}((\lambda_j, \omega(x, \tau))) e^{i(\lambda_j, \varphi)}\Big),$$

(λ_ν, ω), (λ_ν, Ω), and (λ_j, φ) are the scalar products of vectors, and $h_d(t)$ for $d = \varepsilon^\alpha$ is the function defined in Section 4. Note that, by virtue of the finiteness of the function $h_d(t)$, conditions (10.5) are imposed not on all harmonics of the function $a(x,\varphi,\tau,\varepsilon)$, but only on its resonance harmonics. Under these assumptions, according to the results of Section 4, we have

$$\|x_\tau(t,y,\psi,\varepsilon)-\overline{x}_\tau(t,y,\varepsilon)\|\le\sigma\sqrt{\varepsilon},\quad \sigma=\text{const},\tag{10.6}$$

for all $\tau\in[0,L]$, $y\in\mathcal{D}_1$, $\psi\in R^m$, and $\varepsilon\in(0,\varepsilon_0]$. In this estimate, $(x_\tau(t,y,\psi,\varepsilon);\varphi_\tau(t,y,\psi,\varepsilon))$ and $(\overline{x}_\tau(t,y,\varepsilon);\overline{\varphi}_\tau(t,y,\psi,\varepsilon))$ are the solutions of Eqs. (10.1) and the averaged equations (10.4^{I}) and (10.4^{III}) that take the value $(y;\psi)$ for $\tau=t$, and the curve $\overline{x}=\overline{x}_\tau(t,y,\varepsilon)$ lies in $\mathcal{D}$ together with its ρ-neighborhood $\forall(\tau,y,\varepsilon)\in[0,L]\times\mathcal{D}_1\times(0,\varepsilon_0]$.

Theorem 10.1. *Suppose that the following conditions are satisfied:*

(i) conditions (10.2) and (10.5) and the restrictions imposed on a, b, ω, *and* F *are satisfied;*

(ii) the matrices $\dfrac{\partial a_0(x,\tau,\varepsilon)}{\partial x}$ *and* $\dfrac{\partial F(p,\varepsilon)}{\partial p}$ *are uniformly (with respect to* $\varepsilon\in(0,\varepsilon_0]$*) uniformly continuous in* $x\in\mathcal{D}$, $\tau\in[0,L]$, *and* $p=(p_1,\ldots,p_r)\in\mathcal{D}\times\ldots\times\mathcal{D}\equiv\mathcal{D}^r$;

(iii) for every $\varepsilon\in(0,\varepsilon_0]$, *there exists a solution* $\overline{x}=\overline{x}_\tau(\tau_{\nu0},x^0,\varepsilon)$, $x^0=x^0(\varepsilon)$, *of problem (10.4^{I}), (10.4^{II}) that lies in* $\mathcal{D}$ *together with its* ρ-*neighborhood;*

(iv) $\|S^{-1}(\varepsilon)\|\le c_3=\text{const}$ $\forall\varepsilon\in(0,\varepsilon_0]$, *where*

$$S(\varepsilon)=\sum_{j=1}^{r}\frac{\partial F^0}{\partial p_j}\frac{\partial\overline{x}_{\tau_j}(\tau_{\nu0},x^0,\varepsilon)}{\partial x^0},$$

and $\dfrac{\partial F^0}{\partial p_j}$ *denotes the matrix of the first-order partial derivatives of the function* $F(p_1,\ldots,p_r,\varepsilon)$ *with respect to* p_j *for* $p_\mu=\overline{x}_{\tau_\mu}(\tau_{\nu0},x^0,\varepsilon)$, $\mu=\overline{1,r}$.

Then one can find constants $\overline{c}_1>0$ *and* $\varepsilon_1\in(0,\varepsilon_0]$ *such that, for every* $\varepsilon\in(0,\varepsilon_1]$, *problem (10.1), (10.3) has at least one solution* $(x(\tau,\varepsilon);\varphi(\tau,\varepsilon))$ *for which*

$$\|x(\tau,\varepsilon)-\overline{x}_\tau(\tau_{\nu0},x^0,\varepsilon)\|\le\overline{c}_1\sqrt{\varepsilon}\quad\forall(\tau,\varepsilon)\in[0,L]\times(0,\varepsilon_1].\tag{10.7}$$

Proof. We seek a solution of problem (10.1), (10.3) in the form $(x_\tau(\tau_{\nu 0}, x^0 + y, \varphi^0, \varepsilon); \varphi_\tau(\tau_{\nu 0}, x^0 + y, \varphi^0, \varepsilon))$ and determine the unknown parameter $y \in R^n$ from the boundary conditions (10.3):

$$y = -S^{-1}(\varepsilon)\Big\{\Big[F(x_{\tau_1}(\tau_{\nu 0}, x^0 + y, \varphi^0, \varepsilon), \ldots, x_{\tau_r}(\tau_{\nu 0}, x^0 + y, \varphi^0, \varepsilon), \varepsilon)$$
$$- F(\overline{x}_{\tau_1}(\tau_{\nu 0}, x^0 + y, \varepsilon), \ldots, \overline{x}_{\tau_r}(\tau_{\nu 0}, x^0 + y, \varepsilon), \varepsilon)\Big]$$
$$+ \Big[F(\overline{x}_{\tau_1}(\tau_{\nu 0}, x^0 + y, \varepsilon), \ldots, \overline{x}_{\tau_r}(\tau_{\nu 0}, x^0 + y, \varepsilon), \varepsilon) - S(\varepsilon)y\Big]\Big\}$$
$$\equiv M_\varepsilon(y). \tag{10.8}$$

Taking into account the restrictions for F and estimate (10.6), we get

$$\|F(x_{\tau_1}(\tau_{\nu 0}, x^0 + y, \varphi^0, \varepsilon), \ldots, \varepsilon) - F(\overline{x}_{\tau_1}(\tau_{\nu 0}, x^0 + y, \varepsilon), \ldots, \varepsilon)\|$$
$$\le c_1 r \sigma \sqrt{\varepsilon}. \tag{10.9}$$

We now fix arbitrary a positive $\mu \le [2(1 + Lc_1)nre^{2nc_1L}c_3]^{-1} \equiv \Delta$. Then it follows from condition (ii) of Theorem 10.1 that there exists $\delta = \delta(\mu)$ such that, for $\|z\| + \|\widetilde{p}\| < \delta$, we have

$$\Big\|\frac{\partial}{\partial x}a_0(x + z, \tau, \varepsilon) - \frac{\partial}{\partial x}a_0(x, \tau, \varepsilon)\Big\|$$
$$+ \Big\|\frac{\partial}{\partial p}F(p + \widetilde{p}, \varepsilon) - \frac{\partial}{\partial p}F(p, \varepsilon)\Big\| < \mu \tag{10.10}$$

We choose $\delta < \dfrac{1}{n}e^{-nc_1L}\rho$ and rewrite the averaged equations (10.4^{I}) in the form

$$\overline{x}_\tau(\tau_{\nu 0}, x^0 + y, \varepsilon) = x^0 + y + \int_{\tau_{\nu 0}}^{\tau} a_0(\overline{x}_t(\tau_{\nu 0}, x^0 + y, \varepsilon), t, \varepsilon)dt.$$

Differentiating this equality with respect to x^0 and using relation (10.10) and the Gronwall–Bellman inequality, we get

$$\Big\|\frac{\partial}{\partial x^0}(\overline{x}_\tau(\tau_{\nu 0}, x^0 + y, \varepsilon) - \overline{x}_\tau(\tau_{\nu 0}, x^0, \varepsilon))\Big\| \le nLe^{2c_1Ln}\mu$$

for any $(\tau, \varepsilon) \in [0, L] \times (0, \varepsilon_0]$ and $\|y\| < \delta$. Therefore,

$$\overline{x}_\tau(\tau_{\nu 0}, x^0 + y, \varepsilon) = \overline{x}_\tau(\tau_{\nu 0}, x^0, \varepsilon) + \frac{\partial}{\partial x^0}\overline{x}_\tau(\tau_{\nu 0}, x^0, \varepsilon)y + h_1(\tau, y, \varepsilon), \tag{10.11}$$

where

$$h_1(\tau, y, \varepsilon) = \int_0^1 \Big[\frac{\partial}{\partial x^0}\overline{x}_\tau(\tau_{\nu 0}, x^0 + ty, \varepsilon) - \frac{\partial}{\partial x^0}\overline{x}_\tau(\tau_{\nu 0}, x^0, \varepsilon)\Big]\, dty, \qquad (10.12)$$

$$\|h_1(\tau, y, \varepsilon)\| \le nLe^{2c_1Ln}\mu\|y\|$$

for all $\tau \in [0, L]$, $\varepsilon \in (0, \varepsilon_0]$, and $\|y\| < \delta$.

The boundary condition (10.4^{II}) and relations (10.10)–(10.12) yield the representation

$$F(\overline{x}_{\tau_1}(\tau_{\nu 0}, x^0 + y, \varepsilon), \ldots, \varepsilon) = S(\varepsilon)y + h_2(y, \varepsilon), \qquad (10.13)$$

where

$$\|h_2(y, \varepsilon)\| \le nr(1 + Lc_1)e^{2c_1Ln}\mu\|y\| \quad \forall\varepsilon \in (0, \varepsilon_0], \ \ \|y\| < \delta. \qquad (10.14)$$

Thus, it follows from (10.8), (10.9), (10.13), and (10.14) that

$$\|M_\varepsilon(y)\| < c_3\Big(c_1r\sigma\sqrt{\varepsilon} + \frac{1}{c_3 2\Delta}\mu\|y\|\Big) \quad \forall\varepsilon \in (0, \varepsilon_0], \ \ \|y\| < \delta.$$

This implies that $M_\varepsilon(y)$ maps the set $\|y\| \le 2c_1c_3r\sigma\sqrt{\varepsilon} \equiv c\sqrt{\varepsilon}$ into itself, provided that $c\sqrt{\varepsilon} < \delta$. Also note that, for every ε, the vector function $M_\varepsilon(y)$ is continuous with respect to y; therefore, according to the Schauder theorem, there exists a solution $y = y(\varepsilon)$, $\|y(\varepsilon)\| \le c\sqrt{\varepsilon}$, of Eq. (10.8), and, hence, there exists a solution

$$(x(\tau, \varepsilon); \varphi(\tau, \varepsilon)) = (x_\tau(\tau_{\nu 0}, x^0 + y(\varepsilon), \varphi^0, \varepsilon); \varphi_\tau(\tau_{\nu 0}, x^0 + y(\varepsilon), \varphi^0, \varepsilon))$$

of the multipoint problem (10.1), (10.3). Estimate (10.7) with the constant $\overline{c}_1 = \sigma + cne^{c_1nL}$ follows from estimate (10.6) and the inequality

$$\|\overline{x}_\tau(\tau_{\nu 0}, x^0 + y(\varepsilon), \varepsilon) - \overline{x}_\tau(\tau_{\nu 0}, x^0, \varepsilon)\| \le ne^{nc_1L}\|y(\varepsilon)\| \le nce^{nc_1L}\sqrt{\varepsilon}.$$

To complete the proof of the theorem, we impose the condition $\overline{c}_1\sqrt{\varepsilon} < \frac{1}{2}\rho$, which guarantees that the curve $x = x(\tau, \varepsilon)$ lies in $\mathcal{D}$ together with its $\frac{1}{2}\rho$-neighborhood $\forall\tau \in [0, L]$.

Condition (iv) is an essential assumption in Theorem 10.1. In what follows, we consider the case where this condition is not satisfied, namely, we assume that

$$\|S^{-1}(\varepsilon)\| \le K\varepsilon^{-l_1}, \quad l_1 = \text{const} > 0, \ \ K = \text{const} > 0. \qquad (10.15)$$

Theorem 10.2. *Suppose that the following conditions are satisfied:*

(a) conditions (i)–(iii) of Theorem 10.1 and inequality (10.15) are satisfied;

(b) the matrices $\frac{\partial}{\partial x} a_0(x, \tau, \varepsilon)$ *and* $\frac{\partial}{\partial p} F(p, \varepsilon)$ *satisfy the Hölder conditions*

$$\left\| \frac{\partial}{\partial x} a_0(x, \tau, \varepsilon) - \frac{\partial}{\partial x} a_0(\overline{x}, \tau, \varepsilon) \right\| \le M \|x - \overline{x}\|^{l_2}, \quad 0 < l_2 \le 1,$$

$$\left\| \frac{\partial}{\partial p} F(p, \varepsilon) - \frac{\partial}{\partial p} F(\overline{p}, \varepsilon) \right\| \le M \|p - \overline{p}\|^{l_2}$$

for all $x, \overline{x} \in \mathcal{D}$, $p, \overline{p} \in D^r$, $\tau \in [0, L]$, *and* $\varepsilon \in (0, \varepsilon_0]$, *and the constant* M *is independent of* ε;

(c) $l_1 < \dfrac{l_2}{2(1 + l_2)}$.

Then, for sufficiently small $\varepsilon_0 > 0$ *and every* $\varepsilon \in (0, \varepsilon_0]$, *there exists at least one solution* $(x(\tau, \varepsilon); \varphi(\tau, \varepsilon))$ *of problem (10.1), (10.3) that satisfies the inequality*

$$\|x(\tau, \varepsilon) - \overline{x}_\tau(\tau_{\nu 0}, x^0, \varepsilon)\| \le \underline{c}_1 \varepsilon^{\frac{1}{2} - l_1}$$

$$\forall (\tau, \varepsilon) \in [0, L] \times (0, \varepsilon_0], \quad \underline{c}_1 = \text{const.}$$

Proof. We follow the scheme of the proof of Theorem 10.1. To determine a solution

$$(x_\tau(\tau_{\nu 0}, x^0 + y, \varphi^0, \varepsilon); \varphi_\tau(\tau_{\nu 0}, x^0 + y, \varphi^0, \varepsilon))$$

of problem (10.1), (10.3), i.e., to find y, we write equality (10.8) and inequality (10.9). It is easy to verify that the fact that $\frac{\partial}{\partial x} a_0(x, \tau, \varepsilon)$ belongs to the Hölder class guarantees that

$$\left\| \frac{\partial}{\partial x^0} \overline{x}_\tau(\tau_{\nu 0}, x^0 + y, \varepsilon) - \frac{\partial}{\partial x^0} \overline{x}_\tau(\tau_{\nu 0}, x^0, \varepsilon) \right\| \le M_1 \|y\|^{l_2},$$

$$M_1 = M L n^{1 + l_2} e^{n c_1 (2 + l_2) L}$$

for $\|y\| < \left(\frac{\rho}{2n}\right) e^{-n c_1 L}$, $\tau \in [0, L]$, and $\varepsilon \in (0, \varepsilon_0]$. Therefore, for the function $h_1(\tau, y, \varepsilon)$ defined by equality (10.12), the following estimate is true:

$$\|h_1(\tau, y, \varepsilon)\| \le \frac{1}{1 + l_2} M_1 \|y\|^{1 + l_2} \qquad (10.16)$$

Taking into account equality (10.11), estimate (10.16), and the fact that the matrix $\dfrac{\partial}{\partial p}F(p,\varepsilon)$ belongs to the Hölder class, we obtain relation (10.13) in which

$$\|h_2(y,\varepsilon)\| \le M_2\|y\|^{1+l_2}, \quad M_2 = 3rc_1M_1 + rn^{1+l_2}Me^{nc_1(2+l_2)L}.$$

Thus,

$$\|M_\varepsilon(y)\| \le K\varepsilon^{-l_1}\left[rc_1\sigma\sqrt{\varepsilon} + M_2\|y\|^{1+l_2}\right]$$

$$\forall\varepsilon \in (0,\varepsilon_0), \quad \|y\| \le \left(\frac{\rho}{2n}\right)e^{-nc_1L}.$$

The analysis of the last inequality shows that if condition (c) of Theorem 10.2 is satisfied, then $M_\varepsilon(y)$ maps the set

$$\{y\colon \|y\| \le 2Krc_1\sigma\varepsilon^{\frac{1}{2}-l_1}\}$$

into itself for every $\varepsilon \in (0,\varepsilon_0]$, provided that

$$\varepsilon_0 \le \min\left\{\left(\frac{4}{\rho}nc_1r\sigma Ke^{nc_1L}\right)^{\frac{2}{2l_1-1}};\ \left[\frac{1}{2KM_2}\left(\frac{1}{2rc_1\sigma K}\right)^{l_2}\right]^{\frac{2}{l_2-2(1+l_2)l_1}}\right\}.$$

Since the mapping $M_\varepsilon(y)$ is continuous in y, there exists a solution $y = y(\varepsilon)$ of Eq. (10.8) that satisfies the inequality $\|y(\varepsilon)\| \le 2Krc_1\sigma\varepsilon^{\frac{1}{2}-l_1}$. Therefore,

$$(x(\tau,\varepsilon);\varphi(\tau,\varepsilon)) = (x_\tau(\tau_{\nu 0},x^0+y(\varepsilon),\varphi^0,\varepsilon);\varphi_\tau(\tau_{\nu 0},x^0+y(\varepsilon),\varphi^0\varepsilon))$$

is a solution of problem (10.1), (10.3), and

$$\begin{aligned}\|x(\tau,\varepsilon) - \overline{x}_\tau(\tau_{\nu 0},x^0,\varepsilon)\| &\le \|x(\tau,\varepsilon) - \overline{x}_\tau(\tau_{\nu 0},x^0+y(\varepsilon),\varepsilon)\| \\ &\quad + \|\overline{x}_\tau(\tau_{\nu 0},x^0+y(\varepsilon),\varepsilon) - \overline{x}_\tau(\tau_{\nu 0},x^0,\varepsilon)\| \\ &\le \underline{c}_1\varepsilon^{\frac{1}{2}-l_1}, \\ \underline{c}_1 &= \sigma + 2Krc_1\sigma ne^{nc_1L}.\end{aligned}$$

Theorem 10.2 is proved.

For Eqs. (10.1), we now introduce boundary conditions of the form

$$x|_{\tau=\tau_{\nu_0}} = y^0 \in \mathcal{D}, \quad \sum_{j=1}^{r} B_j(\varepsilon)\varphi|_{\tau=\tau_j} = f(x|_{\tau=\tau_1}, \dots, x|_{\tau=\tau_r}, \varepsilon). \qquad (10.17)$$

Here, $B_j(\varepsilon)$ are quadratic m-dimensional matrices and $f(p_1, \dots, p_r, \varepsilon)$ is an m-dimensional vector function.

Theorem 10.3. *Suppose that the following conditions are satisfied:*

(a) condition (i) of Theorem 10.1 is satisfied;

(b) $f(p_1, \dots, p_r, \varepsilon)$ *is continuous in* $p_j \in \mathcal{D}$, $j = \overline{1, r}$, *and*

$$\det \sum_{j=1}^{r} B_j(\varepsilon) \neq 0 \quad \forall \varepsilon \in (0, \varepsilon_0];$$

(c) the curve $\overline{x} = \overline{x}_\tau(\tau_{\nu_0}, y^0, \varepsilon)$ *lies in* $\mathcal{D}$ *together with its* ρ*-neighborhood for* $\tau \in [0, L]$ *and* $\varepsilon \in (0, \varepsilon_0]$.

Then a solution $(x(\tau,\varepsilon); \varphi(\tau,\varepsilon))$ *of problem (10.1), (10.17) exists, and the slow variables* $x(\tau,\varepsilon)$ *of every solution lie in a* $\sigma\sqrt{\varepsilon}$*-neighborhood of the curve* $\overline{x} = \overline{x}_\tau(\tau_{\nu_0}, y^0, \varepsilon)$ $\forall (\tau, \varepsilon) \in [0, L] \times (0, \varepsilon_0]$.

Proof. We represent the fast variables $\varphi(\tau, \varepsilon)$ of the required solution

$$(x(\tau,\varepsilon); \varphi(\tau,\varepsilon)) = (x_\tau(\tau_{\nu_0}, y^0, \psi, \varepsilon); \varphi_\tau(\tau_{\nu_0}, y^0, \psi, \varepsilon)) \qquad (10.18)$$

of the multipoint (10.1), (10.17) in the form

$$\varphi(\tau, \varepsilon) = \psi + \frac{1}{\varepsilon}\theta(\tau, \psi, \varepsilon),$$

$$\theta(\tau, \psi, \varepsilon) = \int_{\tau_{\nu_0}}^{\tau} [\omega(x_t(\tau_{\nu_0}, y^0, \psi, \varepsilon), t) + \varepsilon b(x_\tau(\tau_{\nu_0}, y^0, \psi, \varepsilon), \varphi_t(\tau_{\nu_0}, y^0, \psi, \varepsilon), t, \varepsilon)]dt,$$

$$\|\theta(\tau, \psi, \varepsilon)\| \le c_1 L(1 + \varepsilon) \quad \forall (\tau, \psi, \varepsilon) \in [0, L] \times R^m \times (0, \varepsilon_0].$$

Here, ψ is unknown. To determine ψ, we use the boundary conditions (10.17). As a result, we get

$$\psi = \left(\sum_{j=1}^{r} B_j(\varepsilon)\right)^{-1} \left[\widetilde{f}(\psi,\varepsilon) - \frac{1}{\varepsilon}\sum_{j=1}^{r} B_j(\varepsilon)\theta(\tau_j,\psi,\varepsilon)\right] \equiv T_\varepsilon(\psi), \quad (10.19)$$

where

$$\widetilde{f}(\psi,\varepsilon) = f(x_{\tau_1}(\tau_{\nu 0}, y^0, \psi, \varepsilon), \ldots, x_{\tau_r}(\tau_{\nu 0}, y^0, \psi, \varepsilon), \varepsilon).$$

Taking into account the continuity of $f(p_1, \ldots, p_r, \varepsilon)$ in $p_j \in \mathcal{D},\ j = \overline{1,r}$, condition (c) of Theorem 10.3, and an estimate of the form (10.6), namely

$$\|x(\tau,\varepsilon) - \overline{x}_\tau(\tau_{\nu 0}, y^0, \varepsilon)\| \le \sigma\sqrt{\varepsilon},$$

and choosing $\varepsilon_0 \le \left(\dfrac{\rho}{2\sigma}\right)^2$, we establish the existence of a constant $c(\varepsilon)$ such that $\|\widetilde{f}(\psi,\varepsilon)\| \le c(\varepsilon)\ \ \forall \psi \in R^m,\ \ \varepsilon \in (0,\varepsilon_0]$. Then relation (10.19) yields

$$\|T_\varepsilon(\psi)\| \le \left\| \left(\sum_{j=1}^{r} B_j(\varepsilon)\right)^{-1} \right\| \left[c(\varepsilon) + \frac{1}{\varepsilon} L c_1 (1+\varepsilon) \sum_{j=1}^{r} \|B_j(\varepsilon)\|\right] \equiv \overline{c}(\varepsilon).$$

This inequality, together with the condition of the continuity of the function $T_\varepsilon(\psi)$ with respect to ψ, guarantees the existence of a solution $\psi = \psi(\varepsilon)$, $\|\psi(\varepsilon)\| \le \overline{c}(\varepsilon)$, of Eq. (10.19) and, hence, the existence of a solution (10.18) of problem (10.1), (10.17). Theorem 10.3 is proved.

The linear dependence of the boundary conditions (10.17) on $\varphi|_{\tau=\tau_j}$, $j = \overline{1,r}$, is an essential assumption in Theorem 10.3. Below, we establish sufficient conditions for the solvability of a multipoint problem for a one-frequency system in the case where the boundary conditions contain nonlinearities indicated above.

Consider the case of the one-frequency $(m = 1)$ system (10.1) with the boundary conditions

$$x|_{\tau=\tau_{\nu 0}} = y^0, g(\varphi|_{\tau=\tau_1}, \ldots, \varphi|_{\tau=\tau_r}, \varepsilon) = f(x|_{\tau=\tau_1}, \ldots, x|_{\tau=\tau_r}, \varepsilon). \quad (10.20)$$

Here, $g(q_1, \ldots, q_r, \varepsilon)$ and $f(p_1, \ldots, p_r, \varepsilon)$ are scalar functions of the variables $q_j \in R^m$, $p_j \in \mathcal{D}$, $j = \overline{1,r}$, and $\varepsilon \in (0, \varepsilon_0]$.

Theorem 10.4. *Suppose that the following conditions are satisfied:*

(i) conditions (a) and (c) of Theorem 10.3 are satisfied;

(ii) $f(p_1, \dots, p_r, \varepsilon)$ *and* $g(q_1, \dots, q_r, \varepsilon)$ *are continuous in* $p_j \in \mathcal{D}$ *and* $q_j \in R^m$, $j = \overline{1, r}$;

(iii) for every $N > 0$, *the following limits exist uniformly in* $c^{(j)} = \text{const}$, $|c^{(j)}| \leq N$, $j = \overline{1, r}$:

$$\lim_{t \to \infty} g(t + c^{(1)}, \dots, t + c^{(r)}, \varepsilon) = \infty,$$

$$\lim_{t \to -\infty} g(t + c^{(1)}, \dots, t + c^{(r)}, \varepsilon) = -\infty, \tag{10.21}$$

or

$$\lim_{t \to \infty} g(t + c^{(1)}, \dots, t + c^{(r)}, \varepsilon) = -\infty,$$

$$\lim_{t \to -\infty} g(t + c^{(1)}, \dots, t + c^{(r)}, \varepsilon) = \infty. \tag{10.22}$$

Then a solution $(x(\tau, \varepsilon); \varphi(\tau, \varepsilon))$ *of the multipoint problem (10.1), (10.20) exists, and the slow variables* $x(\tau, \varepsilon)$ *of every solution satisfy the estimate*

$$\|x(\tau, \varepsilon) - \overline{x}_\tau(\tau_{\nu_0}, y^0, \varepsilon)\| < \sigma\sqrt{\varepsilon} \quad \forall (\tau, \varepsilon) \in [0, L] \times (0, \varepsilon_0]. \tag{10.23}$$

Proof. To find solution (10.18) of problem (10.1), (10.20), we rewrite the boundary conditions in the form

$$\widetilde{g}(\psi, \varepsilon) \equiv g\Big(\psi + \frac{1}{\varepsilon}\theta(\tau_1, \psi, \varepsilon), \dots, \psi + \frac{1}{\varepsilon}\theta(\tau_r, \psi, \varepsilon), \varepsilon\Big) - \widetilde{f}(\psi, \varepsilon) = 0,$$

where

$$\widetilde{f}(\psi, \varepsilon) \qquad (|\widetilde{f}(\psi, \varepsilon)| \leq c(\varepsilon) \quad \forall \psi \in R^m)$$

and

$$\theta(\tau, \psi, \varepsilon) \qquad (|\theta(\tau, \psi, \varepsilon)| \leq c_1 L(1 + \varepsilon) \quad \forall \tau \in [0, L], \psi \in R)$$

are defined in the proof of Theorem 10.3. We use condition (iii) of Theorem 10.4 for $N = \Big(\dfrac{1}{\varepsilon} + 1\Big) L c_1$ and consider, e.g., case (10.21). According to condition (iii), there exists $N_1(\varepsilon) > 0$ such that, for $\psi^{(1)} < -N_1(\varepsilon)$ and $\psi^{(2)} > N_1(\varepsilon)$, the following inequalities are true:

$$g\left(\psi^{(1)} + \frac{1}{\varepsilon}\theta(\tau_1, \psi^{(1)}, \varepsilon), \ldots, \psi^{(1)} + \frac{1}{\varepsilon}\theta(\tau_r, \psi^{(1)}, \varepsilon), \varepsilon\right) < -2c(\varepsilon),$$

$$g\left(\psi^{(2)} + \frac{1}{\varepsilon}\theta(\tau_1, \psi^{(2)}, \varepsilon), \ldots, \psi^{(2)} + \frac{1}{\varepsilon}\theta(\tau_r, \psi^{(2)}, \varepsilon), \varepsilon\right) > 2c(\varepsilon).$$

Taking into account that $|\widetilde{f}(\psi, \varepsilon)| < c(\varepsilon),$ we get

$$\widetilde{g}(\psi^{(1)}, \varepsilon) \leq -c(\varepsilon) < 0, \qquad \widetilde{g}(\psi^{(2)}, \varepsilon) \geq c(\varepsilon) > 0.$$

Since (according to the assumptions made) the function $\widetilde{g}(\psi, \varepsilon)$ is continuous in $\psi \in R,$ there exists $\psi = \psi(\varepsilon)$ such that $\widetilde{g}(\psi(\varepsilon), \varepsilon) = 0.$ This yields the existence of solution (10.18) of problem (10.1), (10.20), and estimate (10.20) follows from estimate (10.6). Theorem 10.4 is proved.

11. Boundary-Value Problems with Parameters

In the present section, we study boundary-value problems with parameters for the oscillation system

$$\frac{dx}{d\tau} = a(x, \varphi, \tau), \qquad \frac{d\varphi}{d\tau} = \frac{\omega(\tau)}{\varepsilon} + b(x, \varphi, \tau). \tag{11.1}$$

Most investigations of boundary-value problems with parameters relate to the case where unknown parameters are present only in differential equations. However, for practical purposes, it is also necessary to study problems with parameters in boundary conditions [Luc, Sam9, SaR]. Below, we show that the averaging method can be efficiently applied to the proof of the solvability of boundary-value problems with parameters.

Assume that the functions a and b have continuous bounded partial derivatives with respect to $(x, \varphi, \tau) \in D \times R^m \times [0, L]$ up to the second order inclusive and are almost periodic in $\varphi_\nu,$ $\nu = \overline{1, m},$ and such that

$$[a(x, \varphi, \tau); b(x, \varphi, \tau)] = \sum_{s=0}^{\infty} [a_s(x, \tau); b_s(x, \tau)] e^{i(\lambda_s, \varphi)},$$

where i is the imaginary unit, $\lambda_0 = 0,$ $\lambda_s \neq 0$ for $s \geq 1,$ (λ_s, φ) is the scalar product of vectors $\lambda_s = (\lambda_s^{(1)}, \ldots, \lambda_s^{(m)})$ and $\varphi = (\varphi_1, \ldots, \varphi_m),$ and the functions $c_s = [a_s(x, \tau); b_s(x, \tau)]$ satisfy the inequality

$$\sup\|c_0\| + \sup\left\|\frac{\partial c_0}{\partial \tau}\right\| + \sup\left\|\frac{\partial c_0}{\partial x}\right\| + \sum_{j=1}^{n}\sup\left\|\frac{\partial^2 c_0}{\partial x \partial x_j}\right\|$$

$$+ \sum_{s=1}^{\infty}\Big[\Big(\|\lambda_s\| + \frac{1}{\|\lambda_s\|}\Big)\sup\|c_s\|$$

$$+ \Big(1 + \frac{1}{\|\lambda_s\|}\Big)\Big(\sup\left\|\frac{\partial c_s}{\partial \tau}\right\| + \sup\left\|\frac{\partial c_s}{\partial x}\right\|\Big)$$

$$+ \frac{1}{\|\lambda_s\|}\Big(\sup\left\|\frac{\partial^2 c_s}{\partial \tau \partial x}\right\| + \sum_{j=1}^{n}\sup\left\|\frac{\partial^2 c_s}{\partial x \partial x_j}\right\|\Big)\Big] \le \sigma_1, \qquad (11.2)$$

where the supremum is taken over all $(x, \tau) \in \mathcal{D} \times [0, L]$.

Consider boundary conditions of the form

$$A_1 x|_{\tau=0} + A_2 x|_{\tau=\mu} = C_1, \quad x_n|_{\tau=0} = x_n^0,$$

$$B_1 \varphi|_{\tau=0} + B_2 x|_{\tau=\mu} = C_2, \qquad (11.3)$$

where A_1 and A_2 are $n \times n$ matrices, B_1 and B_2 are $m \times m$ matrices, C_1 and C_2 are n-dimensional and m-dimensional vectors, $\mu \in (0, L)$ is an unknown parameter, x_n is the nth coordinate of the component $x = (x_1, \ldots, x_n)$ of a solution $(x; \varphi)$ of system (11.1), and x_n^0 is a given number.

Problem (11.1), (11.3) is a boundary-value problem with nonfixed right boundary. To solve this problem, i.e., to determine the unknown parameter μ and a solution of system (11.1) that satisfies the boundary conditions (11.3), we use the method of averaging over all fast variables φ. We write the averaged problem

$$\frac{d\overline{x}}{d\tau} = \overline{a}(\overline{x}, \tau), \qquad (11.4^{\mathrm{I}})$$

$$\frac{d\overline{\varphi}}{d\tau} = \frac{\omega(\tau)}{\varepsilon} + \overline{b}(\overline{x}, \tau), \qquad (11.4^{\mathrm{II}})$$

$$A_1 \overline{x}|_{\tau=0} + A_2 \overline{x}|_{\tau=\mu} = C_1, \quad \overline{x}_n|_{\tau=0} = x_n^0, \qquad (11.4^{\mathrm{III}})$$

$$B_1 \overline{\varphi}|_{\tau=0} + B_2 \overline{\varphi}|_{\tau=\mu} = C_2, \qquad (11.4^{\mathrm{IV}})$$

where

$$[\overline{a};\overline{b}] = \lim_{T\to\infty} T^{-m} \int_0^T \dots \int_0^T [a(\overline{x},\varphi,\tau); b(\overline{x},\varphi,\tau)]d\varphi_1 \dots d\varphi_m$$

$$\equiv [a_0(\overline{x},\tau); b_0(\overline{x},\tau)].$$

As in the previous sections, we denote by $(x(\tau,y,\psi,\varepsilon);\ \varphi(\tau,y,\psi,\varepsilon))$ and $(\overline{x}(\tau,y);\overline{\varphi}(\tau,y,\psi,\varepsilon))$, respectively, the solutions of problems (11.1) and (11.4^{I}), (11.4^{II}) that take the value $(y;\psi)$ for $\tau = 0$. By $\mathcal{D}_\rho$, we denote the set of points $y \in \mathcal{D}$ for which the curve $\overline{x} = \overline{x}(\tau,y)$ lies in $\mathcal{D}$ together with its ρ-neighborhood $\forall\tau \in [0,L]$. Assume that the set $\mathcal{D}_{\rho_0}$ is nonempty for certain $\rho_0 > 0$.

Lemma 11.1. *Suppose that the following conditions are satisfied:*

(a) there exists a unique solution $\widetilde{x}^0, \mu^0$ *of the equation*

$$A_1x^0 + A_2\overline{x}(\mu^0,x^0) = C_1$$

such that $x^0 = (\widetilde{x}^0, x_n^0) \in \mathcal{D}_{\rho_0}$ *and* $\mu^0 \in (0,L)$*;*

(b) $\det(B_1 + B_2) \neq 0$.

Then there exists a unique solution $\{\mu^0,\overline{x}(\tau,x^0),\overline{\varphi}(\tau,x^0,\varphi^0,\varepsilon)\}$ *of the averaged problem* (11.4^{I})*–*(11.4^{IV}) *defined* $\forall(\tau,\varepsilon) \in [0,L]\times(0,\varepsilon_0]$.

Proof. It follows from condition (a) that the solution $\overline{x} = \overline{x}(\tau,x^0)$ of Eq. (11.4^{I}) is defined for all $\tau \in [0,L]$ and satisfies the boundary condition (11.4^{III}) for $\mu = \mu^0$. It is easy to verify that this solution is associated with the unique solution

$$\overline{\varphi}(\tau,x^0,\varphi^0,\varepsilon) = \varphi^0 + \frac{1}{\varepsilon}\int_0^\tau [\omega(t) + \varepsilon\overline{b}(\overline{x}(t,x^0),t)]dt, \qquad \tau \in [0,L], \qquad (11.5)$$

of problem (11.4^{II}), (11.4^{IV}), where

$$\varphi^0 = (B_1+B_2)^{-1}\Big\{C_2 - \frac{1}{\varepsilon}B_2\int_0^{\mu_0}[\omega(t) + \varepsilon\overline{b}(\overline{x}(t,x^0),t)]dt\Big\}.$$

Lemma 11.1 is proved.

We now study the problem of the existence of a solution of the original problem (11.1), (11.3) and establish an estimate for its deviation from a solution of the averaged problem (11.4^{I})–(11.4^{IV}). For this purpose, we denote by P the $n \times n$ square matrix

$$P = \left(A_1^1 + A_2 \frac{\partial \overline{x}(\mu^0, x^0)}{\partial \widetilde{x}^0}, A_2 \overline{a}(\overline{x}(\mu^0, x^0), \mu^0)\right).$$

Here, A_1^1 is the $n \times (n-1)$ rectangular matrix whose columns are the first $n-1$ columns of the matrix A.

Theorem 11.1. *Suppose that the following conditions are satisfied:*

(i) $\omega(\tau) = (\omega_1(\tau), \ldots, \omega_m(\tau)) \in C^{m-1+l}_{[0,L]}$ *and the Wronskian of the functions* $\omega_1(\tau), \ldots, \omega_m(\tau)$ *has zeros of multiplicity not higher than* l *on* $[0, L]$;

(ii) the conditions of Lemma 11.1 and inequality (11.2) are satisfied;

(iii) $\det P \neq 0$ *and*

$$\sigma_0 = \sup_{\varphi \in R^m} \|P^{-1} A_2 \widetilde{a}(\overline{x}(\mu^0, x^0), \varphi, \mu^0)\| < 1,$$

where $\widetilde{a}(x, \varphi, \tau) = a(x, \varphi, \tau) - \overline{a}(x, \tau)$.

Then one can find positive constants $\overline{\varepsilon}_0 \leq \varepsilon_0$ *and* σ *such that, for every* $\varepsilon \in (0, \overline{\varepsilon}_0]$, *there exists a solution* $\{\mu(\varepsilon), x(\tau, \varepsilon), \varphi(\tau, \varepsilon)\}$ *of the boundary-value problem (11.1), (11.3) that satisfies the estimates*

$$|\mu(\varepsilon) - \mu^0| + \|x(\tau, \varepsilon) - \overline{x}(\tau, x^0)\| \leq \sigma \varepsilon^\alpha,$$

$$\|\varphi(\tau, \varepsilon) - \overline{\varphi}(\tau, x^0, \varphi^0, \varepsilon)\| \leq \sigma \varepsilon^{\alpha - 1} \quad \forall \tau \in [0, L], \alpha = \frac{1}{m + l}. \tag{11.6}$$

Proof. We seek a solution of problem (11.1), (11.3) in the form

$$\{\mu^0 + h, x(\tau, x^0 + y, \varphi^0 + \psi, \varepsilon), \varphi(\tau, x^0 + y, \varphi^0 + \psi, \varepsilon)\},$$

where $y = (\widetilde{y}, 0) = (y_1, \ldots, y_{n-1}, 0)$, h, and $\psi = (\psi_1, \ldots, \psi_m)$ are unknown parameters. For their determination, we use the boundary conditions (11.3). As a result, we obtain

$$A_1 y + A_2 \overline{x}(\mu^0 + h, x^0 + y) = C_1 - A_1 x^0 - A_2 \Delta x_{\mu^0 + h},$$

$$B_1 \psi + B_2 \overline{\varphi}(\mu^0 + h, x^0 + y, \psi^0 + \psi, \varepsilon) = C_2 - B_1 \varphi^0 - B_2 \Delta \varphi_{\mu^0 + h}, \tag{11.7}$$

where

$$\Delta x_\tau = x(\tau, x^0 + y, \varphi^0 + \psi, \varepsilon) - \overline{x}(\tau, x^0 + y),$$
$$\Delta \varphi_\tau = \varphi(\tau, x^0 + y, \varphi^0 + \psi, \varepsilon) - \overline{\varphi}(\tau, x^0 + y, \varphi^0 + \psi, \varepsilon).$$

Note that if the conditions of Theorem 11.1 are satisfied and $\varepsilon_0 > 0$ is sufficiently small, then it follows from the results of Chapter 1 that the following inequality holds for all $\tau \in [0, L]$, $\|y\| \le \frac{1}{2}\rho_0$, $\psi \in R^m$, and $\varepsilon \in (0, \varepsilon_0]$:

$$\|\Delta x_\tau\| + \|\Delta \varphi_\tau\| + \left\|\frac{\partial}{\partial y}\Delta x_\tau\right\| + \left\|\frac{\partial}{\partial y}\Delta \varphi_\tau\right\| + \left\|\frac{\partial}{\partial \psi}\Delta x_\tau\right\| + \left\|\frac{\partial}{\partial \psi}\Delta \varphi_\tau\right\| \le \underline{\sigma}\varepsilon^\alpha, \tag{11.8}$$

where $\underline{\sigma}$ is a certain positive constant independent of ε.

Note that

$$\overline{x}(\mu^0 + h, x^0 + y) = x(\mu^0, x^0) + \frac{\partial \overline{x}(\mu^0, x^0)}{\partial x^0} y + \overline{a}(\overline{x}(\mu^0, x^0), \mu^0)h + X(x^0, \mu^0, h, y),$$

$$\overline{\varphi}(\mu^0 + h, x^0 + y, \varphi^0 + \psi, \varepsilon) = \varphi^0 + \psi + \frac{1}{\varepsilon}\int\limits_0^{\mu^0+h} [\omega(t) + \varepsilon \overline{b}(\overline{x}(t, x^0 + y), t)]\, dt,$$

$$\|X(x^0, \mu^0, h, y)\| \le \overline{\sigma}_1(\|y\|^2 + h^2), \tag{11.9}$$

$$\left\|\frac{\partial}{\partial y}X(x^0, \mu^0, h, y)\right\| + \left\|\frac{\partial}{\partial h}X(x^0, \mu^0, h, y)\right\| \le \overline{\sigma}_1(\|y\| + |h|),$$

$$\overline{\sigma}_1 = \text{const},$$

Therefore, denoting $(\widetilde{y}, h) = z$, we can rewrite equalities (11.7) in the form

$$z = -P^{-1}A_2[\Delta x_{\mu^0+h} + X(x^0, \mu^0, h, y)] \equiv M(z, \psi, \varepsilon),$$

$$\psi = -(B_1 + B_2)^{-1} B_2 \Big[\int\limits_{\mu^0}^{\mu^0+h} \Big(\frac{\omega(t)}{\varepsilon} + \overline{b}(\overline{x}(t, x^0 + y), t) \Big)\, dt$$

$$+ \int\limits_0^{\mu^0} (\overline{b}(\overline{x}(t, x^0 + y), t) - \overline{b}(\overline{x}(t, x^0), t))dt + \Delta\varphi_{\mu^0+h} \Big]$$

$$\equiv N(z, \psi, \varepsilon). \tag{11.10}$$

It follows from (11.8) and (11.9) that, for every fixed $\psi \in R^m$ and $\varepsilon \in (0, \varepsilon_0]$, $M(z, \psi, \varepsilon)$ maps the set

$$K = \{z\colon z \in R^n,\ \ \|z\| \le \sigma_2 \varepsilon^\alpha\}, \quad \sigma_2 = 2\underline{\sigma}\|P^{-1}A_2\|,$$

into itself for

$$\varepsilon_0 \le (4\overline{\sigma}_1 \underline{\sigma})^{-\frac{1}{\alpha}} \|P^{-1}A_2\|^{-\frac{2}{\alpha}}.$$

In addition, we impose the restriction

$$\varepsilon_0^\alpha \le \frac{1}{\sigma_2} \min\{\mu^0; L - \mu^0\},$$

which guarantees that the condition $\mu^0 + h \in [0, L]$ is satisfied.

Further, we show that $M\colon K \to K$ is a contracting mapping. Using estimate (11.8) and the equality

$$\frac{d}{d\tau} \Delta x_\tau = [\overline{a}(x, \tau) - \overline{a}(\overline{x}, \tau)] + \widetilde{a}(x, \varphi, \tau),$$

from the relation

$$\frac{\partial M}{\partial z} = -P^{-1}A_2 \Big[\Big(\frac{\partial}{\partial \widetilde{y}} \Delta x_{\mu^0+h}; \frac{\partial}{\partial h} \Delta x_{\mu^0+h} \Big) + \frac{\partial}{\partial z} X(x^0, \mu^0, h, y) \Big]$$

we derive the inequality

$$\Big\| \frac{\partial M}{\partial z} \Big\| \le \|P^{-1}A_2\| (\underline{\sigma} + \underline{\sigma}\sigma_1 + \overline{\sigma}_1 \sigma_2) \varepsilon^\alpha$$

$$+ \|P^{-1} A_2 \widetilde{a}(x(\mu^0 + h, x^0 + y, \varphi^0 + \psi, \varepsilon),$$

$$\varphi(\mu^0 + h, x^0 + y, \varphi^0 + \psi, \varepsilon), \mu^0 + h)\|.$$

This inequality, the restriction $\sigma_0 < 1$, and the inequality

$$\|\widetilde{a}(x(\mu^0 + h, x^0 + y, \varphi^0 + \psi, \varepsilon), \varphi, \mu^0 + h) - \widetilde{a}(\overline{x}(\mu^0, x^0, \varphi, \mu^0)\|$$

$$\leq \sigma_1[\underline{\sigma} + \sigma_2(ne^{\sigma_1 L} + 1 + \sigma_1)]\varepsilon^\alpha$$

$$\forall z \in K, \quad \varphi \in R^m, \quad \varepsilon \in (0, \varepsilon_0],$$

[which follows from (11.2), (11.8), and (11.9)] imply that, for all $z \in K$, $\varepsilon \in (0, \varepsilon_0]$, $\varepsilon_0 \leq [(2\overline{\sigma}_2)^{-1}(1 - \sigma_0)]^{\frac{1}{\alpha}}$, and $\psi \in R^m$, the following estimate is true:

$$\left\|\frac{\partial M}{\partial z}\right\| \leq \sigma_0 + \overline{\sigma}_2\varepsilon^\alpha \leq \frac{\sigma_0 + 1}{2} < 1;$$

here,

$$\overline{\sigma}_2 = \|P^{-1}A_2\| \, [\underline{\sigma} + 2\underline{\sigma}\sigma_1 + \overline{\sigma}_1\sigma_2 + \sigma_1\sigma_2(1 + ne^{\sigma_1 L} + \sigma_1)].$$

Thus, $M\colon K \to K$ is a contracting mapping, and, therefore, there exists a unique solution $z = z^0(\psi, \varepsilon) \equiv (\widetilde{y}^0(\psi, \varepsilon), h^0(\psi, \varepsilon)) \in K$ that continuously depends on $(\psi, \varepsilon) \in R^m \times (0, \varepsilon_0]$.

Substituting the value of $z = z^0(\psi, \varepsilon)$ into the second equation in (11.10), we obtain

$$\psi = N\,(z^0(\psi, \varepsilon), \psi, \varepsilon). \tag{11.11}$$

Since the mapping N is continuous in $\psi \in R^m$ and, according to (11.9),

$$\|N(z^0(\psi, \varepsilon), \psi, \varepsilon)\| \leq \sigma_3\varepsilon^{\alpha-1} \quad \forall(\psi, \varepsilon) \in R^m \times (0, \varepsilon_0],$$

where

$$\sigma_3 = \|(B_1 + B_2)^{-1}B_2\|(\underline{\sigma} + \sigma_1\sigma_2 + 2\sigma_1 L + \sigma_2 \max_{[0,L]} \|\omega(\tau)\|),$$

using the Schauder theorem we establish that there exists a solution $\psi = \psi^0(\varepsilon)$, $\|\psi^0(\varepsilon)\| \leq \sigma_3\varepsilon^{\alpha-1}$, of Eq. (11.11). Hence,

$$\begin{aligned} &\{\mu(\varepsilon), x(\tau, \varepsilon), \varphi(\tau, \varepsilon)\} \\ &\qquad = \{\mu^0 + \mu_0(\varepsilon), x(\tau, x^0 + y^0(\varepsilon), \varphi^0 + \psi^0(\varepsilon), \varepsilon), \\ &\qquad\qquad\qquad \varphi(\tau, x^0 + y^0(\varepsilon), \varphi^0 + \psi^0(\varepsilon), \varepsilon)\}, \end{aligned}$$

where $\mu_0(\varepsilon) = h^0(\psi^0(\varepsilon), \varepsilon)$, $y^0(\varepsilon) = (\widetilde{y}^0(\psi^0(\varepsilon), \varepsilon), 0)$, is a solution of the boundary-value problem (11.1), (11.3). Estimates (11.6) with the constant $\sigma = \underline{\sigma} + \sigma_3 + n\sigma_2 e^{\sigma_1 L} \max\{1; \sigma_1 L\}$ follow from estimate (11.8) and the inequalities $\|y^0(\varepsilon)\| + |\mu_0(\varepsilon)| \leq \sigma_2\varepsilon^\alpha$ and $\|\psi^0(\varepsilon)\| \leq \sigma_3\varepsilon^{\alpha-1}$. Theorem 11.1 is proved.

Corollary 1. *Suppose that* $B_2 = 0$ *in Theorem 11.1, i.e., the boundary condition for the fast variables* φ *turns into the initial condition* $\varphi|_{\tau=0} = B_1^{-1}C_2 \equiv \varphi^0$. *Then, in a small neighborhood of the solution* $\{\mu^0, \overline{x}(\tau, x^0), \overline{\varphi}(\tau, x^0, \varphi^0, \varepsilon)\}$ *of the averaged problem (11.4^I)–(11.4^IV), there exists a unique solution* $\{\mu(\varepsilon), x(\tau, \varepsilon), \varphi(\tau, \varepsilon)\}$ *of the boundary-value problem (11.1), (11.3), and this solution satisfies the following inequality for all* $(\tau, \varepsilon) \in [0, L] \times (0, \varepsilon_0]$:

$$|\mu(\varepsilon) - \mu^0| + \|x(\tau, \varepsilon) - \overline{x}(\tau, x^0)\| + \|\varphi(\tau, \varepsilon) - \overline{\varphi}(\tau, x^0, \varphi^0, \varepsilon)\| \le \sigma\varepsilon^\alpha.$$

The method proposed above can be generalized to the case of a multipoint problem that contains unknown parameters $\mu_1, \ldots, \mu_r$ $(2 \le r < n)$ in the boundary conditions. Instead of (11.3), we consider the boundary conditions

$$\sum_{j=1}^{r} A_j x|_{\tau=\mu_j} = C_1, \quad \underset{\sim}{x}|_{\tau=0} = \underset{\sim}{x}^0, \quad \sum_{j=1}^{r} B_j \varphi|_{\tau=\mu_j} = C_2, \tag{11.12}$$

where $0 < \mu_1 < \mu_2 < \ldots < \mu_r < L$, $\underset{\sim}{x} = (x_{n-r+1}, \ldots, x_n)$ is the vector whose coordinates are the last r coordinates of the slow component $x = (x_1, \ldots, x_n)$ of the solution $(x; \varphi)$ of system (11.1), and $\underset{\sim}{x}^0$ is a given r-dimensional vector.

Lemma 11.2. *If the matrix* $\sum_{j=1}^{r} B_j$ *is nondegenerate and there exists a unique solution* $\widetilde{x}^0 = (x_1^0, \ldots, x_{n-r}^0)$, $\mu^0 = (\mu_1^0, \ldots, \mu_r^0)$ *of the equation*

$$\sum_{j=1}^{r} A_j \overline{x}(\mu_j^0, x^0) = C_1$$

that satisfies the conditions $x^0 = (\widetilde{x}^0, \underset{\sim}{x}^0) \in \mathcal{D}_{\rho_0}$, $0 < \mu_1^0 < \mu_2^0 < \ldots < \mu_r^0 < L$, *then there exists a unique solution* $\{\mu^0, \overline{x}(\tau, x^0), \overline{\varphi}(\tau, x^0, \varphi^0, \varepsilon)\}$ *of the averaged problem (11.4^I), (11.4^II), (11.12).*

Proof. It is obviously sufficient to find the $\overline{\varphi}$-component of a solution of the averaged problem. To do this, we use formula (11.5), in which we set

$$\varphi^0 = \Big(\sum_{j=1}^{r} B_j\Big)^{-1} \Big[C_2 - \frac{1}{\varepsilon}\sum_{j=1}^{r} B_j \int_0^{\mu_j^0} (\omega(t) + \varepsilon\overline{b}(\overline{x}(t, x^0), t))\, dt\Big].$$

Lemma 11.2 is proved.

Denote by Q the $n \times n$ square matrix

$$\Big(\sum_{j=1}^{r} A_j \frac{\partial \overline{x}(\mu_j^0, x^0)}{\partial \widetilde{x}^0}, A_1\overline{a}(\overline{x}(\mu_1^0, x^0), \mu_1^0), \dots, A_r\overline{a}(\overline{x}(\mu_r^0, x^0), \mu_r^0)\Big).$$

The proof of the theorem below, in fact, repeats the proof of Theorem 11.1, and, therefore, we present only its formulation.

Theorem 11.2. *Suppose that the following conditions are satisfied:*

(a) condition (i) of Theorem 11.1, inequality (11.2), and the conditions of Lemma 11.2 are satisfied;

(b) the matrix Q is nondegenerate and

$$\sup_{\varphi^{(j)} \in R^m, 1 \le j \le r} \Big\| Q^{-1} \sum_{j=1}^{r} A_j \widetilde{a}(\overline{x}(\mu_j^0, x^0), \varphi^{(j)}, \mu_j^0) \Big\| < 1.$$

Then, for sufficiently small $\varepsilon_0 > 0$ and every $\varepsilon \in (0, \varepsilon_0]$, there exists a solution $\{\mu(\varepsilon), x(\tau, \varepsilon), \varphi(\tau, \varepsilon)\}$ of the multipoint problem (11.1), (11.2) that satisfies the following inequality for any $(\tau, \varepsilon) \in [0, L] \times (0, \varepsilon_0]$:

$$\|\mu(\varepsilon) - \mu^0\| + \|x(\tau, \varepsilon) - \overline{x}(\tau, x^0)\| + \varepsilon \|\varphi(\tau, \varepsilon) - \overline{\varphi}(\tau, x^0, \varphi^0, \varepsilon)\| \le \sigma_4 \varepsilon^\alpha,$$

where the constant σ_4 is independent of ε.

Finally, we consider the case where an unknown scalar parameter $\mu \in R$ enters into the boundary conditions in a linear manner. For the multifrequency system (11.1), we introduce the boundary conditions

$$\begin{aligned} A_1 x|_{\tau=0} + \mu A_2 x|_{\tau=L} &= C_1, \quad x_n|_{\tau=0} = x_n^0, \\ B_1 \varphi|_{\tau=0} + B_2 \varphi|_{\tau=L} &= C_2 \quad (A_2 \ne 0). \end{aligned} \tag{11.13}$$

The solvability of the averaged problem (11.4^{I}), (11.4^{II}), (11.13) follows from the lemma presented below.

Lemma 11.3. *Suppose that the matrix $B_1 + B_2$ is nondegenerate and the equation*

$$A_1 x^0 + \mu^0 A_2 \overline{x}(L, x^0) = C_1$$

has a unique solution μ^0, $\widetilde{x}^0 = (x_1^0, \ldots, x_{n-1}^0)$ *for which* $x^0 = (\widetilde{x}^0, x_n^0) \in \mathcal{D}_{\rho_0}$. *Then, for all* $(\tau, \varepsilon) \in [0, L] \times (0, \varepsilon_0]$, *the unique solution* $\{\mu^0, \overline{x}(\tau, x^0), \overline{\varphi}(\tau, x^0, \varphi^0, \varepsilon)\}$, *where*

$$\varphi^0 = (B_1 + B_2)^{-1}\left[C_2 - \frac{1}{\varepsilon}B_2 \int_0^L (\omega(t) + \varepsilon\overline{b}(\overline{x}(t, x^0), t))\, dt\right],$$

of the boundary-value problem (11.4^I), (11.4^{II}), (11.13) is defined.

Lemma 11.3 can be proved by analogy with Lemmas 11.1 and 11.2.

Further, we choose $h \in R$, $\widetilde{y} \in R^{n-1}$, and $\psi \in R^m$ so that a solution of problem (11.1), (11.13) has the form

$$\{\mu^0 + h, \overline{x}(\tau, x^0 + y, \varphi^0 + \psi, \varepsilon), \varphi(\tau, x^0 + y, \varphi^0 + \psi, \varepsilon)\},$$

where $y = (\widetilde{y}, 0) \in R^n$. Using the boundary conditions (11.13) and Lemma 11.3, we get

$$\begin{aligned} z &= -\overline{P}^{-1} A_2\left[(\Delta x_L + X(x^0, L, 0, y))(\mu^0 + h) + \frac{\partial \overline{x}(L, x^0)}{\partial x^0} yh\right] \\ &\equiv \overline{M}(z, \psi, \varepsilon), \end{aligned} \tag{11.14}$$

$$\begin{aligned} \psi &= -(B_1 + B_2)^{-1} B_2\left[\Delta\varphi_L + \int_0^L (\overline{b}(\overline{x}(t, x^0 + y), t) - \overline{b}(\overline{x}(t, x^0), t))\, dt\right] \\ &\equiv \overline{N}(z, \psi, \varepsilon), \end{aligned} \tag{11.15}$$

where $z = (\widetilde{y}, h)$, $\overline{P}$ is an $n \times n$ matrix of the form

$$\overline{P} = \left(A_1^1 + \mu^0 A_2 \frac{\partial \overline{x}(L, x^0)}{\partial \widetilde{x}^0}, A_2 \overline{x}(L, x^0)\right),$$

and X and A_1^1 are defined in Theorem 11.1.

If $\mu^0 \neq 0$, then the analysis of the inequality

$$\|\overline{M}(z, \psi\varepsilon)\| \leq \|\overline{P}^{-1} A_2\| [(\underline{\sigma}\varepsilon^\alpha + \overline{\sigma}_1 \|z\|^2)(\mu^0 + \|z\|) + n e^{\sigma_1 L}\|z\|^2]$$

[which follows from (11.2), (11.8), (11.9), and (11.14)] shows that, for fixed $\psi \in R^m$, $\varepsilon \in (0, \varepsilon_0]$, and

$$\varepsilon_0 \leq \left[\frac{\sigma_5}{\underline{\sigma}\mu^0}(\underline{\sigma} + \overline{\sigma}_1\sigma_5\mu^0 + \overline{\sigma}_1\sigma_5^2 + n e^{\sigma_1 L}\sigma_5)\right]^{-\frac{1}{\alpha}}, \quad \sigma_5 = 2\underline{\sigma}\mu^0\|\overline{P}^{-1} A_2\|,$$

$\overline{M}(z,\psi,\varepsilon)$ maps the set $\overline{K} = \{z\colon z\in R^n, \|z\| \le \sigma_5\varepsilon^\alpha\}$ into itself. Moreover, it follows from (11.14) that

$$\begin{aligned}&\left\|\frac{\partial}{\partial z}\overline{M}(z,\psi,\varepsilon)\right\|\\ &\qquad \le \|\overline{P}^{-1}A_2\|[\underline{\sigma}(\mu^0+1)+\sigma_5(\underline{\sigma}+\overline{\sigma}_1\mu^0+n^2e^{\sigma_1 L})+2\overline{\sigma}_1\sigma_5^2]\varepsilon^\alpha\\ &\qquad \equiv \sigma_6\varepsilon^\alpha \le \frac{1}{2} \qquad \text{for} \quad \varepsilon_0^\alpha \le (2\sigma_6)^{-1}.\end{aligned}$$

Therefore, for $\mu^0 \ne 0$, there exists a unique solution

$$z = z(\psi,\varepsilon) = (\widetilde{y}(\psi,\varepsilon), h(\psi,\varepsilon)) \in \overline{K}$$

of Eq. (11.14), which can be determined by the method of successive approximations:

$$\begin{gathered}z_{s+1}(\psi,\varepsilon) = \overline{M}(z_s(\psi,\varepsilon),\psi,\varepsilon), \quad s\ge 0,\\ z_0(\psi,\varepsilon)\equiv 0, \quad \lim_{s\to\infty} z_s(\psi,\varepsilon) = z(\psi,\varepsilon).\end{gathered}$$

Differentiating the equality $z_{s+1}(\psi,\varepsilon) = \overline{M}(z_s(\psi,\varepsilon),\psi,\varepsilon)$ with respect to ψ and taking into account estimate (11.8), we get

$$\left\|\frac{\partial}{\partial\psi}z_{s+1}(\psi,\varepsilon)\right\| \le \sigma_7\varepsilon^\alpha\left\|\frac{\partial}{\partial\psi}z_s(\psi,\varepsilon)\right\| + \sigma_8\varepsilon^\alpha, \quad s\ge 0, \tag{11.16}$$

where

$$\begin{gathered}\sigma_7 = \|\overline{P}^{-1}A_2\|[(\mu^0+\sigma_5)(\underline{\sigma}+\overline{\sigma}_1\sigma_5)+\underline{\sigma}+\overline{\sigma}_1\sigma_5^2+n\sigma_5e^{\sigma_1 L}],\\ \sigma_8 = \|\overline{P}^{-1}A_2\|\underline{\sigma}(\mu^0+\sigma_5).\end{gathered}$$

Inequality (11.16) yields

$$\begin{gathered}\left\|\frac{\partial}{\partial\psi}z_{s+1}(\psi,\varepsilon)\right\| \le \frac{\sigma_8}{1-\sigma_7\varepsilon_0^\alpha}\varepsilon^\alpha \equiv \sigma_9\varepsilon^\alpha\\ \forall s\ge 0, \quad \psi\in R^m, \quad \varepsilon\in(0,\varepsilon_0],\end{gathered}$$

provided that $\varepsilon_0^\alpha \le (2\sigma_7)^{-1}$. This is sufficient for the function $z(\psi,\varepsilon)$ to satisfy the Lipschitz condition

$$\|z(\overline{\psi},\varepsilon) - z(\underline{\psi},\varepsilon)\| \le \sigma_9\varepsilon^\alpha\|\overline{\psi}-\underline{\psi}\| \qquad \forall\overline{\psi},\underline{\psi}\in R^m. \tag{11.17}$$

If $\mu^0 = 0$, then it follows from Eq. (11.14) that $z = z(\psi, \varepsilon) \equiv 0$ is its unique solution for small $\|z\|$.

Substituting $z = z(\psi, \varepsilon)$ in (11.15), we obtain the equation

$$\psi = \overline{N}(z(\psi, \varepsilon), \psi, \varepsilon). \tag{11.18}$$

Inequality (11.8) and the restriction $\|y\| \leq \sigma_5 \varepsilon^\alpha$ yield

$$\|\overline{N}(z(\psi, \varepsilon), \psi, \varepsilon)\| \leq \sigma_{10} \varepsilon^\alpha,$$

$$\sigma_{10} = \|(B_1 + B_2)^{-1} B_2\| (\underline{\sigma} + Ln\sigma_1 \sigma_5 e^{\sigma_1 L}),$$

and the Lipschitz condition (11.17) guarantees that the following inequality holds for all $\overline{\psi}, \underline{\psi} \in R^m$ and $\varepsilon \in (0, \varepsilon_0]$:

$$\|\overline{N}(z(\overline{\psi}, \varepsilon), \overline{\psi}, \varepsilon) - \overline{N}(z(\underline{\psi}, \varepsilon), \underline{\psi}, \varepsilon)\| \leq \sigma_{10} \varepsilon^\alpha \|\overline{\psi} - \underline{\psi}\|.$$

If we choose $\varepsilon_0 > 0$ so small that $\sigma_{10} \varepsilon_0^\alpha \leq \dfrac{1}{2}$, then $\overline{N}(z(\psi, \varepsilon), \psi, \varepsilon)$ maps the set $\|\psi\| \leq \sigma_{10} \varepsilon^\alpha$ into itself and is a contracting mapping. Therefore, there exists a unique solution $\psi = \psi(\varepsilon)$ of Eq. (11.18), and, hence, there exists a unique solution $z = z(\varepsilon) = z(\psi(\varepsilon), \varepsilon)$, $\psi = \psi(\varepsilon)$ of system (11.14), (11.15) that satisfies the inequalities

$$\|z(\varepsilon)\| \leq \sigma_5 \varepsilon^\alpha, \quad \|\psi(\varepsilon)\| \leq \sigma_{10} \varepsilon^\alpha \quad \forall \varepsilon \in (0, \varepsilon_0]. \tag{11.19}$$

Thus,

$$\begin{aligned} \{\mu(\varepsilon), x(\tau, \varepsilon), \varphi(\tau, \varepsilon)\} \\ = \{\mu^0 + h(\varepsilon), x(\tau, x^0 + y(\varepsilon), \varphi^0 + \psi(\varepsilon), \varepsilon), \\ \varphi(\tau, x^0 + y(\varepsilon), \varphi^0 + \psi(\varepsilon), \varepsilon)\}, \end{aligned}$$

where $z(\varepsilon) = (\widetilde{y}(\varepsilon), h(\varepsilon))$, $y(\varepsilon) = (\widetilde{y}(\varepsilon), 0)$ is the unique solution of problem (11.1), (11.13) in a small neighborhood of the solution of the averaged problem. Moreover, according to inequalities (11.8) and (11.9), the following estimate holds for all $(\tau, \varepsilon) \in [0, L] \times (0, \varepsilon_0]$:

$$|\mu(\varepsilon) - \mu^0| + \|x(\tau, \varepsilon) - \overline{x}(\tau, x^0)\| + \|\varphi(\tau, \varepsilon) - \overline{\varphi}(\tau, x^0, \varphi^0, \varepsilon)\| \leq \sigma_{11} \varepsilon^\alpha, \tag{11.20}$$

where $\sigma_{11} = \underline{\sigma} + \sigma_{10} + ne^{\sigma_1 L} \sigma_5 \max\{1; L\sigma_1\}$.

Thus, the following statement is true:

Theorem 11.3. *If condition (i) of Theorem 11.1, inequality (11.2), the conditions of Lemma 11.3, and the condition* $\det \overline{P} \neq 0$ *are satisfied, then, for sufficiently small* $\varepsilon_0 > 0$ *and all* $\varepsilon \in (0, \varepsilon_0]$, *in a small neighborhood of the solution* $\{\mu^0, \overline{x}(\tau, x^0), \overline{\varphi}(\tau, x^0, \varphi^0, \varepsilon)\}$ *of the averaged problem (11.4^I), (11.4^{II}), (11.13) the boundary-value problem (11.1), (11.13) has a unique solution* $\{\mu(\varepsilon), x(\tau, \varepsilon), \varphi(\tau, \varepsilon)\}$ *that satisfies inequality (11.20).*

3. INTEGRAL MANIFOLDS

12. Auxiliary Statements

Consider a multifrequency system of the form

$$\frac{dx}{d\tau} = a(x,\tau) + \widetilde{a}(x,\varphi,\tau) + \varepsilon A(x,\varphi,\tau,\varepsilon),$$

$$\frac{d\varphi}{d\tau} = \frac{\omega(\tau)}{\varepsilon} + b(x,\varphi,\tau,\varepsilon), \tag{12.1}$$

where $x \in \mathcal{D} \subset R^n$, $\varphi \in R^m$, $m \geq 2$, $\tau \in R$, $\varepsilon \in (0,\varepsilon_0]$, $\mathcal{D}$ is a bounded domain, and the real vector functions a, $\widetilde{a}$, A, ω, and b are defined and 2π-periodic in each variable φ_ν, $\nu = \overline{1,m}$, on the set $\overline{G} = \mathcal{D} \times R^m \times R \times (0,\varepsilon_0]$. Without loss of generality, we can assume that the function $\widetilde{a}(x,\varphi,\tau)$ averaged with respect to φ over the cube of periods is identically equal to zero [otherwise, it can be included in $a(x,\tau)$ in system (12.1)].

Assume that

$$[a,\widetilde{a},b] \in C^1_\tau(\overline{G},\sigma_1) \cap C^2_{x,\varphi}(\overline{G},\sigma_1),$$

$$\frac{\partial a}{\partial x} \in C^1_\tau(\overline{G},\sigma_1), \quad A \in C^2_{x,\varphi}(\overline{G},\sigma_1), \tag{12.2}$$

$$\sum_{k\neq 0}\left[\|k\|^2 \sup_{\overline{G}}\|c_k\| + \|k\|\left(\sup_{\overline{G}}\left\|\frac{\partial c_k}{\partial \tau}\right\| + \sup_{\overline{G}}\left\|\frac{\partial c_k}{\partial x}\right\|\right)\right] \leq \sigma_1,$$

and $\dfrac{\partial}{\partial \tau}A(x,\varphi,\tau,\varepsilon)$ is continuous in $(x,\varphi,\tau) \in \mathcal{D} \times R^m \times R$ for every fixed $\varepsilon \in (0,\varepsilon_0]$. Here, σ_1 is a certain positive constant, $c_k = c_k(x,\tau,\varepsilon)$ are the Fourier coefficients of the harmonics $\exp\{i(k,\varphi)\}$ in the Fourier expansion of the function $c(x,\varphi,\tau,\varepsilon) = [\widetilde{a}(x,\varphi,\tau); b(x,\varphi,\tau,\varepsilon)]$, i is the imaginary unit, $(k,\varphi) = k_1\varphi_1 + \ldots + k_m\varphi_m$ is the scalar product of vectors $k = (k_1,\ldots,k_m)$

and $\varphi = (\varphi_1, \ldots, \varphi_m)$, $\|k\| = |k_1| + \ldots + |k_m|$, and $C^l_{x,\varphi}(\overline{G}, \sigma_1)$ $(C^l_\tau(\overline{G}, \sigma_1))$ denotes the set of vector functions that have partial derivatives with respect to all variables x and φ (τ) up to the lth order inclusive that are continuous in x, φ, and τ and bounded in $\overline{G}$ by the constant σ_1. Unless otherwise stated, the norm of a matrix is understood as the sum of the absolute values of its elements.

We also impose certain restrictions on the coordinates $\omega_\nu(\tau)$, $\nu = \overline{1, m}$, of the frequency vector $\omega(\tau)$. Assume that the functions

$$\omega_\nu^{(\mu)}(\tau) \equiv \frac{d^\mu}{d\tau^\mu}\omega_\nu(\tau), \quad \nu = \overline{1, m}, \quad \mu = \overline{0, p-1}, \quad p \geq m,$$

are uniformly continuous on the entire axis and

$$\|(W_p^T(\tau)W_p(\tau))^{-1}W_p^T(\tau)\| \leq \sigma_2 = \text{const} \quad \forall \tau \in R, \tag{12.3}$$

where $W_p(\tau)$ and $W_p^T(\tau)$ denote the matrix

$$(\omega_\nu^{(\mu-1)}(\tau))_{\nu,\mu=1}^{m,p}$$

and its transpose, respectively.

Consider the system of equations of the first approximation for slow variables averaged with respect to all angular variables φ

$$\frac{d\overline{x}}{d\tau} = a\,(\overline{x}, \tau), \tag{12.4}$$

and assume that it has a solution $\overline{x} = \overline{x}(\tau)$ defined on the entire numerical straight line and lying in $\mathcal{D}$ together with its ρ-neighborhood.

Lemma 12.1. *If conditions (12.2) and (12.3) are satisfied and* $\varphi = \varphi_\tau^t(\psi, \varepsilon)$ *is a solution of the Cauchy problem*

$$\frac{d\varphi_\tau^t}{dt} = \frac{\omega(t)}{\varepsilon} + b(\overline{x}(t) + Y(\varphi_\tau^t, t, \varepsilon), \varphi_\tau^t, t, \varepsilon), \quad \varphi_\tau^\tau = \psi \in R^m, \tag{12.5}$$

where $Y(\varphi, t, \varepsilon)$ *is continuously differentiable with respect to* $(\varphi, t) \in R^m \times R$ *for every fixed* ε,

$$\left\|\frac{\partial Y}{\partial t} + \frac{\partial Y}{\partial \varphi}\frac{\omega(t)}{\varepsilon}\right\| \leq \overline{d}_1, \qquad \left\|\frac{\partial Y}{\partial \varphi}\right\| \leq d_2\varepsilon^\alpha$$

$$\forall (\varphi, t, \varepsilon) \in R^m \times R \times (0, \varepsilon_0] \equiv G_1,$$

$\overline{d}_1, d_2 = \text{const}$, *and* $\alpha = \dfrac{1}{p}$, *then there exist constants* c_1 *and* c_2 *independent of* ε *and such that*

$$\left\| \frac{\partial}{\partial \psi}(\varphi_\tau^t(\psi, \varepsilon) - \psi) \right\| \le c_1 \varepsilon^\alpha (1 + d_2) e^{c_2 (1+d_2)\varepsilon^\alpha |\tau - t|}(1 + |\tau - t|)$$

for sufficiently small $\varepsilon_0 > 0$ *and all* $(\psi, t, \varepsilon) \in G_1$ *and* $\tau \in R$.

Proof. We rewrite problem (12.5) in the form

$$\varphi_\tau^t - \psi = \int_\tau^t \Big[\frac{\omega(l)}{\varepsilon} + b(\overline{x}(l) + Y(\varphi_\tau^l, l, \varepsilon), \varphi_\tau^l, l, \varepsilon)\Big]\, dl.$$

Then, denoting $z_\tau^t = \dfrac{\partial}{\partial \psi}(\varphi_\tau^t - \psi)$, we obtain

$$z_\tau^t = \int_\tau^t \frac{\partial b}{\partial x}\frac{\partial Y}{\partial \varphi}(z_\tau^l + E_m) dl + \int_\tau^t \frac{\partial b}{\partial \varphi}(z_\tau^l + E_m) dl, \tag{12.6}$$

whence

$$\|z_\tau^t\| \le n\sigma_1 d_2 \varepsilon^\alpha \Big(m|\tau - t| + \Big|\int_\tau^t \|z_\tau^t\|\, dl\Big|\Big)$$

$$+ \sum_{k \ne 0} \Big\| \int_\tau^t B_k(\overline{x}(l) + Y(\varphi_\tau^l, l, \varepsilon), l, \varepsilon)(z_\tau^l + E_m)$$

$$\times \exp\{i(k, \theta_\tau^l)\} \exp\Big\{\frac{i}{\varepsilon}\int_\tau^l (k, \omega(r))\, dr\Big\}\, dl \Big\|. \tag{12.7}$$

Here, E_m is the m-dimensional identity matrix, $B_k(x, \tau, \varepsilon)$ are the Fourier coefficients of the function $\dfrac{\partial}{\partial \varphi} b(x, \varphi, \tau, \varepsilon)$, and $\theta_\tau^t = \varphi_\tau^t - \dfrac{1}{\varepsilon}\displaystyle\int_\tau^t \omega(r) dr$.

First, we consider the case $t \ge \tau + 2$. We represent the segment $[\tau, t]$ as a union of segments, namely

$$[\tau, t] = \bigcup_{s=0}^{q-1} [\tau + s, \tau + s + 1] \cup [\tau + q, t],$$

where q is the integer part of the number $t-\tau-1,\ 1 \le t-(\tau+q) < 2$. Then we represent the integral over $[\tau, t]$ under the summation sign on the right-hand side of (12.7) as the sum of integrals over the segments indicated. Estimating the integral over the segment $[\tau+s, \tau+s+1]$ of unit length by using condition (12.3) and the uniform estimate (1.20), we get

$$\Delta_{s,k} \equiv \Big\| \int\limits_{\tau+s}^{\tau+s+1} B_k(z_\tau^l + E_m) \exp\{i(k,\theta_\tau^l)\} \exp\Big\{\frac{i}{\varepsilon}\int\limits_{\tau}^{l}(k,\omega(r))dr\Big\}dl\Big\|$$

$$\le c_3\varepsilon^\alpha \Big\{ \Big[(1+\sigma_1)(m + \max_{[\tau+s,\tau+s+1]} \|z_\tau^l\|) + \max_{[\tau+s,\tau+s+1]} \Big\|\frac{d}{dl}z_\tau^l\Big\|\Big]$$

$$\times \sup_{\overline{G}} \|B_k(x,\tau,\varepsilon)\| + (m + \max_{[\tau+s,\tau+s+1]} \|z_\tau^l\|) \frac{1}{\|k\|} \sup_{\overline{G}} \Big\|\frac{\partial}{\partial\tau}B_k(x,\tau,\varepsilon)\Big\|$$

$$+ (\sigma_1 + d_2\varepsilon_0^\alpha\sigma_1 + \overline{d}_1)(m + \max_{[\tau+s,\tau+s+1]} \|z_\tau^l\|) \frac{1}{\|k\|} \sup_{\overline{G}} \Big\|\frac{\partial}{\partial x}B_k(x,\tau,\varepsilon)\Big\|\Big\},$$

where c_3 is the constant corresponding to the constant σ_3 in estimate (1.20). Since

$$\frac{dz_\tau^l}{dl} = \Big(\frac{\partial b}{\partial x}\frac{\partial Y}{\partial \varphi} + \frac{\partial b}{\partial \varphi}\Big)(z_\tau^l + E_m),$$

the inequality

$$\max_{[\tau+s,\tau+s+1]} \Big\|\frac{d}{dl}z_\tau^l\Big\| \le (m + nd_2\varepsilon_0^\alpha)(m + \max_{[\tau+s,\tau+s+1]} \|z_\tau^l\|) \tag{12.8}$$

yields the following estimate for $d_2\varepsilon_0^\alpha \le 1$:

$$\Delta_{s,k} \le c_4\varepsilon^\alpha(1 + \max_{[\tau+s,\tau+s+1]} \|z_\tau^l\|)\Big[\sup_{\overline{G}} \|B_k\|$$

$$+ \frac{1}{\|k\|}\Big(\sup_{\overline{G}} \Big\|\frac{\partial}{\partial\tau}B_k\Big\| + \sup_{\overline{G}} \Big\|\frac{\partial}{\partial x}B_k\Big\|\Big)\Big],$$

where $c_4 = mc_3[(1+\sigma_1)(1+\sigma_1(n+m)) + 2\sigma_1 + \overline{d}_1]$.

Further, we consider the differentiable norm

$$\|z\|_1 = \Big(\sum_{i,j=1}^{m} z_{ij}^2\Big)^{\frac{1}{2}}$$

of the matrix $z = (z_{ij})_{i,j=1}^{m}$. It is obvious that

$$\|z\|_1 \le \|z\| \le m^2 \|z\|_1, \quad \Big| \frac{d}{d\tau} \|z\|_1 \Big| \le \Big\| \frac{d}{d\tau} z \Big\|_1, \quad z = z(\tau).$$

Let l_1 and l_2 be, respectively, the maximum point and the minimum point of a continuously differentiable function $\|z_\tau^l\|_1$ of a variable l on the segment $[\tau + s, \tau + s + 1]$. Then the following inequalities hold:

$$\max_{[\tau+s,\tau+s+1]} \|z_\tau^l\| \le m^2 \max_{[\tau+s,\tau+s+1]} \|z_\tau^l\|_1 = m^2 [\|z_\tau^{l_1}\|_1 - \|z_\tau^{l_2}\|_1 + \|z_\tau^{l_2}\|_1]$$

$$= m^2 \Big(\int_{l_1}^{l_2} \frac{d}{dl} \|z_\tau^l\|_1 dl + \|z_\tau^{l_2}\| \Big)$$

$$\le m^2 \int_{\tau+s}^{\tau+s+1} \Big[\Big\| \frac{d}{dl} z_\tau^l \Big\|_1 + \|z_\tau^l\|_1 \Big] dl$$

$$\le \sigma_1 (m + n d_2 \varepsilon_0^\alpha) m^2 \int_{\tau+s}^{\tau+s+1} \|z_\tau^l\| dl + m^3 (m + n d_2 \varepsilon_0^\alpha) \sigma_1,$$

In view of these inequalities, the estimate for Δ_{sk} takes the form

$$\Delta_{s,k} \le \overline{c}_4 \varepsilon^\alpha \Big(1 + \int_{\tau+s}^{\tau+s+1} \|z_\tau^l\| dl \Big)$$

$$\times \Big[\sup_{\overline{G}} \|B_k\| + \frac{1}{\|k\|} \Big(\sup_{\overline{G}} \Big\| \frac{\partial}{\partial \tau} B_k \Big\| + \sup_{\overline{G}} \Big\| \frac{\partial}{\partial x} B_k \Big\| \Big) \Big], \tag{12.9}$$

$$\overline{c}_4 = c_4 (1 + m^3 (m + n)), \quad s = \overline{0, q - 1}.$$

Since the length of the segment $[\tau + q, t]$ is not less than 1 and less then 2, we conclude that the expression

$$\Delta_{q,k} = \Big\| \int_{\tau+q}^{t} B_k(\overline{x}(l) + Y(\varphi_\tau^l, l, \varepsilon), l, \varepsilon)(z_\tau^l + E_m) \exp\{i(k, \varphi_\tau^l)\} dl \Big\|$$

can also be estimated using inequality (1.20). Repeating the scheme of the proof of estimate (12.9), we get

$$\Delta_{q,k} \le \widetilde{c}_4 \varepsilon^\alpha \Big(1 + \int\limits_{\tau+s}^{\tau+s+1} \|z_\tau^l\| dl \Big)$$

$$\times \Big[\sup_{\overline{G}} \|B_k\| + \frac{1}{\|k\|} \Big(\sup_{\overline{G}} \Big\| \frac{\partial}{\partial \tau} B_k \Big\| + \sup_{\overline{G}} \Big\| \frac{\partial}{\partial x} B_k \Big\| \Big) \Big]. \quad (12.10)$$

Combining (12.9) and (12.10) and using condition (12.2) for the Fourier coefficients, we deduce from (12.7) for $t \ge \tau + 2$ that

$$\|z_\tau^t\| \le mn \Big(1 + \int\limits_\tau^t \|z_\tau^l\| dl \Big) \sigma_1 d_2 \varepsilon_0^\alpha + \sigma_1 (\overline{c}_4 + \widetilde{c}_4) \varepsilon^\alpha \Big(\int\limits_\tau^t \|z_\tau^l\| dl + t - \tau \Big) m$$

$$\le c_5 \varepsilon^\alpha (d_2 + 1) \Big[\int\limits_\tau^t \|z_\tau^l\| dl + t - \tau \Big], \quad (12.11)$$

$$c_5 = m\sigma_1 (n + \overline{c}_4 + \widetilde{c}_4).$$

Now let $t \in [\tau, \tau + 2)$. Then equality (12.6) yields

$$\|z_\tau^t\| \le (n\sigma_1 d_2 \varepsilon_0^\alpha + m\sigma_1) \int\limits_\tau^t \|z_\tau^l\| dl + 2mn\sigma_1 d_2 \varepsilon^\alpha$$

$$+ \sum_{k \ne 0} \Big\| \int\limits_\tau^t B_k(\overline{x}(l) + Y(\varphi_\tau^l, l, \varepsilon), l, \varepsilon)$$

$$\times \exp\{i(k, \theta_\tau^l)\} \exp\Big\{ \frac{i}{\varepsilon} \int\limits_\tau^l (k, \omega(r)) dr \Big\} dl \Big\|. \quad (12.12)$$

According to (1.20) and (12.2), the last term on the right-hand side of (12.12) is bounded from above by $c_6 \varepsilon^\alpha$. Therefore, inequality (12.12) for $d_2 \varepsilon_0^\alpha \le 1$ yields

$$\|z_\tau^t\| \le \widetilde{c}_6 \varepsilon^\alpha \quad \forall t \in [\tau, \tau + 2), \qquad \widetilde{c}_6 = e^{2(n+m)\sigma_1} (c_6 + 2mn\sigma_1). \quad (12.13)$$

We return to estimate (12.11) for $t \geq \tau + 2$. If we represent the segment $[\tau, t]$ as the union of the segments $[\tau, \tau + 2]$ and $[\tau + 2, t]$ and use inequality (12.13), then estimate (12.11) takes the form

$$\|z_\tau^t\| \leq c_5(1 + d_2)\varepsilon^\alpha \int\limits_{\tau+2}^{t} \|z_\tau^l\| dl + (1 + d_2)c_5(1 + 4\tilde{c}_6)\varepsilon^\alpha (t - \tau)$$

for $t \geq \tau + 2$, $d_2\varepsilon_0^\alpha \leq 1$, and $\varepsilon_0 < 1$. Since $t - \tau$ increases monotonically as a function of t, we have

$$\frac{\|z_\tau^t\|}{t - \tau} \leq c_5(1 + d_2)\varepsilon^\alpha \int\limits_{\tau+2}^{t} \frac{\|z_\tau^l\|}{l - \tau} dl + (1 + d_2)c_5(1 + 4\tilde{c}_6)\varepsilon^\alpha,$$

Hence, according to the Gronwall–Bellman lemma, we get

$$\|z_\tau^t\| = \left\| \frac{\partial}{\partial \psi}(\varphi_\tau^t - \psi) \right\| \leq c_5(1 + 4\tilde{c}_6)(1 + d_2)\varepsilon^\alpha (t - \tau) e^{c_5(1+d_2)\varepsilon^\alpha (t-\tau)}$$

$\forall t \geq \tau + 2$. Combining the last estimate for $t \geq \tau + 2$ and estimate (12.13) for $\tau \leq t < \tau + 2$, we complete the proof of the lemma for $t \geq \tau$. For $t < \tau$, the proof is analogous.

Lemma 12.2. *Suppose that the conditions of Lemma 12.1 are satisfied. Then one can indicate constants c_7 and c_8 independent of ε and such that*

$$\left\| \frac{\partial}{\partial \tau} \varphi_\tau^t(\psi, \varepsilon) \right\| \leq c_7 \left(1 + \frac{\|\omega(\tau)\|}{\varepsilon} \right) e^{c_8 \varepsilon^\alpha |\tau - t|}$$

for all $(\psi, \tau, \varepsilon) \in G_1$ and $t \in R$.

Proof. The Cauchy problem (12.5) yields

$$\frac{\partial \varphi_\tau^t}{\partial \tau} = -\left(\frac{\omega(\tau)}{\varepsilon} + b(\overline{x}(\tau) + Y(\psi, \tau, \varepsilon), \psi, \tau, \varepsilon) \right)$$

$$+ \int\limits_{\tau}^{t} \left[\frac{\partial b}{\partial x} \frac{\partial Y}{\partial \varphi} + \frac{\partial b}{\partial \varphi} \right] \frac{\partial \varphi_\tau^l}{\partial \tau} dl, \quad \frac{d}{dt} \frac{\partial \varphi_\tau^t}{\partial \tau}$$

$$= \Big[\frac{\partial}{\partial x} b(\overline{x}(t) + Y(\varphi_\tau^t, t, \varepsilon), \varphi_\tau^t, t, \varepsilon) \frac{\partial}{\partial \varphi} Y(\varphi_\tau^t, t, \varepsilon)$$

$$+ \frac{\partial}{\partial \varphi} b(\overline{x}(t) + Y(\varphi_\tau^t, t, \varepsilon), \varphi_\tau^t, t, \varepsilon) \Big] \frac{\partial \varphi_\tau^t}{\partial \tau}. \tag{12.14}$$

The first of these equalities yields

$$\Big\|\frac{\partial\varphi_\tau^t}{\partial\tau}\Big\| \le \Big(\frac{\|\omega(\tau)\|}{\varepsilon}+\sigma_1\Big)+n\sigma_1 d_2\varepsilon^\alpha\Big|\int\limits_\tau^t\Big\|\frac{\partial\varphi_\tau^l}{\partial\tau}\Big\|dl\Big|$$

$$+\Big\|\int\limits_\tau^t\frac{\partial b}{\partial\varphi}\frac{\partial\varphi_\tau^l}{\partial\tau}dl\Big\|. \tag{12.15}$$

If $t\in[\tau,\tau+2)$, then the last term on the right-hand side of (12.15) can be estimated from above by the value

$$m\sigma_1\int\limits_\tau^t\Big\|\frac{\partial\varphi_\tau^l}{\partial\tau}\Big\|\,dl.$$

Hence, using the Gronwall–Bellman inequality, we get

$$\Big\|\frac{\partial\varphi_\tau^t}{\partial\tau}\Big\| \le \Big(\frac{\|\omega(\tau)\|}{\varepsilon}+\sigma_1\Big)\,e^{2\sigma_1(m+n)}\quad \forall t\in[\tau,\tau+2) \tag{12.16}$$

for $d_2\varepsilon_0^\alpha\le 1$. If $t\ge\tau+2$, then we represent the segment $[\tau,t]$ as the union of segments of unit length and the last segment whose length is not less than 1 and less than 2. Then we decompose the integral under the norm sign on the right-hand side of (12.15) into the sum of integrals over the segments indicated. Taking into account inequalities (1.20) and (12.2) and the second equality in (12.14), by analogy with the proof of Lemma 12.1 we get

$$\Big\|\int\limits_\tau^t\frac{\partial b}{\partial\varphi}\,\frac{\partial\varphi_\tau^l}{\partial\tau}\,dl\Big\|\le c_9\varepsilon^\alpha\int\limits_\tau^t\Big\|\frac{\partial\varphi_\tau^l}{\partial\tau}\Big\|\,dl,\quad c_9=\text{const}.$$

The last estimate, together with estimates (12.15) and (12.16), yields

$$\Big\|\frac{\partial\varphi_\tau^t}{\partial\tau}\Big\|\le[1+(n\sigma_1+c_9)2e^{2\sigma_1(m+n)}]\Big(\frac{\|\omega(\tau)\|}{\varepsilon}+\sigma_1\Big)$$

$$+(c_9+n\sigma_1 d_2)\varepsilon^\alpha\int\limits_{\tau+2}^t\Big\|\frac{\partial\varphi_\tau^l}{\partial\tau}\Big\|\,dl.$$

Solving this inequality, we get

$$\left\|\frac{\partial \varphi_\tau^t}{\partial \tau}\right\| \leq [1 + (n\sigma_1 + c_9)2e^{2\sigma_1(m+n)}] \times \left(\frac{\|\omega(\tau)\|}{\varepsilon} + \sigma_1\right) e^{(n\sigma_1 d_2 + c_9)\varepsilon^\alpha (t-\tau)} \quad (12.17)$$

$\forall t \geq \tau + 2$. Estimates (12.16) and (12.17) complete the proof of the lemma for $t \geq \tau$. For $t < \tau$, the lemma is proved by analogy.

Lemma 12.3. *If the functions* $\dfrac{\partial^2}{\partial\psi\partial\psi_\nu} Y(\psi, \tau, \varepsilon)$, $\nu = \overline{1, m}$, *are continuous in* $(\psi, \tau) \in R^m \times R$ *for every* $\varepsilon \in (0, \varepsilon_0]$ *and the conditions of Lemma 12.1 are satisfied, then, for any* $(\psi, \tau, \varepsilon) \in G_1$, $t \in R$, *the following inequality is true:*

$$\sum_{\nu=1}^m \left\|\frac{\partial^2 \varphi_\tau^t(\psi, \varepsilon)}{\partial\psi\partial\psi_\nu}\right\| \leq c_{10}\left(\varepsilon^\alpha + \sum_{\nu=1}^m \sup_{\psi,\tau} \left\|\frac{\partial^2}{\partial\psi\partial\psi_\nu} Y(\psi, \tau, \varepsilon)\right\|\right) \times (1 + |t-\tau|^2) e^{c_{11}\varepsilon^\alpha |t-\tau|}, \quad (12.18)$$

where the constants c_{10} *and* c_{11} *are independent of* ε.

Proof. We prove estimate (12.18) for $t \geq \tau$ (the proof for $t < \tau$ is analogous). Using (12.5), we get

$$\begin{aligned} \frac{\partial^2 \varphi_\tau^t(\psi, \varepsilon)}{\partial\psi\partial\psi_\nu} = \int_\tau^t \Bigg[&\sum_{r=1}^n \frac{\partial^2 b}{\partial x \partial x_r} \frac{\partial Y^{(r)}}{\partial\varphi} \left(\frac{\partial(\varphi_\tau^l - \psi)}{\partial\psi_\nu} + e_\nu\right) \frac{\partial Y}{\partial\varphi} \\ &+ \sum_{\mu=1}^m \frac{\partial^2 b}{\partial x \partial\varphi_\mu} \left(\delta_{\nu\mu} + \frac{\partial(\varphi_{\tau,\mu}^l - \psi_\mu)}{\partial\psi_\nu}\right) \frac{\partial Y}{\partial\varphi} \\ &+ \sum_{r=1}^n \frac{\partial^2 b}{\partial\varphi \partial x_r} \frac{\partial Y^{(r)}}{\partial\varphi} \left(\frac{\partial(\varphi_\tau^l - \psi)}{\partial\psi_\nu} + e_\nu\right) \\ &+ \sum_{\mu=1}^m \frac{\partial^2 b}{\partial\varphi \partial\varphi_\mu} \frac{\partial(\varphi_{\tau,\mu}^l - \psi_\mu)}{\partial\psi_\nu} \end{aligned}$$

$$+\frac{\partial b}{\partial x}\sum_{\mu=1}^{m}\frac{\partial^2 Y}{\partial\varphi\partial\varphi_\mu}\Big(\frac{\partial(\varphi^l_{\tau,\mu}-\psi_\mu)}{\partial\psi_\nu}+\delta_{\nu\mu}\Big)$$

$$\times\left(E_m+\frac{\partial(\varphi^l_\tau-\psi)}{\partial\psi}\right)\Bigg]dl$$

$$+\int_\tau^t\frac{\partial^2 b}{\partial\varphi\partial\varphi_\nu}\frac{\partial(\varphi^l_\tau-\psi)}{\partial\psi}dl+\int_\tau^t\frac{\partial^2 b}{\partial\varphi\partial\varphi_\nu}dl$$

$$+\int_\tau^t\Big(\frac{\partial b}{\partial x}\frac{\partial Y}{\partial\varphi}+\frac{\partial b}{\partial\varphi}\Big)\frac{\partial^2\varphi^l_\tau}{\partial\psi\partial\psi_\nu}\,dl,$$

where $Y=(Y^{(1)},\dots,Y^{(n)})$, $\varphi^t_\tau=(\varphi^t_{\tau,1},\dots,\varphi^t_{\tau,m})$, $\delta_{\nu,\mu}$ is the Kronecker symbol, and e_ν is the unit vector in the space R^m. Using conditions (12.2) and Lemma 12.1, we get

$$\Big\|\frac{\partial^2\varphi^t_\tau}{\partial\psi\partial\psi_\nu}\Big\|\le c_{12}\Big(\varepsilon^\alpha+\sum_{\mu=1}^{m}\sup_{\psi,\tau}\Big\|\frac{\partial^2}{\partial\psi\partial\psi_\mu}Y(\psi,\tau,\varepsilon)\Big\|\Big)$$

$$\times(t-\tau+(t-\tau)^2)e^{c_{13}\varepsilon^\alpha(t-\tau)}+n\sigma_1 d_2\varepsilon^\alpha\int_\tau^t\Big\|\frac{\partial^2\varphi^l_\tau}{\partial\psi\partial\psi_\nu}\Big\|\,dl$$

$$+\Big\|\int_\tau^t\frac{\partial^2 b}{\partial\varphi\partial\varphi_\nu}\,dl\Big\|+\Big\|\int_\tau^t\frac{\partial b}{\partial\varphi}\,\frac{\partial^2\varphi^l_\tau}{\partial\psi\partial\psi_\nu}\,dl\Big\|,\qquad(12.19)$$

where c_{12} and c_{13} are constants independent of ε. The last two terms on the right-hand side of (12.19) can be estimated by analogy with the corresponding integrals in the proof of Lemmas 12.1 and 12.2. Conditions (12.2) for the Fourier coefficients of the function $b(x,\varphi,\tau,\varepsilon)$ and the uniform estimates (1.20) of the oscillation integrals yield the following inequalities for any $t\ge\tau+2$:

$$\Big\|\int_\tau^t\frac{\partial^2 b}{\partial\varphi\partial\varphi_\nu}\,dl\Big\|\le c_{14}\varepsilon^\alpha(t-\tau),$$

$$\left\| \int_{\tau}^{t} \frac{\partial b}{\partial \varphi} \frac{\partial^2 \varphi_\tau^l}{\partial \psi \partial \psi_\nu} \, dl \right\| \leq \bar{c}_{14} \Bigg[\Big(\varepsilon^\alpha + \sum_{\mu=1}^{m} \sup_{\psi,\tau} \Big\| \frac{\partial^2}{\partial \psi \partial \psi_\mu} Y(\psi, \tau, \varepsilon) \Big\| \Big) \times (t - \tau + (t - \tau)^2) e^{c_{15} \varepsilon^\alpha (t-\tau)} + \varepsilon^\alpha \int_{\tau}^{t} \Big\| \frac{\partial^2 \varphi_\tau^l}{\partial \psi \partial \psi_\nu} \Big\| \, dl \Bigg], \tag{12.20}$$

where c_{14}, $\bar{c}_{14}$, and c_{15} are certain constants independent of ε. Combining inequalities (12.19) and (12.20), we obtain

$$\sum_{\nu=1}^{m} \Big\| \frac{\partial^2 \varphi_\tau^t}{\partial \psi \partial \psi_\nu} \Big\| \leq (c_{12} + c_{14} + \bar{c}_{14}) m \Big(\varepsilon^\alpha + \sum_{\nu=1}^{m} \sup_{\psi,\tau} \Big\| \frac{\partial^2}{\partial \psi \partial \psi_\nu} Y(\psi, \tau, \varepsilon) \Big\| \Big) \times (t - \tau + (1 - \tau)^2) e^{\bar{c}_{15} \varepsilon^\alpha (t-\tau)} + (n\sigma_1 d_2 + \bar{c}_{14}) \varepsilon^\alpha \int_{\tau}^{t} \sum_{\nu=1}^{m} \Big\| \frac{\partial^2 \varphi_\tau^l}{\partial \psi \partial \psi_\nu} \Big\| \, dl \tag{12.21}$$

$$\forall t \geq \tau + 2, \qquad \bar{c}_{15} = \max\{c_{13}; c_{15}\}.$$

For $t \in [\tau, \tau + 2)$, relation (12.19) yields

$$\sum_{\nu=1}^{m} \Big\| \frac{\partial^2 \varphi_\tau^t}{\partial \psi \partial \psi_\nu} \Big\| \leq c_{16} \Big(\varepsilon^\alpha + \sum_{\nu=1}^{m} \sup_{\psi,\tau} \Big\| \frac{\partial^2}{\partial \psi \partial \psi_\nu} Y(\psi, \tau, \varepsilon) \Big\| \Big) \tag{12.22}$$

where the constant c_{16} is independent of ε. Therefore, decomposing the integral over $[\tau, t]$ on the right-hand side of (12.21) into the sum of integrals over the segments $[\tau, \tau + 2]$ and $[\tau + 2, t]$ and using estimate (12.22), we deduce from (12.21) that

$$\sum_{\nu=1}^{m} \Big\| \frac{\partial^2 \varphi_\tau^t}{\partial \psi \partial \psi_\nu} \Big\| \leq c_{17} \Big(\varepsilon^\alpha + \sum_{\nu=1}^{m} \sup_{\psi,\tau} \Big\| \frac{\partial^2}{\partial \psi \partial \psi_\nu} Y(\psi, \tau, \varepsilon) \Big\| \Big) \times (t - \tau)^2 e^{(\bar{c}_{15} + n\sigma_1 d_2 + \bar{c}_{14}) \varepsilon^\alpha (t-\tau)} \tag{12.23}$$

$$\forall t \geq \tau + 2, \quad c_{17} = \text{const}.$$

Inequalities (12.22) and (12.23) yield estimate (12.18) for all $t \geq \tau$. Lemma 12.4 is proved.

The methods proposed above can be used for the proof of the following statements:

Lemma 12.4. *If the conditions of Lemma 12.3 are satisfied, then the following estimate holds for any* $(\psi, \tau, \varepsilon) \in G_1$ *and* $t \in R$:

$$\left\| \frac{\partial}{\partial \tau} \frac{\partial}{\partial \psi} \varphi_\tau^t(\psi, \varepsilon) \right\| \le \overline{c}_{10} \varepsilon^{\alpha - 1} (\|\omega(\tau)\| + 1) e^{\overline{c}_{11} \varepsilon^\alpha |t - \tau|},$$

where the constants $\overline{c}_{10}$ *and* $\overline{c}_{11}$ *are independent of* ε.

Lemma 12.5. *Suppose that the following conditions are satisfied:*

(i) conditions (12.2) and (12.3) are satisfied;

(ii) the functions $Y_1(\varphi, \tau, \varepsilon)$ *and* $Y_2(\varphi, \tau, \varepsilon)$ *are twice continuously differentiable with respect to* $(\varphi, \tau) \in R^m \times R$ *for every* $\varepsilon \in (0, \varepsilon_0]$, 2π*-periodic in* φ_ν, $\nu = \overline{1, m}$, *and such that*

$$\|Y_s\| \le d_1 \varepsilon_0^\alpha, \quad \left\| \frac{\partial}{\partial \varphi} Y_s \right\| \le d_2 \varepsilon_0^\alpha, \quad \sum_{\nu=1}^m \left\| \frac{\partial^2}{\partial \varphi \partial \varphi_\nu} Y_s \right\| \le d_3 \varepsilon_0^\alpha,$$

$$\left\| \frac{\partial}{\partial \tau} Y_s + \frac{\partial}{\partial \varphi} Y_s \frac{\omega(\tau)}{\varepsilon} \right\| \le \overline{d}_1 \quad \forall (\varphi, \tau, \varepsilon) \in G_1, \quad s = 1, 2.$$

Then there exist constants c_{18} *and* c_{19} *such that, for all* $(\psi, \tau, \varepsilon) \in G_1$ *and* $t \in R$, *the following estimates are true:*

$$\begin{aligned} &\|\varphi_{\tau,1}^t(\psi, \varepsilon) - \varphi_{\tau,2}^t(\psi, \varepsilon)\| \\ &\qquad \le c_{18}(1 + |t - \tau|) e^{c_{19} \varepsilon_0^\alpha |t - \tau|} \max_{l \in N(\tau, t)} \sup_{\psi \in R^m} \|Y_1(\psi, l, \varepsilon) - Y_2(\psi, l, \varepsilon)\|, \end{aligned}$$

$$\begin{aligned} &\left\| \frac{\partial}{\partial \psi} (\varphi_{\tau,1}^t(\psi, \varepsilon) - \varphi_{\tau,2}^t(\psi, \varepsilon)) \right\| \\ &\qquad \le c_{18}(1 + |t - \tau|^2) e^{c_{19} \varepsilon_0^\alpha |t - \tau|} \Big[\sup_{G_1} \|Y_1(\psi, \tau, \varepsilon) - Y_2(\psi, \tau, \varepsilon)\| \\ &\qquad\qquad + \sup_{G_1} \left\| \frac{\partial}{\partial \psi} (Y_1(\psi, \tau, \varepsilon) - Y_2(\psi, \tau, \varepsilon)) \right\| \Big], \end{aligned} \tag{12.24}$$

where $\varphi^t_{\tau,s}(\psi,\varepsilon)$ *is the solution of the Cauchy problem*

$$\frac{d}{dt}\varphi^t_{\tau,s}(\psi,\varepsilon) = \frac{\omega(t)}{\varepsilon} + b(\overline{x}(t) + Y_s(\varphi^t_{\tau,s}(\psi,\varepsilon),t,\varepsilon),\varphi^t_{\tau,s}(\psi,\varepsilon),t,\varepsilon),$$

$$\varphi^\tau_{\tau,s}(\psi,\varepsilon) = \psi,$$

$N(\tau,t) = [\tau,t]$ *for* $\tau < t$, *and* $N(\tau,t) = [t,\tau]$ *for* $\tau \geq t$.

In what follows, we use the results obtained above in the proof of the existence of the integral manifold of the multifrequency system (12.1) and in the investigation of its properties.

13. Construction of Successive Approximations

Consider a solution $\overline{x} = \overline{x}(\tau)$ of the averaged equations (12.4) that lies in $\mathcal{D}$ together with its ρ-neighborhood $\forall \tau \in R$ and assume that the variational system of equations $\dfrac{dz}{d\tau} = \dfrac{\partial}{\partial x}\overline{a}(\overline{x}(\tau),\tau)z$ is hyperbolic [Pli2]. Without loss of generality, we can rewrite the variational system in the form

$$\frac{dz_+}{d\tau} = H_+(\tau)z_+, \qquad \frac{dz_-}{d\tau} = H_-(\tau)z_-, \tag{13.1}$$

where $z = (z_+, z_-)$, z_+ and z_- are, respectively, n_0-dimensional and $(n - n_0)$-dimensional vectors, and

$$H(\tau) = \operatorname{diag}\left[H_+(\tau), H_-(\tau)\right] = \frac{\partial \overline{a}(\overline{x}(\tau),\tau)}{\partial x}.$$

In this case, the normal fundamental matrices $Q_+(\tau,t)$ and $Q_-(\tau,t)$ of solutions of the first and the second equations in (13.1) satisfy the inequalities

$$\|Q_+(\tau,t)\| \leq Ke^{\gamma(\tau-t)} \quad \forall \tau \leq t,$$

$$\|Q_-(\tau,t)\| \leq Ke^{-\gamma(\tau-t)} \quad \forall \tau \geq t, \tag{13.2}$$

where $K \geq 1$ and $\gamma > 0$ are certain constants. Furthermore, in what follows, we assume that

$$\sigma_0 = \frac{2}{\gamma}K\sup_{\varphi,\tau}\left\|\frac{\partial}{\partial x}\widetilde{a}(\overline{x}(\tau),\varphi,\tau)\right\| < 1. \tag{13.3}$$

Performing the change of variables $y = x - \overline{x}(\tau),\ y = (y_+, y_-),$ we transform Eqs. (12.1) as follows:

$$\frac{dy_+}{d\tau} = H_+(\tau)y_+ + F_+(y,\tau) + \widetilde{a}_+(y+\overline{x}(\tau),\varphi,\tau) + \varepsilon A_+(y+\overline{x}(\tau),\varphi,\tau,\varepsilon),$$

$$\begin{aligned} \frac{dy_-}{d\tau} &= H_-(\tau)y_- + F_-(y,\tau) + \widetilde{a}_-(y+\overline{x}(\tau),\varphi,\tau) \\ &\qquad + \varepsilon A_-(y+\overline{x}(\tau),\varphi,\tau,\varepsilon), \\ \frac{d\varphi}{d\tau} &= \frac{\omega(\tau)}{\varepsilon} + b(y+\overline{x}(\tau),\varphi,\tau,\varepsilon), \end{aligned} \tag{13.4}$$

where $(\widetilde{a}_+, \widetilde{a}_-) = \widetilde{a},\ (A_+, A_-) = A,$

$$\begin{aligned} (F_+, F_-) = F &= a(y+\overline{x}(\tau),\tau) - a(\overline{x}(\tau),\tau) - H(\tau)y \\ &\equiv \int\limits_0^1 \Big[\frac{\partial}{\partial y} a(ly+\overline{x}(\tau),\tau) - H(\tau)\Big]\, dly, \end{aligned}$$

$$\|F(y,\tau)\| \le \frac{1}{2} n^2 \sigma_1 \|y\|^2, \quad \Big\|\frac{\partial}{\partial y} F(y,\tau)\Big\| \le n^2 \sigma_1 \|y\|.$$

Let $Q(\tau,t)$ denote the quadratic n-dimensional matrix

$$Q(\tau,t) = \begin{cases} -\operatorname{diag}(Q_+(\tau,t),0), & \tau < t, \\ \operatorname{diag}(0, Q_-(\tau,t)), & \tau > t. \end{cases}$$

For $\tau \ne t,$ we obviously have

$$\frac{dQ(\tau,t)}{d\tau} = H(\tau)Q(\tau,t), \quad \frac{dQ(\tau,t)}{dt} = -Q(\tau,t)H(t),$$
$$\|Q(\tau,t)\| \le Ke^{-\gamma|\tau-t|}. \tag{13.5}$$

We determine the integral manifold of Eqs. (13.4) by the method of successive approximations as the limit (as $j \to \infty$) of the integral manifolds $y = Y_j(\psi,\tau,\varepsilon),\ (\psi,\tau,\varepsilon) \in G_1,$ of the equations

$$\frac{dy}{d\tau} = H(\tau)y + F(Y_{j-1}(\varphi,\tau,\varepsilon),\tau) + \widetilde{a}(X_{j-1}(\varphi,\tau,\varepsilon),\varphi,\tau) + \varepsilon A(X_{j-1}(\varphi,\tau,\varepsilon),\varphi,\tau,\varepsilon),$$

$$\frac{d\varphi}{d\tau} = \frac{\omega(\tau)}{\varepsilon} + b(X_{j-1}(\varphi,\tau,\varepsilon),\varphi,\tau,\varepsilon), \tag{13.6}$$

where $Y_0 \equiv 0$ and $X_{j-1}(\varphi,\tau,\varepsilon) = \overline{x}(\tau) + Y_{j-1}(\varphi,\tau,\varepsilon)$.

By using the matrix $Q(\tau,t)$, one can determine the integral manifold of Eqs. (13.6) as follows:

$$\begin{aligned} Y_j(\psi,\tau,\varepsilon) = \int\limits_{-\infty}^{\infty} Q(\tau,t)[&F(Y_{j-1}(\varphi^t_{\tau,j}(\psi,\varepsilon),t,\varepsilon),t) \\ &+ \widetilde{a}(X_{j-1}(\varphi^t_{\tau,j}(\psi,\varepsilon),t,\varepsilon),\varphi^t_{\tau,j}(\psi,\varepsilon),t) \\ &+ \varepsilon A(X_{j-1}(\varphi^t_{\tau,j}(\psi,\varepsilon),t,\varepsilon),\varphi^t_{\tau,j}(\psi,\varepsilon),t,\varepsilon)]\,dt, \end{aligned} \tag{13.7}$$

where $\varphi = \varphi^{\tau}_{t,j}(\psi,\varepsilon)$ is a solution of the second equation of system (13.6) that takes the value ψ for $\tau = t$. Indeed, assuming that the order of differentiation and integration on the right-hand side of (13.7) may be changed, we get

$$\begin{aligned} &\frac{\partial Y_j}{\partial\tau} + \frac{\partial Y_j}{\partial\psi}\Big[\frac{\omega(\tau)}{\varepsilon} + \widetilde{b}(\psi,\tau,\varepsilon)\Big] \\ &\quad = H(\tau)Y_j + F(Y_{j-1}(\psi,\tau,\varepsilon),\tau) \\ &\qquad + \widetilde{a}(X_{j-1}(\psi,\tau,\varepsilon),\psi,\tau) + \varepsilon A(X_{j-1}(\psi,\tau,\varepsilon),\psi,\tau,\varepsilon) \\ &\qquad + \int\limits_{-\infty}^{\infty} \frac{\partial B}{\partial\varphi}\Big[\frac{\partial\varphi^t_{\tau,j}}{\partial\tau} + \frac{\partial\varphi^t_{\tau,j}}{\partial\psi}\Big(\frac{\omega(\tau)}{\varepsilon} + \widetilde{b}(\psi,\tau,\varepsilon)\Big)\Big]dt, \end{aligned} \tag{13.8}$$

where $Y_j = Y_j(\psi,\tau,\varepsilon)$, $\widetilde{b}(\psi,\tau,\varepsilon) = b(X_{j-1}(\psi,\tau,\varepsilon),\psi,\tau,\varepsilon)$, and B is the integrand of integral (13.7). The second equation of system (13.6) yields

$$\begin{aligned} &\frac{\partial\varphi^t_{\tau,j}}{\partial\tau} + \frac{\partial\varphi^t_{\tau,j}}{\partial\psi}\left[\frac{\omega(\tau)}{\varepsilon} + \widetilde{b}(\psi,\tau,\varepsilon)\right] \\ &\quad = \int\limits_{\tau}^{t} \frac{\partial\widetilde{b}(\varphi^l_{\tau,j},l,\varepsilon)}{\partial\varphi}\Big\{\frac{\partial\varphi^l_{\tau,j}}{\partial\tau} + \frac{\partial\varphi^l_{\tau,j}}{\partial\psi}\left[\frac{\omega(\tau)}{\varepsilon} + \widetilde{b}(\psi,\tau,\varepsilon)\right]\Big\}\,dl. \end{aligned}$$

Setting

$$f(t)=\frac{\partial\varphi^t_{\tau,j}}{\partial\tau}+\frac{\partial\varphi^t_{\tau,j}}{\partial\psi}\Big[\frac{\omega(\tau)}{\varepsilon}+b(\psi,\tau,\varepsilon)\Big],$$

we deduce from the last equality that

$$\|f(t)\|\le\Big|\int\limits_{\tau}^{t}\|f(l)\|\sup\Big\|\frac{\partial\widetilde{b}(\psi,\tau,\varepsilon)}{\partial\psi}\Big\|\,dl\Big|.$$

Solving this integral inequality, we get $f(t)\equiv 0$. Therefore, for all $(\psi,\tau,\varepsilon)\in G_1$, the following identity is true:

$$\frac{\partial Y_j}{\partial\tau}+\frac{\partial Y_j}{\partial\psi}\Big[\frac{\omega(\tau)}{\varepsilon}+b(X_{j-1},\psi,\tau,\varepsilon)\Big]$$
$$=H(\tau)Y_j+F(Y_{j-1},\tau)+\widetilde{a}(X_{j-1},\psi,\tau)+\varepsilon A(X_{j-1},\psi,\tau,\varepsilon),\quad(13.9)$$

where the values of the functions Y_{j-1}, X_{j-1}, and Y_j are taken at a point (ψ,τ,ε). Hence, $y=Y_j(\psi,\tau,\varepsilon)$ is indeed the integral manifold of system (13.6).

Theorem 13.1. *Suppose that conditions (12.2), (12.3), (13.2), and (13.3) are satisfied. Then, for sufficiently small* $\varepsilon_0>0$, *the functions* $Y_j=Y_j(\psi,\tau,\varepsilon)$, $j=\overline{0,\infty}$, *defined by equality (13.7) are* 2π*-periodic in* ψ_ν, $\nu=\overline{1,m}$, *twice continuously differentiable with respect to* $(\psi,\tau)\in R^m\times R$ *for every fixed* $\varepsilon\in(0,\varepsilon_0]$, *and such that, for any* $(\psi,\tau,\varepsilon)\in G_1$, *the following inequalities are satisfied:*

$$\|Y_j\|\le d_1\varepsilon^\alpha,\quad \Big\|\frac{\partial}{\partial\psi}Y_j\Big\|\le d_2\varepsilon^\alpha,\quad \sum_{\nu=1}^{m}\Big\|\frac{\partial^2}{\partial\psi\partial\psi_\nu}Y_j\Big\|\le d_3\varepsilon^\alpha.\qquad(13.10)$$

If, in addition, the norms $\|\omega(\tau)\|$, $\Big\|\frac{d\omega(\tau)}{d\tau}\Big\|$, *and* $\Big\|\frac{\partial}{\partial\tau}A(x,\varphi,\tau,\varepsilon)\Big\|$ *are uniformly bounded on the set* $\overline{G}$, *then the following inequalities are also true:*

$$\Big\|\frac{\partial}{\partial\tau}Y_j\Big\|\le d_4\varepsilon^{\alpha-1},\quad \Big\|\frac{\partial^2}{\partial\psi\partial\tau}Y_j\Big\|\le d_5\varepsilon^{\alpha-1},\quad \Big\|\frac{\partial^2}{\partial\tau^2}Y_j\Big\|\le d_6\varepsilon^{\alpha-2}.\quad(13.11)$$

Here, $d_1,\ldots,d_6$ *are constants independent of* ε *and* j.

Proof. Consider the sequence $\{Y_j(\psi,\tau,\varepsilon)\}$. Let us prove that it is bounded $\forall(\psi,\tau,\varepsilon)\in G_1$. Denote

$$\theta^t_{\tau,j}=\varphi^t_{\tau,j}-\frac{1}{\varepsilon}\int\limits_\tau^t\omega(r)dr.$$

Since

$$\widetilde{a}(\overline{x}(\tau)+Y_j,\psi,\tau)$$
$$=\widetilde{a}(\overline{x}(\tau),\psi,\tau)+\frac{\partial}{\partial x}\widetilde{a}(\overline{x}(\tau),\psi,\tau)Y_j+\widetilde{A}(\overline{x}(\tau),Y_j,\psi,\tau),$$

where

$$\|\widetilde{A}\|=\left\|\int\limits_0^1\left[\frac{\partial}{\partial x}\widetilde{a}(\overline{x}(\tau)+lY_j,\psi,\tau)-\frac{\partial}{\partial x}\widetilde{a}(\overline{x}(\tau),\psi,\tau)\right]dlY_j\right\|$$
$$\leq\frac{1}{2}n^2\sigma_1\|Y_j\|^2,$$

it follows from (13.7) that

$$\|Y_{j+1}(\psi,\tau,\varepsilon)\|\leq\sigma_0\sup_{\psi,\tau}\|Y_j(\psi,\tau,\varepsilon)\|$$
$$+\frac{2}{\gamma}K\left[\varepsilon\sigma_1+n^2\sigma_1\sup_{\psi,\tau}\|Y_j(\psi,\tau,\varepsilon)\|^2\right]$$
$$+\sum_{k\neq 0}\sum_{s=-\infty}^{\infty}\left\|\int\limits_{\tau+s}^{\tau+s+1}Q(\tau,t)a_k(\overline{x}(t),t)\exp\{i(k,\theta^t_{\tau,j+1})\}\right.$$
$$\left.\times\exp\left\{\frac{i}{\varepsilon}\int\limits_\tau^t(k,\omega(r))dr\right\}dt\right\|. \tag{13.12}$$

Here, $a_k(x,\tau)$ are the Fourier coefficients of the function $\widetilde{a}(x,\varphi,\tau)$. Using conditions (12.2) and (13.5) and the uniform estimate (1.20) of the oscillation integral, we estimate the last term on the right-hand side of (13.12) from above by the value

$$\sigma_3 K \sum_{k\neq 0} \sum_{s=-\infty}^{\infty} \Big\{ \Big[\sup_{\overline{G}} \|a_k\| + \frac{1}{\|k\|} \Big(\sup_{\overline{G}} \Big\| \frac{\partial a_k}{\partial \tau} \Big\| + \sup_{\overline{G}} \Big\| \frac{\partial a_k}{\partial x} \Big\| \Big) \Big]$$

$$\times (1 + \sigma_1 + n\sigma_1) \max_{[\tau+s,\tau+s+1]} e^{-\gamma|t-\tau|} \Big\} \varepsilon^\alpha \leq \sigma_4 \varepsilon^\alpha,$$

$$\sigma_4 = 2K\sigma_1\sigma_3(1 + \sigma_1 + n\sigma_1)\frac{e^\gamma}{1 - e^{-\gamma}}.$$

Then relation (13.12) yields

$$\sup_{\psi,\tau} \|Y_{j+1}\| \leq \sigma_0 \sup_{\psi,\tau} \|Y_j\| + \frac{2}{\gamma} K n^2 \sigma_1 \sup_{\psi,\tau} \|Y_j\|^2 + \Big(\sigma_4 + \frac{2}{\gamma} K\sigma_1\Big)\varepsilon^\alpha,$$

which leads to the following estimate in view of the fact that $Y_0 \equiv 0$ and $\sigma_0 < 1$:

$$\sup_{\psi,\tau} \|Y_j(\psi, \tau, \varepsilon)\| \leq \frac{1}{2} d_1 \varepsilon^\alpha < d_1 \varepsilon^\alpha \quad \forall j \geq 0, \quad \varepsilon \in (0, \varepsilon_0], \tag{13.13}$$

where $\varepsilon_0 \leq (nd_1)^{-\frac{2}{\alpha}}$ and $d_1 = \frac{1}{\gamma}(4K\sigma_1 + \gamma\sigma_4)(1 - \sigma_0)^{-1}$. Note that it is necessary to impose the restriction $d_1 \varepsilon_0^\alpha \leq \rho$. If this condition is satisfied, then $X_j(\psi, \tau, \varepsilon) = \overline{x}(\tau) + Y_j(\psi, \tau, \varepsilon)$ lies in the $\frac{1}{2}\rho$-neighborhood of the curve $x = \overline{x}(\tau)$ $\forall(\psi, \tau, \varepsilon) \in G_1$, i.e., in the course of the construction of successive approximations, we do not leave the domain of definition of the right-hand side of system (12.1).

Let us prove that the sequence $\Big\{\frac{\partial}{\partial \psi} Y_j(\psi, \tau, \varepsilon)\Big\}$ is also uniformly bounded in G_1 by the value $d_2 \varepsilon^\alpha$, where d_2 is a constant independent of ε and $j \geq 0$. Denote by $A_k(\overline{x}(\tau), \tau)$ the $m \times n$ rectangular matrix

$$A_k(\overline{x}(\tau), \tau) = (a_k^{(\mu)}(\overline{x}(\tau), \tau) k_\nu)_{\mu,\nu=1}^{n,m},$$

$$a_k(x, \tau) = (a_k^{(1)}(x, \tau), \ldots, a_k^{(n)}(x, \tau)).$$

It is obvious that

$$\|A_k\| \leq \|a_k\| \cdot \|k\|,$$

$$\Big\|\frac{\partial}{\partial \tau} A_k\Big\| \leq \Big\|\frac{\partial a_k}{\partial \tau}\Big\| \cdot \|k\|, \quad \Big\|\frac{\partial}{\partial x} A_k\Big\| \leq n\Big\|\frac{\partial a_k}{\partial x}\Big\| \cdot \|k\|.$$

Consider the inequality

$$\Big\|\frac{\partial}{\partial\psi}Y_1(\psi,\tau,\varepsilon)\Big\|$$

$$\leq \varepsilon n\sigma_1 K\int\limits_{-\infty}^{\infty} e^{-\gamma|t-\tau|}\Big(m+\Big\|\frac{\partial}{\partial\psi}(\varphi_{\tau,1}^t-\psi)\Big\|\Big)d\tau$$

$$+\sum_{k\neq 0}\sum_{s=-\infty}^{\infty}\Big\|\int\limits_{s+\tau}^{s+\tau+1} Q(\tau,t)A_k(\overline{x}(t),t)\Big(E_m+\frac{\partial}{\partial\psi}(\varphi_{\tau,1}^t-\psi)\Big)$$

$$\times \exp\{i(k,\theta_{\tau,1}^t)\}\,\exp\Big\{\frac{i}{\varepsilon}\int\limits_{\tau}^{t}(k,\omega(r))\,dr\Big\}\,dt\Big\|, \tag{13.14}$$

which follows from (13.7) for $j=1$. Estimating the last term on the right-hand side of (13.14) [denote it by Δ] using inequalities (1.20) and (12.2), we get

$$\Delta \leq (1+\sigma_1+n\sigma_1)n\sigma_1 K\sigma_3\varepsilon^{\alpha}$$

$$\times\sum_{s=-\infty}^{\infty}\Big(m+\max_{[\tau+s,\tau+s+1]}\|z_\tau^t\|+\max_{[\tau+s,\tau+s+1]}\Big\|\frac{d}{dt}z_\tau^t\Big\|\Big)$$

$$\times \max_{[\tau+s,\tau+s+1]} e^{-\gamma|t-\tau|}, \tag{13.15}$$

where $z_\tau^t = \dfrac{\partial}{\partial\psi}(\varphi_{\tau,1}^t-\psi)$. Further, we use Lemma 12.1 for the function $Y = Y_0(\psi,\tau,\varepsilon)\equiv 0$. Since $\Big\|\dfrac{\partial}{\partial\psi}Y_0(\psi,\tau,\varepsilon)\Big\| \leq d_2\varepsilon^{\alpha}$ (the constant $d_2>0$ is fixed in what follows) and

$$\Big\|\frac{\partial Y_0}{\partial\tau}+\frac{\partial Y_0}{\partial\psi}\,\frac{\omega(\tau)}{\varepsilon}\Big\| \leq \overline{d}_1 = \sigma_1\Big(3+nd_1+\frac{1}{2}n^2d_1^2\Big),$$

$$(1+|t-\tau|)e^{c_2(1+d_2)\varepsilon^{\alpha}|t-\tau|} \leq \sigma_5 e^{\frac{\gamma}{2}|t-\tau|},\quad \sigma_5=\max\Big\{1;\frac{4}{\gamma}\Big\},$$

for $c_2(1+d_2)\varepsilon_0^\alpha \leq \frac{1}{4}\gamma$ and $d_2\varepsilon_0^\alpha \leq 1,$ inequality (12.8) and Lemma 12.1 imply that, for all $s \geq 0,$ the following relation is true:

$$\mu \equiv m + \max_{[\tau+s,\tau+s+1]} \|z_\tau^t\| + \max_{[\tau+s,\tau+s+1]} \left\| \frac{d}{dt} z_\tau^t \right\| \leq \sigma_5^{(1)} + \sigma_5^{(2)} e^{\frac{\gamma}{2}(s+1)},$$

$$\sigma_5^{(1)} = m + (m+n)\sigma_1, \quad \sigma_5^{(2)} = (1+(m+n)\sigma_1)2c_1\sigma_5.$$

For $s < 0,$ we obviously have $\mu \leq \sigma_5^{(1)} + \sigma_5^{(2)} e^{-\frac{\gamma}{2}s}.$ Taking this into account, we can rewrite inequality (13.15) in the form

$$\Delta \leq (1+\sigma_1+n\sigma_1)n\sigma_1 K 2\sigma_3 \Big(\sigma_5^{(1)} \frac{1}{1-e^{-\gamma}} + \sigma_5^{(2)} \frac{e^{\frac{\gamma}{2}}}{1-e^{-\frac{\gamma}{2}}}\Big)\varepsilon^\alpha$$

$$\equiv \sigma_6 \varepsilon^\alpha. \tag{13.16}$$

Inequalities (13.14) and (13.16) yield

$$\left\| \frac{\partial}{\partial \psi} Y_1(\psi,\tau,\varepsilon) \right\| \leq \varepsilon n \sigma_1 K \Big(\frac{2m}{\gamma} + \frac{8}{\gamma} c_1 \sigma_5\Big) + \sigma_6 \varepsilon^\alpha < d_2 \varepsilon^\alpha$$

$$\forall (\psi,\tau,\varepsilon) \in G_1.$$

Note that, for $\tau \in [-T,T]$ and $N > T,$ we have

$$\left\| \int_N^\infty Q(\tau,t) \Big[F(Y_0,t) + \widetilde{a}(\overline{x}(t)+Y_0,\varphi_{\tau,1}^t,t) + \varepsilon A(\overline{x}(t)+Y_0,\varphi_{\tau,1}^t,t,\varepsilon)\Big]\, dt \right\|$$

$$\leq \frac{1}{\gamma} K \Big(\frac{1}{2} n^2 \sigma_1 d_1^2 \varepsilon_0^\alpha + \sigma_1 + \varepsilon_0 \sigma_1\Big) e^{-\gamma(N-T)},$$

$$\Big\| \int_N^\infty \frac{\partial}{\partial \psi} \Big\{ Q(\tau,t)[F(Y_0,t) + \widetilde{a}(\overline{x}(t)+Y_0,\varphi_{\tau,1}^t,t)$$

$$+ \varepsilon A(\overline{x}(t)+Y_0,\varphi_{\tau,1}^t,t,\varepsilon)] \Big\} dt \Big\|$$

$$\leq K\sigma_1(n^2 d_1 d_2 \varepsilon_0^{2\alpha} + n d_2 \varepsilon_0^\alpha + m$$

$$+ n d_2 \varepsilon_0^{1+\alpha} + \varepsilon_0 m)\Big(\frac{m}{\gamma} + \frac{4}{\gamma} c_1 \sigma_5\Big) e^{-\frac{\gamma}{2}(N-T)},$$

and the corresponding inequalities with the integration interval $[N, \infty)$ replaced by $(-\infty, -N]$ are also true. Taking this into account, we conclude that integral (13.7) for $j = 1$ and the integral obtained from (13.7) by differentiation with respect to ψ under the integral sign are uniformly convergent on the set $(\psi, \tau, \varepsilon) \in R^m \times [-T, T] \times (0, \varepsilon_0]$. By virtue of the smoothness of the right-hand side of Eqs. (12.1) and the arbitrariness of $T > 0$, this implies that the functions $Y_1(\psi, \tau, \varepsilon)$ and $\frac{\partial}{\partial \psi} Y_1(\psi, \tau, \varepsilon)$ are continuous in $(\psi, \tau) \in R^m \times R$ for every fixed $\varepsilon \in (0, \varepsilon_0]$. Using (13.5) and Lemma 12.2, we can similarly establish the uniform convergence (for $(\psi, \tau, \varepsilon) \in R^m \times [-T, T] \times [\underline{\varepsilon}_0, \varepsilon_0]$, where $T > 0$ and $\underline{\varepsilon}_0 \in (0, \varepsilon_0)$ are arbitrary) of the integral obtained from (13.7) by differentiation with respect to τ under the integral sign. Therefore, $\frac{\partial}{\partial \tau} Y_1(\psi, \tau, \varepsilon)$ is also continuous in $(\psi, \tau) \in R^m \times R$ for every $\varepsilon \in (0, \varepsilon_0]$. Moreover, the uniform convergence of the corresponding integrals enables us to change the order of integration and differentiation with respect to ψ and τ. As a result, we establish that the function $Y_1(\psi, \tau, \varepsilon)$ satisfies identity (13.9) $\forall (\psi, \tau, \varepsilon) \in G_1$ and the inequality

$$\left\| \frac{\partial Y_1}{\partial \tau} + \frac{\partial Y_1}{\partial \psi} \frac{\omega(\tau)}{\varepsilon} \right\| \leq \sigma_1 \Big(3 + n d_1 + \frac{1}{2} n^2 d_1^2 \Big) = \overline{d}_1.$$

We now assume that, for all $j = \overline{2, l-1}$, $l > 2$, the functions $Y_j = Y_j(\psi, \tau, \varepsilon)$ are continuously differentiable with respect to $(\psi, \tau) \in R^m \times R$ for every $\varepsilon \in (0, \varepsilon_0]$ and satisfy identity (13.9) and the inequalities

$$\left\| \frac{\partial Y_j}{\partial \psi} \right\| \leq d_2 \varepsilon^\alpha, \quad \left\| \frac{\partial Y_j}{\partial \tau} + \frac{\partial Y_j}{\partial \psi} \frac{\omega(\tau)}{\varepsilon} \right\| \leq \overline{d}_1 \quad \forall (\psi, \tau, \varepsilon) \in G_1. \qquad (13.17)$$

Let us prove that $Y_l(\psi, \tau, \varepsilon)$ is also continuously differentiable with respect to ψ and τ for every fixed ε and satisfies (13.9) and (13.17) for $j = l$. It follows from (13.7) that

$$\begin{aligned} \left\| \frac{\partial}{\partial \psi} Y_l(\psi, \tau, \varepsilon) \right\| \leq K \int_{-\infty}^{\infty} e^{-\gamma |t - \tau|} \Big\{ & \varepsilon m \sigma_1 + (n+m) n \sigma_1 d_1 \varepsilon^\alpha \\ & + \Big[n^2 \sigma_1 d_1 \varepsilon^\alpha + \varepsilon n \sigma_1 + \sup_{\varphi, \tau} \left\| \frac{\partial \widetilde{a}(\overline{x}(\tau), \varphi, \tau)}{\partial x} \right\| \Big] \\ & \times \sup_{\psi, \tau} \left\| \frac{\partial}{\partial \psi} Y_{l-1}(\psi, \tau, \varepsilon) \right\| \Big\} \Big(1 + \left\| \frac{\partial}{\partial \psi} (\varphi_{\tau, l}^t - \psi) \right\| \Big) \, dt \end{aligned}$$

$$+\sum_{k\neq 0}\sum_{s=-\infty}^{\infty}\Big\| \int_{\tau+s}^{\tau+s+1} Q(\tau,t)A_k(\overline{x}(t),t)$$

$$\times\Big(E_m+\frac{\partial}{\partial\psi}(\varphi^t_{\tau,l}-\psi)\Big)\exp\{i(k,\varphi^t_{\tau,l})\}\,dt\Big\|. \tag{13.18}$$

According to estimate (1.20) and Lemma 12.1, the last term on the right-hand side of (13.18) satisfies inequalities (13.15) and (13.16), and, therefore, it is bounded from above by the value $\sigma_6\varepsilon^\alpha$. Then (13.18) can be rewritten in the form

$$\sup_{\psi,\tau}\Big\|\frac{\partial}{\partial\psi}Y_l(\psi,\tau,\varepsilon)\Big\| \le \Big[\frac{2}{\gamma}K(m+(n+m)nd_1)\sigma_1(1+4c_1\sigma_5)+\sigma_6\Big]\varepsilon^\alpha$$

$$+\Big[\frac{2}{\gamma}Kn\sigma_1(1+nd_1)(1+4c_1\sigma_5)\varepsilon_0^\alpha+\sigma_0$$

$$+2\sigma_0c_1(1+d_2)\sigma_5\varepsilon_0^\alpha\Big]\sup_{\psi,\tau}\Big\|\frac{\partial}{\partial\psi}Y_{l-1}(\psi,\tau,\varepsilon)\Big\|.$$

Taking $\varepsilon_0>0$ so small that

$$\frac{2}{\gamma}Kn\sigma_1(1+nd_1)(1+4c_1\sigma_5)\varepsilon_0^\alpha \le \frac{1-\sigma_0}{4},$$

$$2\sigma_0c_1(1+d_2)\sigma_5\varepsilon_0^\alpha \le \frac{1-\sigma_0}{4},$$

we get

$$\sup_{\psi,\tau}\Big\|\frac{\partial}{\partial\psi}Y_l(\psi,\tau,\varepsilon)\Big\| \le \frac{1+\sigma_0}{2}\sup_{\psi,\tau}\Big\|\frac{\partial}{\partial\psi}Y_{l-1}(\psi,\tau,\varepsilon)\Big\|+\sigma_7\varepsilon^\alpha,$$

where

$$\sigma_7=\frac{2}{\gamma}K(m+(n+m)nd_1)\sigma_1(1+4c_1\sigma_5)+\sigma_6,$$

and the constant $\frac{1}{2}(1+\sigma_0)$ is less than 1 according to condition (13.3). The last inequality yields

$$\Big\|\frac{\partial}{\partial\psi}Y_l(\psi,\tau,\varepsilon)\Big\| \le \frac{2\sigma_7}{1-\sigma_0}\varepsilon^\alpha \equiv d_2\varepsilon^\alpha \quad \forall(\psi,\tau,\varepsilon)\in G_1.$$

As in the case $j=1$, one can easily verify that the improper integral on the right-hand side of (13.7) for $j=l$ and the integrals obtained from it by differentiation with respect to ψ and τ under the integral sign are uniformly convergent on the set

$$\psi \in R^m, \quad \tau \in [-T, T], \quad \varepsilon \in [\underline{\varepsilon}_0, \varepsilon_0], \tag{13.19}$$

where $T > 0$ and $\underline{\varepsilon}_0 \in (0, \varepsilon_0]$ are arbitrary constants. Therefore, the function $Y_l(\psi, \tau, \varepsilon)$ is continuously differentiable with respect to ψ and τ for every fixed ε on set (13.19) and satisfies identity (13.9) with $j = l$ for all ψ, τ, and ε from set (13.19). Since T and $\underline{\varepsilon}_0$ are arbitrary, we get relation (13.9) with $j = l$ for all $(\psi, \tau, \varepsilon) \in G_1$ and the inequality

$$\left\| \frac{\partial Y_l}{\partial \tau} + \frac{\partial Y_l}{\partial \psi} \frac{\omega(\tau)}{\varepsilon} \right\| \le \bar{d}_1.$$

Thus, by induction, we establish that $Y_j(\psi, \tau, \varepsilon)$, $j = \overline{0, \infty}$, are continuously differentiable with respect to $(\psi, \tau) \in R^m \times R$ for every $\varepsilon \in (0, \varepsilon_0]$ and satisfy the inequality

$$\left\| \frac{\partial}{\partial \psi} Y_j(\psi, \tau, \varepsilon) \right\| \le d_2 \varepsilon^\alpha \quad \forall (\psi, \tau, \varepsilon) \in G_1, \;\; j \ge 0.$$

By analogy, using the methods proposed above and Lemmas 12.3 and 12.4, we prove the continuity of the functions $\dfrac{\partial^2}{\partial \psi \partial \psi_\nu} Y_j(\psi, \tau, \varepsilon)$ and $\dfrac{\partial^2}{\partial \tau \partial \psi} Y_j(\psi, \tau, \varepsilon)$, $j \ge 0$, in $(\psi, \tau) \in R^m \times R$ for every $\varepsilon \in (0, \varepsilon_0]$ and the estimate

$$\sum_{\nu=1}^{m} \left\| \frac{\partial^2}{\partial \psi \partial \psi_\nu} Y_j(\psi, \tau, \varepsilon) \right\| \le d_3 \varepsilon^\alpha \quad \forall (\psi, \tau, \varepsilon) \in G_1, \quad j \ge 0,$$

where the constant d_3 is independent of ε and j. By virtue of the smoothness conditions (12.2) and the properties of the functions $Y_j(\psi, \tau, \varepsilon)$ established above, identity (13.9) yields the continuity of the functions $\dfrac{\partial^2}{\partial \psi \partial \tau} Y_j(\psi, \tau, \varepsilon)$ and $\dfrac{\partial^2}{\partial \tau^2} Y_j(\psi, \tau, \varepsilon)$ in ψ and τ for every ε. Hence, each of the functions $Y_j(\psi, \tau, \varepsilon)$ is twice continuously differentiable with respect to $(\psi, \tau) \in R^m \times R$ for every fixed $\varepsilon \in (0, \varepsilon_0]$.

Now let

$$\|\omega(\tau)\| + \left\| \frac{d}{d\tau} \omega(\tau) \right\| + \left\| \frac{\partial}{\partial \tau} A(x, \varphi, \tau, \varepsilon) \right\| \le \sigma_1 \quad \forall (x, \varphi, \tau, \varepsilon) \in \overline{G}.$$

Then identity (13.9) yields estimates (13.11) in which

$$d_4 = \sigma_1\Big(d_2 + (d_2 + nd_1)\varepsilon_0 + \frac{1}{2}n^2 d_1^2 \varepsilon_0^{1+\alpha} + \varepsilon_0^{1-\alpha} + \varepsilon_0^{2-\alpha}\Big),$$

$$d_5 = \sigma_1(d_3 + m(1+\varepsilon_0)\varepsilon_0^{1-\alpha} + nd_2\varepsilon_0^2 + (2nd_2 + md_2 + d_3)\varepsilon_0 + n^2 d_2(d_1 + d_2)\varepsilon_0^{1+\alpha}),$$

$$d_6 = \sigma_1[d_5 + (d_5 + d_2 + n(1 + nd_1\varepsilon_0^{\alpha})d_4)\varepsilon_0 + (2+n^2)\varepsilon_0^{2-\alpha} + (nd_1 + d_2)\varepsilon_0^{2-\alpha} + (nd_1 + d_2)\varepsilon_0^2 + n(\varepsilon_0^{1-\alpha} + d_4)(1 + d_2\varepsilon_0^{\alpha} + \varepsilon_0)\varepsilon_0].$$

Finally, we prove that each function $Y_j(\psi, \tau, \varepsilon)$, $j \geq 0$, is 2π-periodic in each component ψ_ν, $\nu = \overline{1,m}$, of the vector ψ. Indeed, the right-hand side of (12.1) is 2π-periodic in φ_ν, $\nu = \overline{1,m}$. If we impose the condition that $Y_{l-1}(\psi, \tau, \varepsilon)$ is also periodic in ψ_ν with period 2π, then the function $\varphi = \varphi_{\tau,l}^t(\psi, \varepsilon)$ that is the solution of the Cauchy problem

$$\frac{d\varphi_{\tau,l}^t}{dt} = \frac{\omega(t)}{\varepsilon} + b(\overline{x}(t) + Y_{l-1}(\varphi_{\tau,l}^t, t, \varepsilon), \varphi_{\tau,l}^t, t, \varepsilon), \quad \varphi_{\tau,l}^{\tau} = \psi \tag{13.20}$$

can be represented in the form

$$\varphi_{\tau,l}^t(\psi, \varepsilon) = \psi + \widetilde{\varphi}_{\tau,l}^t(\psi, \varepsilon),$$

where $\widetilde{\varphi}_{\tau,l}^t(\psi, \varepsilon)$ is 2π-periodic in ψ_ν, $\nu = \overline{1,m}$. Let e_ν be the unit vector in the space R^m. Then, taking into account that

$$\varphi_{\tau,l}^t(\psi + 2\pi e_\nu, \varepsilon) = \psi + 2\pi e_\nu + \widetilde{\varphi}_{\tau,l}^t(\psi, \varepsilon) = \varphi_{\tau,l}^t(\psi, \varepsilon) + 2\pi e_\nu,$$

we deduce from equality (13.7) for $j = l$ that

$$Y_l(\psi + 2\pi e_\nu, \tau, \varepsilon) = Y_l(\psi, \tau, \varepsilon), \quad \nu = \overline{1,m}.$$

Since $Y_0(\psi + 2\pi e_\nu, \tau, \varepsilon) = Y_0(\psi, \tau, \varepsilon) \equiv 0$, this implies that $Y_j(\psi, \tau, \varepsilon)$ are 2π-periodic functions with respect to ψ_ν, $\nu = \overline{1,m}$, for all $j \geq 0$. Theorem 13.1 is proved.

Remark 1. If we assume, in addition, that the right-hand side of system (12.1) is continuous in all variables $(x, \varphi, \tau, \varepsilon) \in \overline{G}$, then the functions $Y_j(\psi, \tau, \varepsilon)$ are also continuous in $(\psi, \tau, \varepsilon) \in G_1$ for all $j \geq 0$. Indeed, since $Y_0(\psi, \tau, \varepsilon) \equiv 0$ is continuous in G_1, it follows from problem (13.20) that the function $\varphi^t_{\tau,1}(\psi, \varepsilon)$ is continuous in $(\psi, \tau, \varepsilon)$. Then the uniform convergence of the improper integral (13.7) guarantees the continuity of $Y_1(\psi, \tau, \varepsilon)$ in G_1. By analogy, one can establish that $Y_j(\psi, \tau, \varepsilon)$ is continuous for $j > 1$.

14. Existence of Integral Manifold

Below, we show that the sequence $\{X_j(\psi, \tau, \varepsilon)\}$, $X_j(\psi, \tau, \varepsilon) = \overline{x}(\tau) + Y_j(\psi, \tau, \varepsilon)$, constructed in the previous section converges to the integral manifold $x = X(\psi, \tau, \varepsilon)$ of system (12.1).

Theorem 14.1. *If conditions (12.2), (12.3), (13.2), and (13.3) are satisfied, then, for sufficiently small* $\varepsilon_0 > 0$, *the following assertions are true:*

(a) there exists an integral manifold $x = X(\psi, \tau, \varepsilon)$ *of system (12.1) that lies in a* $d_1\varepsilon^\alpha$*-neighborhood of the curve* $x = \overline{x}(\tau)$ $\forall(\psi, \tau, \varepsilon) \in G_1$;

(b) the function $X(\psi, \tau, \varepsilon)$ *is* 2π*-periodic in* $\psi_\nu, \nu = \overline{1, m}$, *and continuously differentiable with respect to* ψ *and* τ *for every fixed* $\varepsilon \in (0, \varepsilon_0]$, *and its matrix of partial derivatives with respect to* ψ *satisfies the inequality*

$$\left\| \frac{\partial}{\partial \psi} X(\psi, \tau, \varepsilon) \right\| \leq d_2 \varepsilon^\alpha$$

for all $(\psi, \tau, \varepsilon) \in G_1$ *and the Lipschitz condition with respect to the variables* ψ:

$$\left\| \frac{\partial X(\psi, \tau, \varepsilon)}{\partial \psi} - \frac{\partial X(\overline{\psi}, \tau, \varepsilon)}{\partial \psi} \right\| \leq d_3 \varepsilon^\alpha \|\psi - \overline{\psi}\|$$

$$\forall(\psi, \tau, \varepsilon) \in G_1, \quad \overline{\psi} \in R^m;$$

(c) on the integral manifold, system (12.1) takes the form

$$\frac{d\varphi}{d\tau} = \frac{\omega(\tau)}{\varepsilon} + b(X(\varphi, \tau, \varepsilon), \varphi, \tau, \varepsilon).$$

Proof. Consider the sequence $\{Y_j(\psi, \tau, \varepsilon)\}$. Let us prove that it converges uniformly on the set G_1 to a certain function $Y(\psi, \tau, \varepsilon)$. For this purpose, we establish an estimate of the norm $\|Y_{j+1} - Y_j\|$. It follows from (13.7) that

$$Y_{j+1}(\psi, \tau, \varepsilon) - Y_j(\psi, \tau, \varepsilon)$$
$$= \int_{-\infty}^{+\infty} Q(\tau, t)\{[F_j - F_{j-1}] + [\widetilde{a}_j - \widetilde{a}_{j-1}] + \varepsilon[A_j - A_{j-1}]\}dt, \quad (14.1)$$

where

$$F_l = F(Y_l, t), \quad \widetilde{a}_l = \widetilde{a}(\overline{x}(t) + Y_l, \varphi^t_{\tau,l+1}, t),$$
$$A_l = A(\overline{x}(t) + Y_l, \varphi^t_{\tau,l+1}, t, \varepsilon), \quad Y_l = Y_l(\varphi^t_{\tau,l+1}, t, \varepsilon), \quad l = j, j-1.$$

Further, we represent the difference $\widetilde{a}_j - \widetilde{a}_{j-1}$ in the form

$$\widetilde{a}_j - \widetilde{a}_{j-1}$$
$$= [\widetilde{a}(\overline{x}(t), \varphi^t_{\tau,j+1}, t) - \widetilde{a}(\overline{x}(t), \varphi^t_{\tau,j}, t)] + \frac{\partial}{\partial x}\widetilde{a}(\overline{x}(t), \varphi^t_{\tau,j}, t)[Y_j - Y_{j-1}]$$
$$+ \int_0^1 \Big[\frac{\partial}{\partial x}\widetilde{a}(\overline{x}(t) + rY_j, \varphi^t_{\tau,j+1}, t) - \frac{\partial}{\partial x}\widetilde{a}(\overline{x}(t) + rY_{j-1}, \varphi^t_{\tau,j}, t)\Big] dr Y_j$$
$$+ \int_0^1 \Big[\frac{\partial}{\partial x}\widetilde{a}(\overline{x}(t) + rY_{j-1}, \varphi^t_{\tau,j}, t) - \frac{\partial}{\partial x}\widetilde{a}(\overline{x}(t), \varphi^t_{\tau,j}, t)\Big] dr [Y_j - Y_{j-1}]. \quad (14.2)$$

Using the smoothness conditions (12.2) and estimate (13.13), we obtain

$$\|F_j - F_{j-1}\| \le n^2\sigma_1 d_1 \varepsilon^\alpha \|Y_j - Y_{j-1}\|,$$

$$\varepsilon\|A_j - A_{j-1}\| \le \varepsilon\sigma_1(n+m)(\|Y_j - Y_{j-1}\| + \|\varphi^t_{\tau,j+1} - \varphi^t_{\tau,j}\|),$$

$$\Big\|\int_0^1 \Big[\frac{\partial}{\partial x}\widetilde{a}(\overline{x}(t) + rY_j, \varphi^t_{\tau,j+1}, t) - \frac{\partial}{\partial x}\widetilde{a}(\overline{x}(t) + rY_{j-1}, \varphi^t_{\tau,j}, t)\Big] dr Y_j\Big\|$$
$$\le n\sigma_1(n+m) d_1 \varepsilon^\alpha (\|Y_j - Y_{j-1}\| + \|\varphi^t_{\tau,j+1} - \varphi^t_{\tau,j}\|),$$

$$\Big\|\int_0^1 \Big[\frac{\partial}{\partial x}\widetilde{a}(\overline{x}(t)+rY_{j-1},\varphi^t_{\tau,j},t)-\frac{\partial}{\partial x}\widetilde{a}(\overline{x}(t),\varphi^t_{\tau,j},t)\Big]dr(Y_j-Y_{j-1})\Big\|$$

$$\leq n^2\sigma_1 d_1\varepsilon^\alpha\|Y_j-Y_{j-1}\|. \tag{14.3}$$

We now estimate the difference $Y_j - Y_{j-1}$ as follows:

$$\|Y_j-Y_{j-1}\| \leq \|Y_j(\varphi^t_{\tau,j+1},t,\varepsilon)-Y_j(\varphi^t_{\tau,j},t,\varepsilon)\|$$

$$+\|Y_j(\varphi^t_{\tau,j},t,\varepsilon)-Y_{j-1}(\varphi^t_{\tau,j},t,\varepsilon)\| \leq d_2\varepsilon^\alpha\|\varphi^t_{\tau,j+1}-\varphi^t_{\tau,j}\|$$

$$+\sup_{G_1}\|Y_j(\psi,\tau,\varepsilon)-Y_{j-1}(\psi,\tau,\varepsilon)\|. \tag{14.4}$$

Since, according to Lemma 12.5, we have

$$\|\varphi^t_{\tau,j+1}-\varphi^t_{\tau,j}\| \leq \sigma_8 e^{\frac{1}{2}\gamma|t-\tau|}\sup_{G_1}\|Y_j(\psi,\tau,\varepsilon)-Y_{j-1}(\psi,\tau,\varepsilon)\| \tag{14.5}$$

for $c_{19}\varepsilon_0^\alpha \leq \dfrac{1}{4}\gamma$ and $\sigma_8 = c_{18}\max\Big\{1;\dfrac{4}{\gamma}\Big\}$, combining (14.1)–(14.5) we get

$$\|Y_{j+1}(\psi,\tau,\varepsilon)-Y_j(\psi,\tau,\varepsilon)\|$$

$$\leq(\sigma_0+\sigma_9\varepsilon^\alpha)\sup_{G_1}\|Y_j(\psi,\tau,\varepsilon)-Y_{j-1}(\psi,\tau,\varepsilon)\|$$

$$+\Big\|\int_{-\infty}^{\infty}Q(\tau,t)[\widetilde{a}(\overline{x}(t),\varphi^t_{\tau,j+1},t)-\widetilde{a}(\overline{x}(t),\varphi^t_{\tau,j},t)]dt\Big\|, \tag{14.6}$$

where

$$\sigma_9=\frac{2}{\gamma}K\{2n^2\sigma_1 d_1+\sigma_1(n+m)(nd_1+1)(1+2\sigma_8)$$

$$+2\sigma_1\sigma_8 d_2[n+2n^2d_1+(n+m)(1+nd_1)]\}.$$

To estimate the integral on the right-hand side of the last inequality (denote it by I_j), we represent it in the form of the sum of integrals over segments of unit length and use estimate (1.20). As a result, we get

$$\|I_j\| \leq \sum_{k\neq 0}\sum_{s=-\infty}^{\infty} \sigma_3 K\varepsilon^{\alpha} \max_{[\tau+s,\tau+s+1]} e^{-\gamma|t-\tau|}\Big\{\Big[(1+n\sigma_1)\sup_{\overline{G}}\|a_k\|$$

$$+\frac{1+\sigma_1}{\|k\|}\sup_{\overline{G}}\Big\|\frac{\partial a_k}{\partial x}\Big\| + \sup_{\overline{G}}\Big\|\frac{\partial a_k}{\partial \tau}\Big\|\Big]$$

$$\times \max_{[\tau+s,\tau+s+1]} |\exp\{i(k,\theta^t_{\tau,j+1})\} - \exp\{i(k,\theta^t_{\tau,j})\}|$$

$$+\frac{1}{\|k\|}\sup_{\overline{G}}\|a_k\| \max_{[\tau+s,\tau+s+1]}\Big|\Big(k,\frac{d}{dt}\theta^t_{\tau,j+1}\Big)\exp\{i(k,\theta^t_{\tau,j+1})\}$$

$$-\Big(k,\frac{d}{dt}\theta^t_{\tau,j}\Big)\exp\{i(k,\theta^t_{\tau,j})\}\Big|\Big\}, \tag{14.7}$$

where $a_k = a_k(x,\tau)$. Taking into account the inequalities

$$|\exp\{i(k,\theta^t_{\tau,j+1})\} - \exp\{i(k,\theta^t_{\tau,j})\}| \leq \|k\|\|\varphi^t_{\tau,j+1} - \varphi^t_{\tau,j}\|,$$

$$\Big|\exp\{i(k,\theta^t_{\tau,j+1})\}\frac{d}{dt}(k,\theta^t_{\tau,j+1}) - \exp\{i(k,\theta^t_{\tau,j})\}\frac{d}{dt}(k,\theta^t_{\tau,j})\Big|$$

$$\leq \|k\|^2\sup_{\overline{G}}\|b\|\|\varphi^t_{\tau,j+1} - \varphi^t_{\tau,j}\| + \|k\|\Big[\sup_{\overline{G}}\Big\|\frac{\partial b}{\partial x}\Big\|\|Y_j(\varphi^t_{\tau,j+1},t,\varepsilon)$$

$$-Y_{j-1}(\varphi^t_{\tau,j},t,\varepsilon)\| + \sup_{\overline{G}}\Big\|\frac{\partial b}{\partial \varphi}\Big\|\|\varphi^t_{\tau,j+1} - \varphi^t_{\tau,j}\|\Big] \tag{14.8}$$

and estimates (14.4), (14.5), and (14.7), we obtain

$$\|I_j\| \leq \sigma_{10}\varepsilon^{\alpha}\sup_{G_1}\|Y_j(\psi,\tau,\varepsilon) - Y_{j-1}(\psi,\tau,\varepsilon)\|,$$

where

$$\sigma_{10} = 2K\sigma_1\sigma_3[(1+n\sigma_1)(1+\sigma_8) + \sigma_1(1+m+nd_2)]\frac{e^{\frac{\gamma}{2}}}{1-e^{-\frac{\gamma}{2}}}.$$

Combining the last inequality with (14.6), for $j \geq 1$ and $\varepsilon_0 \leq [2(1-\sigma_0)^{-1}(\sigma_9+\sigma_{10})]^{-\frac{1}{\alpha}}$ we get

$$\sup_{G_1}\|Y_{j+1}(\psi,\tau,\varepsilon) - Y_j(\psi,\tau,\varepsilon)\|$$

$$\leq \frac{1+\sigma_0}{2}\sup_{G_1}\|Y_j(\psi,\tau,\varepsilon) - Y_{j-1}(\psi,\tau,\varepsilon)\|. \tag{14.9}$$

Since the constant $\frac{1}{2}(1+\sigma_0)$ is less than 1 and $\|Y_1(\psi,t,\varepsilon)\| \le d_1\varepsilon_0^\alpha$, it follows from (14.9) that the sequence $\{Y_j(\psi,\tau,\varepsilon)\}$ is uniformly convergent on the set G_1. Therefore, the function

$$Y(\psi,\tau,\varepsilon) = \lim_{j\to\infty} Y_j(\psi,\tau,\varepsilon)$$

is 2π-periodic in ψ_ν, $\nu = \overline{1,m}$, continuous in (ψ,τ) for every fixed ε, and such that $\|Y(\psi,\tau,\varepsilon)\| \le d_1\varepsilon^\alpha$ $\forall(\psi,\tau,\varepsilon) \in G_1$.

To prove the convergence of the sequence $\left\{\frac{\partial}{\partial\psi}Y_j(\psi,\tau,\varepsilon)\right\}$, we consider the equality

$$\begin{aligned}
&\frac{\partial}{\partial\psi}(Y_{j+1}(\psi,\tau,\varepsilon) - Y_j(\psi,\tau,\varepsilon))\\
&\quad = \int_{-\infty}^{\infty} Q(\tau,t)\Big[\Big(\frac{\partial F_j}{\partial y} + \frac{\partial \widetilde{a}_j}{\partial x} + \varepsilon\frac{\partial A_j}{\partial x}\Big)\frac{\partial Y_j}{\partial \varphi} + \varepsilon\frac{\partial A_j}{\partial \varphi}\Big]\frac{\partial}{\partial\psi}(\varphi^t_{\tau,j+1} - \varphi^t_{\tau,j})dt\\
&\qquad + \int_{-\infty}^{+\infty} Q(\tau,t)\Big\{\Big[\Big(\frac{\partial F_j}{\partial y} - \frac{\partial F_{j-1}}{\partial y}\Big) + \Big(\frac{\partial \widetilde{a}_j}{\partial x} - \frac{\partial \widetilde{a}_{j-1}}{\partial x}\Big)\\
&\qquad + \varepsilon\Big(\frac{\partial A_j}{\partial x} - \frac{\partial A_{j-1}}{\partial x}\Big)\Big]\frac{\partial Y_j}{\partial \varphi} + \varepsilon\Big(\frac{\partial A_j}{\partial \varphi} - \frac{\partial A_{j-1}}{\partial \varphi}\Big)\\
&\qquad + \Big(\frac{\partial F_{j-1}}{\partial y} + \frac{\partial \widetilde{a}_{j-1}}{\partial x} + \varepsilon\frac{\partial A_{j-1}}{\partial x}\Big)\Big(\frac{\partial Y_j}{\partial \varphi} - \frac{\partial Y_{j-1}}{\partial \varphi}\Big)\Big\}\\
&\qquad\qquad \times \Big[\frac{\partial}{\partial\psi}(\varphi^t_{\tau,j} - \psi) + E_m\Big]dt\\
&\qquad + \int_{-\infty}^{+\infty} Q(\tau,t)\Big(\frac{\partial \widetilde{a}_j}{\partial \varphi} - \frac{\partial \widetilde{a}_{j-1}}{\partial \varphi}\Big)\Big[\frac{\partial}{\partial\psi}(\varphi^t_{\tau,j} - \psi) + E_m\Big]dt\\
&\qquad + \int_{-\infty}^{+\infty} Q(\tau,t)\frac{\partial \widetilde{a}_j}{\partial \varphi}\frac{\partial}{\partial\psi}(\varphi^t_{\tau,j+1} - \varphi^t_{\tau,j})dt,
\end{aligned}$$

which follows from (13.7). Using Lemmas 12.1 and 12.5 and following the proof of inequality (14.9), we get

$$\sup_{G_1}\Big\|\frac{\partial}{\partial\psi}(Y_{j+1}(\psi,\tau,\varepsilon)-Y_j(\psi,\tau,\varepsilon))\Big\|$$

$$\le(\sigma_0+\sigma_{11}\varepsilon_0^\alpha)\sup_{G_1}\Big\|\frac{\partial}{\partial\psi}(Y_j(\psi,\tau,\varepsilon)-Y_{j-1}(\psi,\tau,\varepsilon))\Big\|$$

$$+\sigma_{12}\sup_{G_1}\|Y_j(\psi,\tau,\varepsilon)-Y_{j-1}(\psi,\tau,\varepsilon)\|, \tag{14.10}$$

where the constants σ_{11} and σ_{12} are independent of ε and j. Since $\sigma_0<1$, we deduce from the last estimate (by choosing $\varepsilon_0>0$ sufficiently small) that the sequence $\Big\{\dfrac{\partial}{\partial\psi}Y_j(\psi,\tau,\varepsilon)\Big\}$ converges uniformly on the set G_1 to the function $\dfrac{\partial}{\partial\psi}Y(\psi,\tau,\varepsilon)$, and, according to (13.10), the following inequality is true:

$$\Big\|\frac{\partial}{\partial\psi}Y(\psi,\tau,\varepsilon)\Big\|\le d_2\varepsilon^\alpha\quad\forall(\psi,\tau,\varepsilon)\in G_1.$$

Also note that the function $\dfrac{\partial}{\partial\psi}Y(\psi,\tau,\varepsilon)$ is continuous in ψ and τ for every $\varepsilon\in(0,\varepsilon_0]$, and the Lipschitz condition with respect to ψ follows from the last inequality in (13.10).

Now consider the sequence $\Big\{\dfrac{\partial}{\partial\tau}Y_j(\psi,\tau,\varepsilon)\Big\}$. It follows from (13.9) that

$$\begin{aligned}\frac{\partial}{\partial\tau}(Y_{j+1}-Y_j)=H(\tau)(Y_{j+1}-Y_j)&+[F(Y_j,\tau)-F(Y_{j-1},\tau)]\\&+[\widetilde a(X_j,\psi,\tau)-\widetilde a(X_{j-1},\psi,\tau)]\\&+\varepsilon[A(X_j,\psi,\tau,\varepsilon)-A(X_{j-1},\psi,\tau,\varepsilon)]\\&-\frac{\partial}{\partial\psi}(Y_{j+1}-Y_j)\Big[\frac{\omega(\tau)}{\varepsilon}+b(X_j,\psi,\tau,\varepsilon)\Big]\\&-\frac{\partial Y_j}{\partial\psi}[b(X_j,\psi,\tau,\varepsilon)-b(X_{j-1},\psi,\tau,\varepsilon)],\end{aligned}$$

whence

$$\left\| \frac{\partial}{\partial \tau} Y_{j+1}(\psi, \tau, \varepsilon) - \frac{\partial}{\partial \tau} Y_j(\psi, \tau, \varepsilon) \right\|$$

$$\leq n\sigma_1 \sup_{G_1} \| Y_{j+1}(\psi, \tau, \varepsilon) - Y_j(\psi, \tau, \varepsilon) \|$$

$$+ (1 + \varepsilon_0 + d_2 \varepsilon_0^\alpha + n d_1 \varepsilon_0^\alpha) n \sigma_1 \sup_{G_1} \| Y_j(\psi, \tau, \varepsilon) - Y_{j-1}(\psi, \tau, \varepsilon) \|$$

$$+ \left(\frac{\|\omega(\tau)\|}{\varepsilon} + \sigma_1 \right) \sup_{G_1} \left\| \frac{\partial}{\partial \psi} Y_{j+1}(\psi, \tau, \varepsilon) - \frac{\partial}{\partial \psi} Y_j(\psi, \tau, \varepsilon) \right\|.$$

Since the sequences $\{Y_j(\psi, \tau, \varepsilon)\}$ and $\left\{ \frac{\partial}{\partial \psi} Y_j(\psi, \tau, \varepsilon) \right\}$ are uniformly convergent on the set G_1, the last inequality yields the uniform convergence of the sequence $\left\{ \frac{\partial}{\partial \tau} Y_j(\psi, \tau, \varepsilon) \right\}$ on the set

$$\psi \in R^m, \quad \tau \in [-T, T], \quad \varepsilon \in [\underline{\varepsilon}_0, \varepsilon_0], \tag{14.11}$$

where $T > 0$ and $\underline{\varepsilon}_0 \in (0, \varepsilon_0)$ are arbitrary. Therefore,

$$\lim_{j \to \infty} \frac{\partial Y_j(\psi, \tau, \varepsilon)}{\partial \tau} = \frac{\partial Y(\psi, \tau, \varepsilon)}{\partial \tau} \tag{14.12}$$

for all $(\psi, \tau, \varepsilon)$ from set (14.11). By virtue of the arbitrariness of T and $\underline{\varepsilon}_0$, we obtain equality (14.12) for all $(\psi, \tau, \varepsilon) \in G_1$. It is clear that the function $\frac{\partial}{\partial \tau} Y(\psi, \tau, \varepsilon)$ is continuous in $(\psi, \tau) \in R^m \times R$.

Passing to the limit as $j \to \infty$ in Eq. (13.9), we get

$$\frac{\partial X}{\partial \tau} + \frac{\partial X}{\partial \psi} \left[\frac{\omega(\tau)}{\varepsilon} + b(X, \psi, \tau, \varepsilon) \right]$$

$$= a(X, \tau) + \widetilde{a}(X, \psi, \tau) + \varepsilon A(X, \psi, \tau, \varepsilon), \tag{14.13}$$

where $X = X(\psi, \tau, \varepsilon) = \overline{x}(\tau) + Y(\psi, \tau, \varepsilon)$.

Further, we consider the Cauchy problem

$$\frac{d\varphi}{d\tau} = \frac{\omega(\tau)}{\varepsilon} + b(X(\varphi, \tau, \varepsilon), \varphi, \tau, \varepsilon), \quad \varphi|_{\tau = \tau_0} = \psi \in R^m, \quad \tau_0 \in R.$$

The smoothness conditions enable one to extend the solution $\varphi = \varphi_{\tau_0}^{\tau}(\psi, \varepsilon)$ of the Cauchy problem for all $\tau \in R$. Using (14.13), one can easily verify that the function $x^{\tau}_{\tau_0}(\psi,\varepsilon) = X(\varphi^{\tau}_{\tau_0}(\psi,\varepsilon),\tau,\varepsilon)$ satisfies the following equation for all $\tau \in R$:

$$\frac{dx^{\tau}_{\tau_0}}{d\tau} = a(x^{\tau}_{\tau_0},\tau) + \widetilde{a}(x^{\tau}_{\tau_0},\varphi^{\tau}_{\tau_0},\tau) + \varepsilon A(x^{\tau}_{\tau_0},\varphi^{\tau}_{\tau_0},\tau,\varepsilon).$$

Therefore, by definition [MiLy], $x = X(\psi,\tau,\varepsilon)$ is the integral manifold of system (12.1). The properties of the function $X(\psi,\tau,\varepsilon)$ follow from the properties of $\overline{x}(\tau)$ and $Y(\psi,\tau,\varepsilon)$. Theorem 14.1 is proved.

Corollary 1. *If the conditions of Theorem 14.1 are satisfied and the functions* $A(x,\varphi,\tau,\varepsilon)$ *and* $b(x,\varphi,\tau,\varepsilon)$ *are continuous in the collection of variables on the set* $\overline{G}$, *then* $X(\psi,\tau,\varepsilon)$ *is continuous on* G_1.

Indeed, it follows from Remark 1 (Section 13) that each function $Y_j(\psi,\tau,\varepsilon)$, $j \geq 0$, is continuous on the set G_1. Since the sequence $\{Y_j(\psi,\tau,\varepsilon)\}$ converges uniformly on G_1, the limit function $Y(\psi,\tau,\varepsilon)$ and, hence, $X(\psi,\tau,\varepsilon)$ are continuous on G_1.

Corollary 2. *If the conditions of Theorem 14.1 are satisfied and* $\|\omega(\tau)\|$, $\left\|\frac{d}{d\tau}\omega(\tau)\right\|$, *and* $\left\|\frac{\partial}{\partial\tau}A(x,\varphi,\tau,\varepsilon)\right\|$ *are uniformly bounded by a constant* σ_1 *for any* $(x,\varphi,\tau,\varepsilon) \in \overline{G}$, *then*

$$\left\|\frac{\partial}{\partial\tau}X(\psi,\tau,\varepsilon)\right\| \leq (d_4+\sigma_1)\varepsilon^{\alpha-1} \quad \forall(\psi,\tau,\varepsilon) \in G_1,$$

and the matrices $\frac{\partial}{\partial\tau}X$ *and* $\frac{\partial}{\partial\psi}X$ *satisfy the Lipschitz conditions*

$$\left\|\frac{\partial}{\partial\tau}X(\overline{\psi},\overline{\tau},\varepsilon) - \frac{\partial}{\partial\tau}X(\psi,\tau,\varepsilon)\right\|$$
$$\leq d_5\varepsilon^{\alpha-1}\|\psi-\overline{\psi}\| + (d_6+\sigma_1(1+n\sigma_1))\varepsilon^{\alpha-2}|\tau-\overline{\tau}|,$$

$$\left\|\frac{\partial}{\partial\psi}X(\overline{\psi},\overline{\tau},\varepsilon) - \frac{\partial}{\partial\psi}X(\psi,\tau,\varepsilon)\right\| \leq d_3\varepsilon^{\alpha}\|\psi-\overline{\psi}\| + d_5\varepsilon^{\alpha-1}|\tau-\overline{\tau}|$$

for any $\tau,\overline{\tau} \in R$, $\psi,\overline{\psi} \in R^m$, *and* $\varepsilon \in (0,\varepsilon_0]$.

To prove this fact, it suffices to use inequalities (13.11) and the smoothness condition for the function $\overline{x}(\tau)$.

15. Conditional Asymptotic Stability of Integral Manifold

In this section, we establish the conditional asymptotic stability of the integral manifold $x = X(\psi, \tau, \varepsilon)$ of system (12.1) with respect to a certain set of initial data for slow variables. In the theorem presented below, we denote by $(x^{\tau}_{\tau_0}(y, \psi, \varepsilon); \varphi^{\tau}_{\tau_0}(y, \psi, \varepsilon))$ the solution of system (12.1) that takes the value $(y; \psi)$ for $\tau = \tau_0$ and by n_0 the integer number defined in Section 13.

Theorem 15.1. *Suppose that the conditions of Theorem 14.1 are satisfied. Then, for sufficiently small $\varepsilon_0 > 0$ and any $(\psi, \tau_0, \varepsilon) \in G_1$, in a certain neighborhood of the point $\overline{x}(\tau_0)$ there exist an $(n-n_0)$-dimensional manifold S_+ and an n_0-dimensional manifold S_- such that, for $\tau \in [\tau_0, \infty)$ $(\tau \in (-\infty, \tau_0])$, the solution $(x^{\tau}_{\tau_0}(y, \psi, \varepsilon); \varphi^{\tau}_{\tau_0}(y, \psi, \varepsilon))$ of system (12.1) is defined for all $y \in S_+$ $(y \in S_-)$, and the slow variables $x^{\tau}_{\tau_0}(y, \psi, \varepsilon)$ tend exponentially to the integral manifold $x = X(\psi, \tau, \varepsilon)$ as $\tau \to +\infty$ $(\tau \to -\infty)$ for $y \in S_+$ $(y \in S_-)$.*

Proof. We construct a sequence $\{Z_j(\psi, \tau, \varepsilon, \tau_0, d)\}$ using the recurrence formula

$$
\begin{aligned}
&Z_{j+1}(\psi, \tau, \varepsilon, \tau_0, d) \\
&\quad = \overline{Q}(\tau, \tau_0)d + \int_{\tau_0}^{\infty} Q(\tau, t)[F(Z_j, t) + \widetilde{a}(\overline{x}(t) + Z_j, \overline{\varphi}^t_{\tau,j+1}, t) \\
&\qquad\qquad + \varepsilon A(\overline{x}(t) + Z_j, \overline{\varphi}^t_{\tau,j+1}, t, \varepsilon)]dt, \quad Z_0 \equiv 0, \qquad (15.1)
\end{aligned}
$$

where

$$
Z_j = Z_j(\overline{\varphi}^t_{\tau,j+1}, t, \varepsilon, \tau_0, d), \quad \overline{Q}(\tau, \tau_0) = \operatorname{diag}(0, Q_-(\tau, \tau_0)),
$$

d is a constant n-dimensional vector whose first n_0 coordinates are equal to zero, $\varepsilon \in (0, \varepsilon_0]$, $\tau \geq \tau_0$, and $\overline{\varphi}^t_{\tau,j+1} = \overline{\varphi}^t_{\tau,j+1}(\psi, \varepsilon, d)$ is a solution of the Cauchy problem

$$
\frac{d}{dt}\overline{\varphi}^t_{\tau,j+1} = \frac{\omega(t)}{\varepsilon} + b(\overline{x}(t) + Z_j, \overline{\varphi}^t_{\tau,j+1}, t, \varepsilon), \quad \overline{\varphi}^{\tau}_{\tau,j+1} = \psi. \qquad (15.2)
$$

Taking into account that $Q(\tau, \tau_0) = \overline{Q}(\tau, \tau_0)$ for $\tau > \tau_0$, we deduce from (15.1) that

$$
\begin{aligned}
&\|Z_{j+1}(\psi,\tau,\varepsilon,\tau_0,d)\| \\
&\qquad \le K\|d\| + \sigma_0 \sup_{\psi,\tau} \|Z_j(\psi,\tau,\varepsilon,\tau_0,d)\| \\
&\qquad\qquad + \frac{2}{\gamma}K[\varepsilon\sigma_1 + n^2\sigma_1 \sup_{\psi,\tau}\|Z_j(\psi,\tau,\varepsilon,\tau_0,d)\|^2] \\
&\qquad\qquad + \sum_{k\neq 0}\Big[\sum_{s=-q}^{\infty}\Big\| \int_{\tau+s}^{\tau+s+1} Q(\tau,t)a_k(\overline{x}(t),t)\exp\{i(k,\overline{\varphi}^t_{\tau,j+1})\}dt\Big\| \\
&\qquad\qquad + \Big\|\int_{\tau_0}^{\tau-q} Q(\tau,t)a_k(\overline{x}(t),t)\exp\{i(k,\overline{\varphi}^t_{\tau,j+1})\}dt\Big\|\Big]. \qquad (15.3)
\end{aligned}
$$

Here, q is the integer part of the number $\tau-\tau_0$. Using conditions (12.2) and (13.5) and relation (1.20), we estimate the last term on the right-hand side of inequality (15.3) from above by the value $\sigma_4\varepsilon^\alpha$, where σ_4 is the constant defined in Section 13. Thus, inequality (15.3) yields

$$
\begin{aligned}
&\sup_{\psi,\tau}\|Z_{j+1}\| \\
&\qquad \le K\|d\| + \sigma_0\sup_{\psi,\tau}\|Z_j\| + \frac{2}{\gamma}Kn^2\sigma_1\sup_{\psi,\tau}\|Z_j\|^2 + \Big(\sigma_4 + \frac{2}{\gamma}K\sigma_1\Big)\varepsilon_0^\alpha,
\end{aligned}
$$

which, for $\varepsilon_0 \le (nd_1)^{-\frac{2}{\alpha}}$ and $\|d\| \le \Big(\sigma_4K^{-1} + \dfrac{4}{\gamma}\sigma_1\Big)\varepsilon_0^\alpha$, leads to the estimate

$$
\|Z_j(\psi,\tau,\varepsilon,\tau_0,d)\| \le 2\frac{4K\sigma_1+\gamma\sigma_4}{\gamma(1-\sigma_0)}\varepsilon_0^\alpha = d_1\varepsilon_0^\alpha \qquad (15.4)
$$

for all

$$
j\ge 0, \quad (\psi,\tau,\varepsilon,d)\in R^m\times[\tau_0,\infty)\times(0,\varepsilon_0]\times L \equiv G_2,
$$

$$
L = \Big\{d\colon d\in R^n,\ \|d\| \le \Big(\sigma_4K^{-1}+\frac{4}{\gamma}\sigma_1\Big)\varepsilon_0^\alpha\Big\},
$$

where the first n_0 coordinates of the vector d are equal to zero.

The inequality $\|Q(\tau,t)\| \le Ke^{-\gamma|t-\tau|}$ and condition (12.2) guarantee that the integral on the right-hand side of (15.1) converges uniformly for any $\psi\in R^m$,

$\tau \in [\tau_0, T]$, $\varepsilon \in (0, \varepsilon_0]$, and $d \in L$ $(T > \tau_0$ is arbitrary). Since $Z_0 \equiv 0$ and $\varphi^t_{\tau,1} = \overline{\varphi}^t_{\tau,1}$, using Lemmas 12.1–12.4 we establish the estimates

$$\sup_{G_2}\Big\|\frac{\partial}{\partial\psi}Z_1\Big\| \le d_2\varepsilon_0^\alpha, \quad \sum_{\nu=1}^m \sup_{G_2}\Big\|\frac{\partial^2}{\partial\psi\partial\psi_\nu}Z_1\Big\| \le d_3\varepsilon_0^\alpha$$

and the uniform (for all $\psi \in R^m$, $\tau \in [\tau_0, T]$, $\varepsilon \in [\underline{\varepsilon}_0, \varepsilon_0]$, and $d \in L$, where $T > \tau_0$ and $\underline{\varepsilon}_0 \in (0, \varepsilon_0)$ are arbitrary) convergence of the integrals obtained from (15.1) for $j = 0$ by differentiation with respect to ψ and τ under the integral sign. Moreover, by direct differentiation, one can verify that

$$\begin{aligned}\frac{\partial Z_1}{\partial\tau} + \frac{\partial Z_1}{\partial\psi}\Big[\frac{\omega(\tau)}{\varepsilon} + b(\overline{x}(\tau), \psi, \tau, \varepsilon)\Big]& \\ = H(\tau)Z_1 + \widetilde{a}(\overline{x}(\tau), \psi, \tau) &+ \varepsilon A(\overline{x}(\tau), \psi, \tau, \varepsilon), \\ Z_1 = Z_1(\psi, \tau, \varepsilon, \tau_0, d)&,\end{aligned}$$

and prove that Z_1 is 2π-periodic in ψ_ν, $\nu = \overline{1, m}$, and twice continuously differentiable with respect to ψ and τ for fixed ε, τ_0, and d.

Using the method of mathematical induction, by analogy with Section 13 we obtain the inequalities

$$\Big\|\frac{\partial}{\partial\psi}Z_j\Big\| \le d_2\varepsilon_0^\alpha, \quad \sum_{\nu=1}^m \Big\|\frac{\partial^2}{\partial\psi\partial\psi_\nu}Z_j\Big\| \le d_3\varepsilon_0^\alpha \tag{15.5}$$

and the identity

$$\begin{aligned}\frac{\partial Z_{j+1}}{\partial\tau} + \frac{\partial Z_{j+1}}{\partial\tau}\Big[\frac{\omega(\tau)}{\varepsilon} + b(\overline{x}(\tau) + Z_j, \psi, \tau, \varepsilon)\Big]& \\ = H(\tau)Z_{j+1} + F(Z_j, \tau) + \widetilde{a}(\overline{x}(\tau) + Z_j, \psi, \tau)& \\ + \varepsilon A(\overline{x}(\tau) + Z_j, \psi, \tau, \varepsilon)&\end{aligned} \tag{15.6}$$

for all $(\psi, \tau, \varepsilon, d) \in G_2$ and $j \ge 0$. Furthermore, the functions $Z_j = Z_j(\psi, \tau, \varepsilon, \tau_0, d)$ are periodic in ψ_ν, $\nu = \overline{1, m}$, with period 2π and twice continuously differentiable with respect to ψ and τ for fixed ε, τ_0, and d. Moreover, according to Lemma 12.5, they satisfy estimates (14.9) and (14.10) with Y_j and G_1 replaced by Z_j and G_2, respectively. This implies that the sequences $\{Z_j\}$ and $\left\{\frac{\partial}{\partial\psi}Z_j\right\}$ converge uniformly on the set G_2, and equality (15.6) yields the uniform convergence of the sequence $\left\{\frac{\partial}{\partial\tau}Z_j\right\}$ on the set

$$\psi \in R^m, \quad \tau \in [\tau_0, T], \quad \varepsilon \in [\underline{\varepsilon_0}, \varepsilon_0], \quad d \in L$$

for arbitrary $T > \tau_0$ and $\underline{\varepsilon_0} \in (0, \varepsilon_0)$. Passing to the limit as $j \to \infty$ in (15.6), for any $(\psi, \tau, \varepsilon, d) \in G_2$ we get

$$\frac{\partial Z}{\partial \tau} + \frac{\partial Z}{\partial \psi}\Big[\frac{\omega(\tau)}{\varepsilon} + b(\overline{x}(\tau) + Z, \psi, \tau, \varepsilon)\Big]$$
$$= H(\tau)Z + F(Z,\tau) + \widetilde{a}(\overline{x}(\tau) + Z, \psi, \tau) + \varepsilon A(\overline{x}(\tau) + Z, \psi, \tau, \varepsilon), \quad (15.7)$$

where

$$Z = Z(\psi, \tau, \varepsilon, \tau_0, d) = \lim_{j\to\infty} Z_j(\psi, \tau, \varepsilon, \tau_0, d).$$

Let $\overline{\varphi}^{\tau}_{\tau_0} = \overline{\varphi}^{\tau}_{\tau_0}(\psi, \varepsilon, d)$ denote a solution of the Cauchy problem

$$\frac{d\overline{\varphi}^{\tau}_{\tau_0}}{d\tau} = \frac{\omega(\tau)}{\varepsilon} + b(\overline{x}(\tau) + Z(\overline{\varphi}^{\tau}_{\tau_0}, \tau, \varepsilon, \tau_0, d), \overline{\varphi}^{\tau}_{\tau_0}, \tau, \varepsilon),$$
$$\overline{\varphi}^{\tau_0}_{\tau_0} = \psi \in R^m.$$

It now follows from (15.7) that $(\overline{x}^{\tau}_{\tau_0}(\psi, \varepsilon, d); \overline{\varphi}^{\tau}_{\tau_0}(\psi, \varepsilon, d))$, where $\overline{x}^{\tau}_{\tau_0}(\psi, \varepsilon, d) = \overline{x}(\tau) + Z(\overline{\varphi}^{\tau}_{\tau_0}(\psi, \varepsilon, d), \tau, \varepsilon, \tau_0, d)$, is a solution of system (12.1) for $\tau \geq \tau_0$, and

$$\|\overline{x}^{\tau}_{\tau_0}(\psi, \varepsilon, d) - \overline{x}(\tau)\| \leq d_1\varepsilon_0^{\alpha} \quad \forall(\psi, \tau, \varepsilon, d) \in G_2. \quad (15.8)$$

Thus, $\overline{x}(\tau_0) + Z(\psi, \tau_0, \varepsilon, \tau_0, d) \equiv S_+$ (for fixed ψ, τ_0, and ε) is an $(n - n_0)$-dimensional manifold that possesses the following property: every solution of system (12.1) with initial data from the set $S_+ \times R^m$ is defined for all $\tau \geq \tau_0$, and, according to (15.8), its slow variables are uniformly bounded.

To construct the manifold S_-, it is necessary to consider the following sequence instead of the sequence $\{Z_j\}$ defined by (15.1) and (15.2):

$$\widetilde{Z}_{j+1}(\psi,\tau, \varepsilon, \tau_0, d)$$
$$= \widetilde{Q}(\tau, \tau_0)d + \int_{-\infty}^{\tau} Q(\tau, t)[F(\widetilde{Z}_j, t) + \widetilde{a}(\overline{x}(t) + \widetilde{Z}_j, \widetilde{\varphi}^{t}_{\tau,j+1}, t)$$
$$+ \varepsilon A(\overline{x}(t) + \widetilde{Z}_j, \widetilde{\varphi}^{t}_{\tau,j+1}, t, \varepsilon)]dt, \quad \widetilde{Z}_0 \equiv 0,$$

where $\tau \le \tau_0$,

$$\widetilde{Z}_j = \widetilde{Z}_j(\widetilde{\varphi}^t_{\tau,j+1}, t, \varepsilon, \tau_0, d), \quad \widetilde{Q}(\tau, \tau_0) = \operatorname{diag}(-Q_+(\tau, \tau_0), 0),$$

d is a constant n-dimensional vector the last $n - n_0$ coordinates of which are equal to zero, and $\widetilde{\varphi}^t_{\tau,j+1} = \widetilde{\varphi}^t_{\tau,j+1}(\psi, \varepsilon, d)$ is a solution of the Cauchy problem

$$\frac{d\widetilde{\varphi}^t_{\tau,j+1}}{dt} = \frac{\omega(t)}{\varepsilon} + b(\overline{x}(t) + \widetilde{Z}_j, \widetilde{\varphi}^t_{\tau,j+1}, t, \varepsilon), \quad \widetilde{\varphi}^\tau_{\tau,j+1} = \psi.$$

Then

$$S_- = \overline{x}(\tau_0) + \lim_{j\to\infty} \widetilde{Z}_j(\psi, \tau_0, \varepsilon, \tau_0, d).$$

We now prove the second part of the theorem. Taking into account relations (13.7) and the equality

$$Q_-(\tau, \tau_0)Q_-(\tau_0, t) = Q_-(\tau, t), \quad t \le \tau_0 \le \tau,$$

one can represent the function $Y_{j+1}(\psi, \tau, \varepsilon)$ for $\tau \ge \tau_0$ in the form

$$Y_{j+1}(\psi, \tau, \varepsilon) = \overline{Q}(\tau, \tau_0)y_{j+1} + \int_{\tau_0}^{\infty} Q(\tau, t)[F(Y_j, t) + \widetilde{a}(\overline{x}(t) + Y_j, \varphi^t_{\tau,j+1}, t)$$

$$+ \varepsilon A(\overline{x}(t) + Y_j, \varphi^t_{\tau,j+1}, t, \varepsilon)]dt, \tag{15.9}$$

where $Y_j = Y_j(\varphi^t_{\tau,j+1}, t, \varepsilon)$, $y_{j+1} = y_{j+1}(\psi, \tau, \varepsilon)$ is the n-dimensional vector the first n_0 coordinates of which are equal to zero and the other coordinates coincide with the vector

$$\int_{-\infty}^{\tau_0} Q_-(\tau_0, t)[F_-(Y_j, t) + \widetilde{a}_-(\overline{x}(t) + Y_j, \varphi^t_{\tau,j+1}, t)$$

$$+ \varepsilon A_-(\overline{x}(t) + Y_j, \varphi^t_{\tau,j+1}, t, \varepsilon)]dt.$$

Here, we preserve the notation of Section 13. According to inequalities (1.20), (12.2), and (13.3), for any $(\psi, \tau, \varepsilon) \in G_1$ we have

$$\|y_{j+1}(\psi, \tau, \varepsilon)\| \le \Big[\frac{1}{\gamma}(n^2 d_1^2 \sigma_1 \varepsilon_0^\alpha + \sigma_1)K + \frac{1}{2}\sigma_0 d_1 + \sigma_4\Big]\varepsilon^\alpha < d_1 \varepsilon^\alpha.$$

Now consider the inequality

$$\|Z_{j+1}(\psi,\tau,\varepsilon,\tau_0,d)-Y_j(\psi,\tau,\varepsilon)\|$$
$$\leq Ke^{-\gamma(\tau-\tau_0)}\|d-y_{j+1}(\psi,\tau,\varepsilon)\|$$
$$+\left\|\int\limits_{\tau_0}^{-\infty} Q(\tau,t)\{[F_j-\underline{F}_j]+[\widetilde{a}_j-\underline{\widetilde{a}}_j]+\varepsilon[A_j-\underline{A}_j]\}dt\right\|, \quad (15.10)$$

which follows from (15.1) and (15.9). In this inequality, F_j, $\widetilde{a}_j$, and A_j have the same meaning as in (14.1), and

$$\underline{F}_j=F(Z_j,t),\quad \underline{\widetilde{a}}_j=\widetilde{a}(\overline{x}(t)+Z_j,\overline{\varphi}^t_{\tau,j+1},t),$$
$$\underline{A}_j=A(\overline{x}(t)+Z_j,\overline{\varphi}^t_{\tau,j+1},t,\varepsilon),\quad Z_j=Z_j(\overline{\varphi}^t_{\tau,j+1},t,\varepsilon,\tau_0,d).$$

For the difference $\widetilde{a}_j-\underline{\widetilde{a}}_j$, we use a representation of the form (14.2) with Z_j instead of Y_{j-1}. Then, taking into account conditions (12.2), (13.13), and (15.5), we deduce from (15.10) the following inequality:

$$\|Z_{j+1}(\psi,\tau,\varepsilon,\tau_0,d)-Y_j(\psi,\tau,\varepsilon)\|$$
$$\leq 2d_1\varepsilon_0^\alpha Ke^{-\gamma(\tau-\tau_0)}+K\int\limits_{\tau_0}^{\infty}e^{-\gamma|\tau-t|}\Big\{\Big[(md_1+3nd_1+1)n\sigma_1\varepsilon_0^\alpha$$
$$+\sup_{\psi,\tau}\left\|\frac{\partial\widetilde{a}(\overline{x}(\tau),\psi,\tau)}{\partial x}\right\|\Big]\|Z_j-Y_j\|$$
$$+\varepsilon_0^\alpha(m+n(m+n)d_1)\sigma_1\|\overline{\varphi}^t_{\tau,j+1}-\varphi^t_{\tau,j+1}\|\Big\}dt$$
$$+\sum_{k\neq0}\Big[\Big\|\int\limits_{\tau_0}^{\tau-q}Q(\tau,t)a_k(\overline{x}(t),t)(\exp\{i(k,\overline{\theta}^t_{\tau,j+1})\}$$
$$-\exp\{i(k,\theta^t_{\tau,j+1})\})\exp\Big\{\frac{i}{\varepsilon}(k,\int\limits_{\tau}^{t}\omega(r)dr)\Big\}dt\Big\|$$

$$+\sum_{s=-q}^{\infty}\Big\|\int\limits_{\tau+s}^{\tau+s+1} Q(\tau,t)a_k(\overline{x}(t),t)(\exp\{i(k,\overline{\theta}^t_{\tau,j+1})\}$$

$$-\exp\{i(k,\theta^t_{\tau,j+1})\})\exp\Big\{\frac{i}{\varepsilon}\Big(k,\int\limits_{\tau}^{t}\omega(r)dr\Big)\Big\}dt\Big\|\Big], \tag{15.11}$$

where q is the integer part of the number $\tau-\tau_0$ and

$$\overline{\theta}^t_{\tau,j+1}=\overline{\varphi}^t_{\tau,j+1}-\frac{1}{\varepsilon}\int\limits_{\tau}^{t}\omega(r)dr.$$

Denote

$$M_j=\sup_{\tau\in[\tau_0,\infty)}[e^{\frac{\gamma}{l}(\tau-\tau_0)}p_j(\tau,\varepsilon)],\quad l=\begin{cases}2, & \sigma_0=0,\\ \Big(1+\dfrac{2\sigma_0}{1-\sigma_0}\Big)^{\frac{1}{2}}, & \sigma_0>0,\end{cases}$$

$$p_j(\tau,\varepsilon)=\sup_{\psi\in R^m}\|Z_j(\psi,\tau,\varepsilon,\tau_0,d)-Y_j(\psi,\tau,\varepsilon)\|.$$

It is clear that

$$p_j(\tau,\varepsilon)\le M_je^{\frac{\gamma}{l}(\tau-\tau_0)}\quad\forall\tau\ge\tau_0,$$

$$\int\limits_{\tau_0}^{\infty}e^{-\gamma\|\tau-t\|}p_j(t,\varepsilon)dt\le\frac{2l^2}{\gamma(l^2-1)}M_je^{-\frac{\gamma}{l}(\tau-\tau_0)}. \tag{15.12}$$

To estimate $\|\overline{\varphi}^t_{\tau,j+1}-\varphi^t_{\tau,j+1}\|\equiv\Delta\varphi_{j+1}$, we use Lemma 12.5. As a result, we get

$$\Delta\varphi_{j+1}\le\overline{\sigma}_8e^{\frac{\gamma}{2}\frac{l-1}{l}|t-\tau|}\max_{\xi\in N(\tau,t)}p_j(\xi,\varepsilon)$$

for $c_{19}\varepsilon_0^{\alpha}\le\dfrac{\gamma(l-1)}{4l}$ and $\overline{\sigma}_8=c_{18}\max\Big\{1;\dfrac{4l}{\gamma(l-1)}\Big\}$. If $t<\tau$, then

$$\max_{\xi\in[t,\tau]}p_j(\xi,\varepsilon)\le e^{-\frac{\gamma}{l}(t-\tau_0)}\max_{\xi\in[t,\tau]}[e^{-\frac{\gamma}{l}(\xi-\tau_0)}p_j(\xi,\varepsilon)]\le M_je^{-\frac{\gamma}{l}(t-\tau_0)};$$

if $t\ge\tau$, then

$$\max_{\xi\in[\tau,t]}p_j(\xi,\varepsilon)\le M_je^{-\frac{\gamma}{l}(\tau-\tau_0)}.$$

Taking this arguments into account, we obtain

$$\|\overline{\varphi}^t_{\tau,j+1} - \varphi^t_{\tau,j+1}\| \le \overline{\sigma}_8 e^{\frac{\gamma}{2}\frac{l-1}{l}|t-\tau| - \frac{\gamma}{l}(\min\{\tau;t\}-\tau_0)},$$

$$\int\limits_{\tau_0}^{\infty} e^{-\gamma|\tau-t|}\|\overline{\varphi}^t_{\tau,j+1} - \varphi^t_{\tau,j+1}\|dt \le \overline{\sigma}_8 \frac{4l^2}{\gamma(l^2-1)} M_j e^{-\frac{\gamma}{l}(\tau-\tau_0)}. \tag{15.13}$$

Then it follows from (15.12) and (15.13) that

$$\int\limits_{\tau_0}^{\infty} e^{-\gamma|\tau-t|}\|Z_j(\overline{\varphi}^t_{\tau,j+1}, t, \varepsilon, \tau_0, d) - Y_j(\varphi^t_{\tau,j+1}, t, \varepsilon)\|dt$$

$$\le \frac{2l^2}{\gamma(l^2-1)}(1 + 2\overline{\sigma}_8 d_2 \varepsilon_0^\alpha) M_j e^{-\frac{\gamma}{l}(\tau-\tau_0)}. \tag{15.14}$$

According to (1.20) and (14.8), each of the integrals over the segments $[\tau+s, \tau+s+1]$ and $[\tau_0, \tau-q]$ on the right-hand side of (15.11) can be estimated from above by the value

$$\begin{aligned}&\sigma_3 K(1 + n\sigma_1 + \sigma_1 + m\sigma_1)\varepsilon_0^\alpha \max_t e^{-\gamma|t-\tau|}\\ &\quad\times \left\{\|k\| \sup_{\overline{G}}\|a_k\| + \sup_{\overline{G}}\left\|\frac{\partial a_k}{\partial \tau}\right\| + \sup_{\overline{G}}\left\|\frac{\partial a_k}{\partial x}\right\|\right\}\\ &\quad\times \left[\max_t \|\overline{\varphi}^t_{\tau,j+1} - \varphi^t_{\tau,j+1}\|(1 + d_2\varepsilon_0^\alpha) + \max_t p(t,\varepsilon)\right],\end{aligned} \tag{15.15}$$

where $a_k = a_k(x,\tau)$ and the maximum with respect to t is taken over all $t \in [\tau+s, \tau+s+1]$ or $t \in [\tau_0, \tau-q]$, depending on which integral is considered. Therefore, taking into account conditions (12.2) for the Fourier coefficients of the function $\widetilde{a}(x,\varphi,\tau)$ and inequalities (15.12), (15.13) and (15.15), we can estimate the last of the three terms on the right-hand side of (15.11) by the value

$$\sigma_{13}\varepsilon_0^\alpha M_j e^{-\frac{\gamma}{l}(\tau-\tau_0)},$$

where

$$\sigma_{13} = 4e^\gamma\left[1 - e^{\frac{\gamma(l-1)}{2l}}\right]^{-1} K\sigma_3(2 + d_2\varepsilon_0^\alpha)(1 + \overline{\sigma}_8)(1 + \sigma_1(1+n+m))\sigma_1.$$

Thus, with regard for inequalities (15.13) and (15.14), inequality (15.11) takes the form

$$M_{j+1} \le 2Kd_1\varepsilon_0^\alpha + \left[\sigma_{14}\varepsilon_0^\alpha + \sigma_0\frac{l^2}{l^2-1}\right]M_j, \quad j \ge 0. \tag{15.16}$$

Here,

$$\sigma_{14} = \sigma_{13} + \frac{2l^2}{\gamma(l^2-1)} K[2(m+(n+m)n)\sigma_1\overline{\sigma}_8 + n\sigma_1(1+md_1+3nd_1)(1+2\overline{\sigma}_8 d_2)] + \gamma K^{-1}\overline{\sigma}_8 d_2 \sigma_0.$$

Since $\sigma_0 \frac{l^2}{l^2-1} \leq \frac{1+\sigma_0}{2}$ and $M_0 = 0$, for $\sigma_{14}\varepsilon_0^\alpha \leq \frac{1-\sigma_0}{4}$ relation (15.16) yields

$$M_j \leq \frac{8}{1-\sigma_0} K d_1 \varepsilon_0^\alpha \quad \forall j \geq 0,$$

or

$$\|Z_j(\psi,\tau,\varepsilon,\tau_0,d) - Y_j(\psi,\tau,\varepsilon)\| \leq \frac{8}{1-\sigma_0} K d_1 \varepsilon_0^\alpha e^{-\frac{\gamma}{l}(\tau-\tau_0)} \tag{15.17}$$

for all $(\psi,\tau,\varepsilon,d) \in G_2$ and $j \geq 0$. Passing to the limit as $j \to \infty$ in (15.17), we get

$$\|Z(\psi,\tau,\varepsilon,\tau_0,d) - Y(\psi,\tau,\varepsilon)\| \leq \frac{8}{1-\sigma_0} K d_1 \varepsilon_0^\alpha e^{-\frac{\gamma}{l}(\tau-\tau_0)}$$

for $\psi \in R^m$, $\tau \geq \tau_0$, $\varepsilon \in (0,\varepsilon_0]$, and $d \in L$. Hence, as $\tau \to \infty$, the slow variables $\overline{x}_{\tau_0}^\tau(\psi,\varepsilon,d) = \overline{x}(\tau) + Z(\overline{\varphi}_{\tau_0}^\tau(\psi,\varepsilon,d),\tau,\varepsilon,\tau_0,d)$ of the solution $(\overline{x}_{\tau_0}^\tau(\psi,\varepsilon,d);\ \overline{\varphi}_{\tau_0}^\tau(\psi,\varepsilon,d))$ of system (12.1) tend exponentially to the curve $x = \overline{x}(\tau) + Y(\overline{\varphi}_{\tau_0}^\tau(\psi,\varepsilon,d),\tau,\varepsilon)$, which lies on the integral manifold $x = X(\psi,\tau,\varepsilon)$.

By analogy, one can establish that, as $\tau \to -\infty$, the slow variables of every solution of system (12.1) with initial data from the set $S_- \times R^m$ tend exponentially to the integral manifold. Theorem 15.1 is proved.

Remark 2. Inequality (15.8) can be regarded as an error estimate of the averaging method on the semiaxis $[\tau_0,\infty)$ under the condition $\overline{x}_{\tau_0}^{\tau_0}(\psi,\varepsilon,d) \in S_+$.

Remark 3. Theorem 15.1 remains true for $n_0 = 0$. In this case, the integral manifold $x = X(\psi,\tau,\varepsilon)$ of system (12.1) is asymptotically stable for all initial values of the slow variable x from a certain small neighborhood of the point $X(\psi,\tau_0,\varepsilon)$.

16. Smoothness of Integral Manifold

In Sections 12–15, we have proved the existence of the integral manifold $x = X(\psi, \tau, \varepsilon)$ of the system of $n+m$ differential equations

$$\frac{dx}{d\tau} = a(x,\tau) + \widetilde{a}(x,\varphi,\tau) + \varepsilon A(x,\varphi,\tau,\varepsilon),$$

$$\frac{d\varphi}{d\tau} = \frac{\omega(\tau)}{\varepsilon} + b(x,\varphi,\tau,\varepsilon). \tag{16.1}$$

In the present section, we study the problem of the smoothness of the function $X(\psi,\tau,\varepsilon)$. Assume that the following conditions are satisfied:

(a) the functions a, $\widetilde{a}$, A, ω, and b are $l \geq 2$ times continuously differentiable with respect to $(x,\varphi,\tau) \in \mathcal{D} \times R^m \times R \equiv G_3$ for every $\varepsilon \in (0,\varepsilon_0]$, and all their partial derivatives are uniformly bounded in $\overline{G} = G_3 \times (0,\varepsilon_0]$ by a constant c_1 independent of ε;

(b) the following relation is true:

$$\sum_{k\neq 0}\Big[\|k\|^l \sup_{\overline{G}}\|c_k\| + \|k\|^{l-1}\Big(\sup_{\overline{G}}\Big\|\frac{\partial c_k}{\partial \tau}\Big\| + \sup_{\overline{G}}\Big\|\frac{\partial c_k}{\partial x}\Big\|\Big)\Big] \leq c_1, \tag{16.2}$$

where $c_k = c_k(x,\tau,\varepsilon)$ are the Fourier coefficients of the function $[\widetilde{a}(x,\varphi,\tau); b(x,\varphi,\tau,\varepsilon)]$.

Theorem 16.1. *Suppose that conditions (a) and (b) are satisfied and relations (12.1), (13.2), and (13.3) are true. Then there exist constants* $\varepsilon_1 > 0$ *and* $c_2 > 0$ *such that, for all* $(\psi,\tau,\varepsilon) \in G_1 = R^m \times R \times (0,\varepsilon_0]$, $\varepsilon_0 \leq \varepsilon_1$, *the function* $X(\psi,\tau,\varepsilon)$ *is* $l-1$ *times continuously differentiable with respect to* ψ *and* τ *for every fixed* ε,

$$\Big\|D_\psi^s \frac{\partial^q}{\partial \tau^q} X(\psi,\tau,\varepsilon)\Big\| \leq c_2 \varepsilon^{\alpha - q} \quad \forall(\psi,\tau,\varepsilon) \in G_1, \;\; 1 \leq s+q \leq l-1, \tag{16.3}$$

and the derivatives of the $(l-1)$*th order satisfy the Lipschitz condition with respect to the variables* ψ *and* τ. *Here,* D_ψ^s *is an arbitrary partial derivative of order* s *with respect to* ψ.

It follows from Theorem 16.1 that the smoothness of the function $X(\psi,\tau,\varepsilon)$ decreases as compared with the smoothness of the right-hand side of (16.1). Under the conditions imposed on system (16.1), this situation is typical of the theory of integral manifolds, which is confirmed, e.g., by the analysis carried out in [Sam4].

Prior to the proof of Theorem 16.1, we prove the lemma presented below, in which $Y_j(\psi, \tau, \varepsilon)$ are the functions defined by (13.7), and $\varphi^t_{\tau,j+1}(\psi, \varepsilon)$ is a solution of the Cauchy problem

$$\frac{d}{dt}\varphi^t_{\tau,j+1} = \frac{\omega(t)}{\varepsilon} + b(\overline{x}(t) + Y_j(\varphi^t_{\tau,j+1}, t, \varepsilon), \varphi^t_{\tau,j+1}, t, \varepsilon), \quad \varphi^\tau_{\tau,j+1} = \psi. \tag{16.4}$$

Lemma 16.1. *If, for certain* $j \geq 0$, *the function* $Y_j(\psi, \tau, \varepsilon)$ *is* $l \geq 2$ *times continuously differentiable with respect to* $(\psi, \tau) \in R^m \times R$ *for every* $\varepsilon \in (0, \varepsilon_0]$ *and such that*

$$\left\| D^s_\psi \frac{\partial^q}{\partial \tau^q} Y_j(\psi, \tau, \varepsilon) \right\| \leq d_{s,q} \varepsilon^{\alpha - q} \quad \forall (\psi, \tau, \varepsilon) \in G_1, \quad 0 \leq s + q \leq l,$$

then one can find sufficiently large constants $d_{s,q}$ *and a sufficiently small constant* $\varepsilon_0 = \varepsilon_0(d_{s,q}) > 0$ *such that the function* $Y_{j+1}(\psi, \tau, \varepsilon)$ *is* l *times continuously differentiable with respect to* ψ *and* τ *for every fixed* $\varepsilon \in (0, \varepsilon_0]$ *and such that*

$$\left\| D^s_\psi \frac{\partial^q}{\partial \tau^q} Y_{j+1}(\psi, \tau, \varepsilon) \right\| \leq d_{s,q} \varepsilon^{\alpha - q} \tag{16.5}$$

for all $(\psi, \tau, \varepsilon) \in G_1$ *and* $0 \leq s + q \leq l$.

Proof. For $l = 2$, the statement of the lemma follows from Theorem 13.1. Therefore, we assume that $l > 2$. According to the theorems on the existence of a solution of the Cauchy problem and its differentiability with respect to initial data, for all $t \in R$ the function $\varphi^t_{\tau,j+1}(\psi, \varepsilon)$ has l continuous partial derivatives with respect to $(\psi, \tau) \in R^m \times R$ for every fixed $\varepsilon \in (0, \varepsilon_0]$. On the basis of problem (16.4), we consider the derivatives of the function $\varphi^t_{\tau,j+1}$ with respect to ψ. According to Lemmas 12.1 and 12.3, we have

$$\left\| \frac{\partial}{\partial \psi}(\varphi^t_{\tau,j+1} - \psi) \right\| \leq c_0^{(1)} \varepsilon^\alpha e^{\overline{\gamma}|t-\tau|}, \quad \left\| \frac{\partial}{\partial \psi}\varphi^t_{\tau,j+1} \right\| \leq c_0^{(1)} e^{\overline{\gamma}|t-\tau|},$$

$$\left\| D^2_\psi \varphi^t_{\tau,j+1} \right\| \leq c_0^{(2)} \varepsilon^\alpha e^{2\overline{\gamma}|t-\tau|}, \tag{16.6}$$

where

$$\overline{\gamma} = \frac{\gamma}{2l}, \quad c_0^{(1)} = c_1(1 + md_{1,0}) \max\Big\{1; \frac{2}{\gamma}\Big\}, \quad c_0^{(1)} = m + \underline{c}_0^{(1)},$$

$$c_0^{(2)} = c_{10}(1 + m^2 d_{2,0}) \max\Big\{1; \frac{2}{\gamma}\Big\}, \quad \varepsilon_0^\alpha \le \min\Big\{\frac{\overline{\gamma}}{c_{11}}; \frac{\overline{\gamma}}{2c_2(1 + md_{1,0})}\Big\},$$

and c_1, c_2, c_{10}, and c_{11} are the constants defined in Lemmas 12.1 and 12.3.

Assume that, for all $p = \overline{2, s-1}$, $s \le l$, the following inequalities are true:

$$\|D_\psi^p \varphi_{\tau,j+1}^t\| \le c_0^{(p)} \varepsilon^\alpha e^{p\overline{\gamma}|t-\tau|}, \quad (\psi, \tau, \varepsilon) \in G_1, \quad t \in R, \tag{16.7}$$

where the constants $c_0^{(p)}$ depend on $d_{0,0}, d_{1,0}, \ldots, d_{p,0}$. Then the functions $Y_j = Y_j(\varphi_{\tau,j+1}^t, t, \varepsilon)$ satisfy the estimate

$$\begin{aligned}
\|D_\psi^p Y_j\| &\le \sum_{\nu=1}^{p} \|D^\nu_{\varphi_{\tau,j+1}^t} Y_j\| \sum_\beta c_{\nu\beta} \|D_\psi \varphi_{\tau,j+1}^t\|^{\beta_1} \ldots \|D_\psi^p \varphi_{\tau,j+1}^t\|^{\beta_p} \\
&\le \varepsilon^\alpha \sum_{\nu=1}^{p} d_{\nu,0} \sum_\beta c_{\nu\beta} (c_0^{(1)})^{\beta_1} \ldots (c_0^{(p)})^{\beta_p} e^{\overline{\gamma}p|t-\tau|} \equiv \varepsilon^\alpha M_p e^{\overline{\gamma}p|t-\tau|}.
\end{aligned}$$

For $p \ge 2$, an analogous estimate is also true for $u = (u_1, \ldots, u_{n+m}) = (\overline{x}(t) + Y_j, \varphi_{\tau,j+1}^t)$, namely

$$\begin{aligned}
\|D_\psi^p u\| &\le \|D_\psi^p Y_j\| + \|D_\psi^p \varphi_{\tau,j+1}^t\| \\
&\le \varepsilon^\alpha [M_p + c_0^{(p)}] e^{\overline{\gamma}p|t-\tau|} \equiv \varepsilon^\alpha \overline{c}_0^{(p)} e^{\overline{\gamma}p|t-\tau|}.
\end{aligned} \tag{16.8}$$

If $p = 1$, then

$$\begin{aligned}
\|D_\psi u\| &\le (\|D_{\varphi_{\tau,j+1}^t} Y_j\| + 1) \|D_\psi \varphi_{\tau,j+1}^t\| \\
&\le (md_{1,0} + 1) c_0^{(1)} e^{\overline{\gamma}|t-\tau|} \equiv \overline{c}_0^{(1)} e^{\overline{\gamma}|t-\tau|}.
\end{aligned} \tag{16.9}$$

Further, differentiating equality (16.4) $s = s_1 + \ldots + s_m$ times with respect to the variables ψ, we obtain

$$\frac{d}{dt}\frac{\partial^s \varphi^t_{\tau,j+1}}{\partial\psi_1^{s_1}\dots\partial\psi_m^{s_m}}$$

$$= \frac{\partial b}{\partial u}\frac{\partial^s u}{\partial\psi_1^{s_1}\dots\partial\psi_m^{s_m}} + F_{s,j}$$

$$+ \sum_{p_1+\ldots+p_{n+m}=s}\frac{\partial^s b}{\partial u_1^{p_1}\dots\partial u_{n+m}^{p_{n+m}}}\sum\prod_{\nu=1}^{m}\prod_{\mu=1}^{m+n}\left(\frac{\partial u_\mu}{\partial\psi_\nu}\right)^{\beta_\nu^{(\mu)}}. \tag{16.10}$$

Here, the symbol $\sum$ in the third term on the right-hand side denotes summation over all $\beta_\nu^{(\mu)}$ that satisfy the conditions

$$\sum_{\mu=1}^{n+m}\beta_\nu^{(\mu)} = s_\nu, \quad \nu = \overline{1,m}, \qquad \sum_{\nu=1}^{m}\beta_\nu^{(\mu)} = p_\mu, \quad \mu = \overline{1,m+n},$$

and $F_{s,j}$ satisfies the inequality

$$\|F_{s,j}\| \le \sum_{p=2}^{s-1}\|D_u^p b\|\sum_\beta c_{p\beta}\|D_\psi u\|^{\beta_1}\dots\|D_\psi^{s-1}u\|^{\beta_{s-1}},$$

where at least one of the numbers $\beta_2,\dots,\beta_{s-1}$ is not equal to zero. Since the partial derivatives of the function $b(x,\varphi,\tau,\varepsilon)$ with respect to all variables x_k and φ_ν, $k=\overline{1,n}$, $\nu=\overline{1,m}$, up to the order l inclusive are bounded by a constant c_1 and inequalities (16.8) and (16.9) are satisfied, we have

$$\|F_{s,j}\| \le \varepsilon^\alpha\sum_{p=2}^{s-1}c_1\sum_\beta c_{p\beta}(\overline{c}_0^{(1)})^{\beta_1}\dots(\overline{c}_0^{(s-1)})^{\beta_{s-1}}\times e^{\overline{\gamma}s|t-\tau|}$$

$$\equiv \varepsilon^\alpha\sigma_1^{(s)}e^{\overline{\gamma}s|t-\tau|}. \tag{16.11}$$

We represent the first term on the right-hand side of (16.10) in the form

$$\frac{\partial b}{\partial u}\frac{\partial^s u}{\partial\psi_1^{s_1}\dots\partial\psi_m^{s_m}} = \frac{\partial b}{\partial x}\frac{\partial^s Y_j}{\partial\psi_1^{s_1}\dots\partial\psi_m^{s_m}} + \frac{\partial b}{\partial\varphi}L_\tau^t$$

$$= \left(\frac{\partial b}{\partial x}\frac{\partial Y_j}{\partial\varphi} + \frac{\partial b}{\partial\phi}\right)L_\tau^t + \Phi_{s,j}, \tag{16.12}$$

where

$$Y_j = Y_j(\varphi^t_{\tau,j+1},t,\varepsilon), \quad L_\tau^t = \frac{\partial^s\varphi^t_{\tau,j+1}}{\partial\psi_1^{s_1}\dots\partial\psi_m^{s_m}},$$

$$\|\Phi_{s,j}\| \le \left\|\frac{\partial b}{\partial x}\right\| \sum_{p=2}^{s} \|D^p_{\varphi^t_{\tau,j+1}} Y_j\| \sum_{\beta} c_{p\beta} \|D_\psi \varphi^t_{\tau,j+1}\|^{\beta_1} \ldots \|D_\psi^{s-1} \varphi^t_{\tau,j+1}\|^{\beta_{s-1}}$$

$$\le n c_1 \sum_{p=2}^{s} d_{p,0} \varepsilon^\alpha \sum_{\nu} c_{p\beta} (c_0^{(1)})^{\beta_1} \ldots (c_0^{(s-1)})^{\beta_{s-1}} e^{\overline{\gamma} s|t-\tau|}$$

$$\equiv \varepsilon^\alpha \sigma_2^{(s)} e^{\overline{\gamma} s|t-\tau|}. \qquad (16.13)$$

Moreover, taking into account that

$$\left|\frac{\partial u_\mu}{\partial \psi_\nu}\right| \le \left\|\frac{\partial}{\partial \varphi} Y_j^{(\mu)}\right\| \left\|\frac{\partial}{\partial \psi_\nu} \varphi^t_{\tau,j+1}\right\| \le m d_{1,0} c_0^{(1)} \varepsilon^\alpha e^{\overline{\gamma}|t-\tau|} \quad \text{for} \quad \mu = \overline{1,n}$$

and

$$\frac{\partial u_{n+\mu}}{\partial \psi_\nu} = \frac{\partial}{\partial \psi_\nu} (\varphi^{t,\mu}_{\tau,j+1} - \psi_\mu) + \delta_{\nu\mu} \quad \text{for} \quad \mu = \overline{1,m},$$

where $\delta_{\nu\mu}$ is the Kronecker symbol, $\varphi^t_{\tau,j+1} = (\varphi^{t,1}_{\tau,j+1}, \ldots, \varphi^{t,m}_{\tau,j+1})$, and $Y_j = (Y_j^{(1)}, \ldots, Y_j^{(n)})$, we deduce from condition (a) and inequality (16.6) that

$$\sum_{p_1+\ldots+p_{n+m}=s} \frac{\partial^s b}{\partial u_1^{p_1} \ldots \partial u_{n+m}^{p_{n+m}}} \sum \prod_{\nu=1}^{m} \prod_{\mu=1}^{m+n} \left(\frac{\partial u_\mu}{\partial \psi_\nu}\right)^{\beta_\nu^{(\mu)}}$$

$$= \frac{\partial^s b}{\partial \varphi_1^{s_1} \ldots \partial \varphi_m^{s_m}} + R_{s,j}. \qquad (16.14)$$

Here,

$$\|R_{s,j}\|$$

$$\le \sum_{p_1+\ldots+p_{n+m}=s} c_1 \sum \prod_{\nu=1}^{m} \prod_{\mu=1}^{m+n} \left(\max\{1 + \underline{c}_0^{(1)}; m d_{1,0} c_0^{(1)}\}\right)^{\beta_\nu^{(\mu)}} \varepsilon^\alpha e^{\overline{\gamma} s|t-\tau|}$$

$$\equiv \varepsilon^\alpha \sigma_3^{(s)} e^{\overline{\gamma} s|t-\tau|}. \qquad (16.15)$$

Thus, combining (16.12) and (16.14), we can rewrite Eq. (16.10) in the form

$$\frac{d}{dt} L^t_\tau = \left(\frac{\partial b}{\partial x} \frac{\partial Y_j}{\partial \varphi} + \frac{\partial b}{\partial \varphi}\right) L^t_\tau + \frac{\partial^s b}{\partial \varphi_1^{s_1} \ldots \partial \varphi_m^{s_m}} + F_{s,j} + \Phi_{s,j} + R_{s,j}, \quad (16.16)$$

where the functions $F_{s,j}$, $\Phi_{s,j}$, and $R_{s,j}$ satisfy inequalities (16.11), (16.13), and (16.15), respectively. For $s \geq 2$, equation (16.16) yields

$$\|L_\tau^t\| \leq nmc_1 d_{1,0}\varepsilon^\alpha \Big|\int_\tau^t \|L_\tau^\xi\| d\xi\Big| + \Big\|\int_\tau^t \frac{\partial^s b}{\partial\varphi_1^{s_1}\dots\partial\varphi_m^{s_m}} d\xi\Big\|$$

$$+ \Big\|\int_\tau^t \frac{\partial b}{\partial\varphi} L_\tau^\xi d\xi\Big\| + \frac{1}{\overline{\gamma}s}(\sigma_1^{(s)} + \sigma_2^{(s)} + \sigma_3^{(s)})e^{\overline{\gamma}s|t-\tau|}\varepsilon^\alpha. \quad (16.17)$$

Since

$$\frac{\partial^s b}{\partial\varphi_1^{s_1}\dots\partial\varphi_m^{s_m}}$$

$$= \sum_{k\neq 0} b_k(\overline{x}(\xi) + Y_j(\varphi_{\tau,j+1}^\xi, \xi, \varepsilon), \xi, \varepsilon) i^s k_1^{s_1}\dots k_m^{s_m} \exp\{i(k, \varphi_{\tau,j+1}^\xi)\},$$

$$\sup_{\overline{G}} \|b_k k_1^{s_1}\dots k_m^{s_m}\| \leq \|k\|^s \sup_{\overline{G}} \|b_k\|,$$

it follows from the condition for Fourier coefficients (16.2) and the estimate for oscillation integrals (1.20) that

$$\Big\|\int_\tau^t \frac{\partial^s b}{\partial\varphi_1^{s_1}\dots\partial\varphi_m^{s_m}} d\xi\Big\| \leq \varepsilon^\alpha\sigma_3 c_1[1 + 3c_1 + md_{1,0}c_1(1+2n)](1 + |t-\tau|)$$

$$\leq \varepsilon^\alpha \sigma_4 e^{\overline{\gamma}s|t-\tau|}, \quad (16.18)$$

$$\sigma_4 = \sigma_3 c_1[1 + 3c_1 + md_{1,0}c_1(1+2n)] \max\Big\{1; \frac{1}{\overline{\gamma}s}\Big\}.$$

Then, for $t \in [\tau, \tau+2)$, inequality (16.17) yields

$$\|L_\tau^t\| \leq \sigma_4^{(s)}\varepsilon^\alpha \leq \varepsilon^\alpha \sigma_4^{(s)} e^{\overline{\gamma}s|t-\tau|}, \quad (16.19)$$

$$\sigma_4^{(s)} = \Big[\frac{1}{\overline{\gamma}s}(\sigma_1^{(s)} + \sigma_2^{(s)} + \sigma_3^{(s)}) + \sigma_4\Big] \exp\{2(\overline{\gamma}s + (mnd_{1,0} + m)c_1)\}.$$

If $t \geq \tau + 2$, then we represent the third term on the right-hand side of inequality (16.17) in the form

$$\Big\| \int_{\tau}^{t} \frac{\partial b}{\partial \varphi} L_{\tau}^{\xi} d\xi \Big\|$$

$$\leq \sum_{k \neq 0} \Bigg[\sum_{q=0}^{\overline{q}-1} \Big\| \int_{\tau+q}^{\tau+q+1} B_k L_{\tau}^{\xi} \exp\{i(k, \theta_{\tau,j+1}^{\xi})\} \exp\Big\{ \frac{i}{\varepsilon} \int_{\tau}^{\xi} (k, \omega(r)) dr \Big\} d\xi \Big\|$$

$$+ \Big\| \int_{\tau+\overline{q}}^{t} B_k L_{\tau}^{\xi} \exp\{i(k, \theta_{\tau,j+1}^{\xi})\} d\xi \Big\| \Bigg], \qquad (16.20)$$

where $\overline{q}$ is the integer part of the number $t - \tau - 1$, $1 \leq t - (\tau + \overline{q}) < 2$, and $B_k = B_k(\overline{x}(\xi) + Y_j(\varphi_{\tau,j+1}^{\xi}, \xi, \varepsilon), \xi, \varepsilon)$ are the Fourier coefficients of the function $\dfrac{\partial b}{\partial \varphi}$. By analogy with the proof of Lemma 12.1, we estimate each of the integrals over the segments $[\tau + q, \tau + q + 1]$ and $[\tau + \overline{q}, t]$ with regard for inequality (1.20). As a result, we establish that the integral over the segment $[\tau+q, \tau+q+1]$ does not exceed the value

$$\Big[\sup_{\overline{G}} \|B_k\| + \frac{1}{\|k\|} \Big(\sup_{\overline{G}} \Big\| \frac{\partial}{\partial \tau} B_k \Big\| + \sup_{\overline{G}} \Big\| \frac{\partial}{\partial x} B_k \Big\| \Big) \Big] \sigma_5^{(s)} \varepsilon^{\alpha}$$

$$\times \Big(\max_{[\tau+q, \tau+q+1]} \|L_{\tau}^{\xi}\| + \int_{\tau+q}^{\tau+q+1} e^{s\overline{\gamma}(\xi - \tau)} d\xi \Big), \qquad (16.21)$$

where

$$\sigma_5^{(s)} = \sigma_3 \Big[1 + (m+1)c_1 + c_1 m d_{1,0}(2+n) + 2nc_1 d_{0,0} + \frac{1}{\overline{\gamma} s} (\sigma_1^{(s)} + \sigma_2^{(s)} + \sigma_3^{(s)}) \Big] e^{2\overline{\gamma} s}.$$

The integral over the segment $[\tau + \overline{q}, t]$ can also be estimated by a value of the form (16.21) with the only difference that the maximum of $\|L_{\tau}^{\xi}\|$ over $\xi \in [\tau + q, \tau + q + 1]$ must be replaced by the corresponding maximum over $\xi \in [\tau + \overline{q}, t]$, and the integral of the exponent over the segment $[\tau + q, \tau + q + 1]$ must be replaced by the corresponding integral over $\xi \in [\tau + \overline{q}, t]$.

We estimate the maximum of the function $\|L_\tau^\xi\|$ on the segments $[\tau+q, \tau+q+1]$ and $[\tau+\overline{q}, t]$ by analogy with the estimation of the maximum of the function $\|z_\tau^l\|$ in the proof of Lemma 12.1, namely

$$\max_{[\tau+q,\tau+q+1]} \|L_\tau^\xi\| \leq \sigma_6^{(s)} \int\limits_{\tau+q}^{\tau+q+1} [\|L_\tau^\xi\| + e^{s\overline{\gamma}(\xi-\tau)}]d\xi,$$

$$\max_{[\tau+\overline{q},t]} \|L_\tau^\xi\| \leq \sigma_6^{(s)} \int\limits_{\tau+\overline{q}}^{t} [\|L_\tau^\xi\| + e^{s\overline{\gamma}(\xi-\tau)}]d\xi,$$

$$\sigma_6^{(s)} = m\Big[1 + mc_1(1+nd_{1,0}) + \frac{1}{\overline{\gamma}s}(\sigma_1^{(s)} + \sigma_2^{(s)} + \sigma_3^{(s)})\Big].$$

Thus, using (16.2), (16.18), (16.20), and (16.21), we can rewrite inequality (16.17) for $t \geq \tau + 2$ in the form

$$\|L_\tau^t\| \leq \sigma_7^{(s)} \varepsilon^\alpha \Big(\int\limits_\tau^t \|L_\tau^\xi\| d\xi + e^{\overline{\gamma}s(t-\tau)}\Big),$$

where

$$\sigma_7^{(s)} = \max\Big\{(mnd_{1,0} + \sigma_5^{(s)}\sigma_6^{(s)})c_1;$$
$$\sigma_4 + \frac{1}{\overline{\gamma}s}\big(\sigma_1^{(s)} + \sigma_2^{(s)} + \sigma_3^{(s)} + c_1\sigma_5^{(s)}(1+\sigma_6^{(s)})\big)\Big\}.$$

The last inequality, together with inequality (16.19), yields

$$\Big\|\frac{\partial^s \varphi_{\tau,j+1}^t}{\partial\psi_1^{s_1}\ldots\partial\psi_m^{s_m}}\Big\| \leq c_0^{(s)} e^{s\overline{\gamma}|\tau-t|}\varepsilon^\alpha, \quad c_0^{(s)} = \max\Big\{\sigma_4^{(s)}; \frac{2s\sigma_7^{(s)}}{2s-1}\Big\} \tag{16.22}$$

for all $t \geq \tau$ and $s \geq 2$. By analogy, we establish estimate (16.22) for $t < \tau$. Hence, by induction, for all $(\psi, \tau, \varepsilon) \in G_1$, $t \in R$, and $s = \overline{2, l}$ we get

$$\|D_\psi^s \varphi_{\tau,j+1}^t(\psi,\varepsilon)\| \leq c_0^{(s)} \varepsilon^\alpha e^{\overline{\gamma}s|\tau-t|}, \tag{16.23}$$

where the constants $c_0^{(s)}$ depend on $d_{0,0}, \ldots, d_{s,0}$.

We now prove that inequalities (16.6) and (16.23) yield estimate (16.5) for $q = 0$. If $s = 0, 1, 2$, then relation (16.5) follows from Theorem 13.1. Assume that estimate (16.5) holds for $p = \overline{0, s-1}$, $s \geq 3$, $s \leq l$, and $q = 0$. For $s = s_1 + \ldots + s_m$, we consider

$$D^s_\psi Y_{j+1}(\psi, \tau, \varepsilon)$$

$$= \int_{-\infty}^{\infty} Q(\tau, t)[D^s_\psi F(Y_j, t) + D^s_\psi \widetilde{a}(\overline{x}(t) + Y_j, \varphi^t_{\tau,j+1}, t)$$

$$+ \varepsilon D^s_\psi A(\overline{x}(t) + Y_j, \varphi^t_{\tau,j+1}, t, \varepsilon)]dt, \quad (16.24)$$

where

$$D^s_\psi = \frac{\partial^s}{\partial \psi_1^{s_1} \ldots \partial \psi_m^{s_m}}, \quad Y_j = Y_j(\varphi^t_{\tau,j+1}, t, \varepsilon).$$

Since

$$\|D^s_\psi F(Y_j, t)\| \leq \sum_{\nu=2}^{s} \|D^\nu_{Y_j} F(Y_j, t)\| \sum_\beta c_{\nu\beta} \|D_\psi Y_j\|^{\beta_1} \ldots \|D^{s-1}_\psi Y_j\|^{\beta_{s-1}}$$

$$+ \left\| \frac{\partial}{\partial Y_j} F(Y_j, t) \right\| \cdot \|D^s_\psi Y_j\|$$

and

$$\|D^p_\psi Y_j\| \leq \varepsilon^\alpha M_p e^{\overline{\gamma} p |t-\tau|}, \quad M_p = M_p(d_{0,0}, \ldots, d_{p,0}),$$

$$\left\| \frac{\partial}{\partial Y_j} F(Y_j, t) \right\| \leq n^2 c_1 d_{0,0} \varepsilon^\alpha,$$

the following estimate holds for all $(\psi, \tau, \varepsilon) \in G_1$ and $t \in R$:

$$\|D^s_\psi F(Y_j, t)\| \leq \varepsilon^{2\alpha} \Big[\sum_{\nu=2}^{s} c_1 \sum_\beta c_{\nu\beta} M_1^{\beta_1} \ldots M_{s-1}^{\beta_{s-1}} + n^2 c_1 d_{0,0} M_s \Big] e^{\overline{\gamma} s |t-\tau|}$$

$$\equiv \varepsilon^{2\alpha} c_1^{(s)} e^{\overline{\gamma} s |t-\tau|}, \quad (16.25)$$

where the constant $c_1^{(s)}$ depends on $d_{0,0}, \ldots, d_{s,0}$.

By analogy, using inequalities (16.8) and (16.9), we obtain

$$\|D^s_\psi A\| \leq \sum_{\nu=1}^{s} c_1 \sum_\beta c_{\nu\beta} (\overline{c}_0^{(1)})^{\beta_1} \ldots (\overline{c}_0^{(s)})^{\beta_s} e^{\overline{\gamma} s |t-\tau|} \equiv c_2^{(s)} e^{\overline{\gamma} s |t-\tau|}, \quad (16.26)$$

where $c_2^{(s)} = c_2^{(s)}(d_{0,0}, \dots, d_{s,0})$. Further, we consider the second term in the square brackets on the right-hand side of equality (16.24) and represent it in the form

$$D_\psi^s \widetilde{a} = \frac{\partial \widetilde{a}}{\partial u} D_\psi^s u + \sum_{p_1+\dots+p_{n+m}=s} \frac{\partial^s \widetilde{a}}{\partial u_1^{p_1} \dots \partial u_{n+m}^{p_{n+m}}} \sum \prod_{\nu=1}^{m} \prod_{\mu=1}^{m+n} \Big(\frac{\partial u_\mu}{\partial \psi_\nu}\Big)^{\beta_\nu^{(\mu)}} + \widetilde{F}_{s,j}. \tag{16.27}$$

Here, the symbol $\sum$ denotes summation over all $\beta_\nu^{(\mu)}$ that satisfy the conditions

$$\sum_{\mu=1}^{m+n} \beta_\nu^{(\mu)} = s_\nu, \quad \nu = \overline{1,m}, \qquad \sum_{\nu=1}^{m} \beta_\nu^{(\mu)} = p_\mu, \quad \mu = \overline{1, m+n},$$

and $\widetilde{F}_{s,j}$ satisfies the inequality

$$\begin{aligned} \|\widetilde{F}_{s,j}\| &\leq \sum_{p=2}^{s-1} \|D_u^p \widetilde{a}\| \sum_\beta c_{p\beta} \|D_\psi u\|^{\beta_1} \dots \|D_\psi^{s-1} u\|^{\beta_{s-1}} \\ &\leq \varepsilon^\alpha \sum_{p=2}^{s-1} c_1 \sum_\beta c_{p\beta} (\overline{c}_0^{(1)})^{\beta_1} \dots (\overline{c}_0^{(s-1)})^{\beta_{s-1}} e^{\overline{\gamma} s|t-\tau|} \\ &\equiv \varepsilon^\alpha \sigma_1^{(s)} e^{\overline{\gamma} s|t-\tau|}, \end{aligned} \tag{16.28}$$

where at least one of the numbers $\beta_2, \dots, \beta_{s-1}$ is not equal to zero and $\sigma_1^{(s)} = \sigma_1^{(s)}(d_{0,0}, \dots, d_{s-1,0})$.

The second term on the right-hand side of (16.27) (denote it by v) admits a representation of the form (16.14), namely

$$v = D_\varphi^s \widetilde{a} + \widetilde{R}_{s,j}, \tag{16.29}$$

where

$$\|\widetilde{R}_{s,j}\| \leq \varepsilon^\alpha \sigma_3^{(s)} e^{\overline{\gamma} s|t-\tau|}, \quad \sigma_3^{(s)} = \sigma_3^{(s)}(d_{0,0}, \dots, d_{s-1,0}). \tag{16.30}$$

It remains to transform the first term on the right-hand side of (16.27). It is obvious that

$$\frac{\partial \widetilde{a}}{\partial u} D_\psi^s u = \Big(\frac{\partial \widetilde{a}}{\partial x}\frac{\partial Y_j}{\partial \varphi} + \frac{\partial \widetilde{a}}{\partial \varphi}\Big) D_\psi^s \varphi_{\tau,j+1}^t + \widetilde{\Phi}_{s,j}$$

$$+ \frac{\partial a}{\partial x} \sum_{p_1+\ldots+p_m=s} \frac{\partial^s Y_j}{\partial \varphi_1^{p_1}\ldots\partial\varphi_m^{p_m}} \sum \prod_{\mu=1}^m \prod_{\nu=1}^m \Big(\frac{\partial \varphi_\mu}{\partial \psi_\nu}\Big)^{\beta_\nu^{(\mu)}}$$

where the symbol $\sum$ denotes summation over all $\beta_\nu^{(\mu)}$ that satisfy the conditions

$$\sum_{\mu=1}^m \beta_\nu^{(\mu)} = s_\nu, \quad \nu = \overline{1,m}, \qquad \sum_{\nu=1}^m \beta_\nu^{(\mu)} = p_\mu, \quad \mu = \overline{1,m},$$

and $\widetilde{\Phi}_{s,j}$ satisfies the inequality

$$\|\widetilde{\Phi}_{s,j}\| \le \varepsilon^\alpha c_3^{(s)} e^{\overline{\gamma} s|t-\tau|}, \tag{16.31}$$

$$c_3^{(s)} = c_3^{(s)}(d_{0,0},\ldots,d_{s-1,0}) = nc_1 \sum_{p=2}^{s-1} d_{p,0} \sum_\beta c_{p\beta}(c_0^{(1)})^{\beta_1}\ldots(c_0^{(s-1)})^{\beta_{s-1}}.$$

Since

$$\frac{\partial \widetilde{a}}{\partial x} \sum_{p_1+\ldots+p_m=s} \frac{\partial^s Y_j}{\partial \varphi_1^{p_1}\ldots\partial\varphi_m^{p_m}} \sum \prod_{\mu=1}^m \prod_{\nu=1}^m \Big(\frac{\partial}{\partial \psi_\nu}(\varphi_\mu - \psi_\mu) + \delta_{\nu\mu}\Big)^{\beta_\nu^{(\mu)}}$$

$$= \frac{\partial \widetilde{a}}{\partial x} D_\psi^s Y_j + \widetilde{N}_{s,j},$$

where

$$\|\widetilde{N}_{s,j}\| \le \varepsilon^\alpha c_4^{(s)} e^{\overline{\gamma} s|t-\tau|} \sum_{p_1+\ldots+p_m=s} \sup_{\psi,\tau} \Big\| \frac{\partial^s Y_j(\psi,\tau,\varepsilon)}{\partial \psi_1^{p_1}\ldots\partial\psi_m^{p_m}} \Big\|, \tag{16.32}$$

$$c_4^{(s)} = nc_1 \max_{p_1+\ldots+p_m=s} \sum \prod_{\mu=1}^m \prod_{\nu=1}^m (\max\{1+\overline{c}_0^{(1)}; md_{1,0}c_0^{(1)}\})^{\beta_\nu^{(\mu)}},$$

$$c_4^{(s)} = c_4^{(s)}(d_{0,0}, d_{1,0}),$$

we get

$$\frac{\partial \widetilde{a}}{\partial u} D_\psi^s u = \Big(\frac{\partial \widetilde{a}}{\partial x}\frac{\partial Y_j}{\partial \varphi} + \frac{\partial \widetilde{a}}{\partial \varphi}\Big) D_\psi^s \varphi_{\tau,j+1}^t + \frac{\partial \widetilde{a}}{\partial x} D_\varphi^s Y_j + \widetilde{\Phi}_{s,j} + \widetilde{N}_{s,j}. \tag{16.33}$$

Hence, taking into account equalities (16.27), (16.29), and (16.33) and estimates (16.25), (16.26), (16.28), and (16.30)–(16.32), for $s \geq 3$ we deduce the following inequality from (16.24):

$$\|D_\psi^s Y_{j+1}(\psi,\tau,\varepsilon)\|$$

$$\leq \varepsilon^\alpha K\Big[c_4^{(s)} \sum_{p_1+\ldots+p_m=s} \sup_{\psi,\tau}\Big\|\frac{\partial^s Y_j(\psi,\tau,\varepsilon)}{\partial\psi_1^{p_1}\ldots\partial\psi_m^{p_m}}\Big\| + c_1^{(s)}\varepsilon^\alpha$$

$$+ c_2^{(s)}\varepsilon^{1-\alpha} + \sigma_1^{(s)} + \sigma_3^{(s)} + c_3^{(s)}\Big]\int\limits_{-\infty}^{\infty} e^{(-\gamma+\overline{\gamma}s)|t-\tau|}dt$$

$$+ K\int\limits_{-\infty}^{\infty} e^{-\gamma|t-\tau|}\Big\|\frac{\partial\widetilde{a}}{\partial x}\Big\|\Big\|\frac{\partial Y_j}{\partial\varphi}\Big\|\|D_\psi^s\varphi_{\tau,j+1}^t\|dt$$

$$+ K\int\limits_{-\infty}^{\infty} e^{-\gamma|t-\tau|}\Big[\sup_{\varphi,\tau}\Big\|\frac{\partial\widetilde{a}(\overline{x}(\tau),\varphi,\tau)}{\partial x}\Big\|$$

$$+ \Big\|\frac{\partial}{\partial x}\widetilde{a}(\overline{x}(\tau)+Y_j,\varphi_{\tau,j+1}^t,t) - \frac{\partial}{\partial x}\widetilde{a}(\overline{x}(t),\varphi_{\tau,j+1}^t,t,\varepsilon)\Big\|\Big]dt$$

$$+ \Big\|\int\limits_{-\infty}^{\infty} Q(\tau,t)D_\varphi^s\widetilde{a}(\overline{x}(t)+Y_j,\varphi_{\tau,j+1}^t,t)dt\Big\|$$

$$+ \Big\|\int\limits_{-\infty}^{\infty} Q(\tau,t)\frac{\partial}{\partial\varphi}\widetilde{a}(\overline{x}(t)+Y_j,\varphi_{\tau,j+1}^t,t)D_\psi^s\varphi_{\tau,j+1}^t dt\Big\|, \qquad (16.34)$$

where $Y_j = Y_j(\varphi_{\tau,j+1}^t,t,\varepsilon)$, $\overline{\gamma} = \dfrac{\gamma}{2l}$, and the constants $\sigma_1^{(s)}$, $\sigma_3^{(s)}$, $c_3^{(s)}$, and $c_4^{(s)}$ are independent of $d_{s,0}$. We choose $\varepsilon_0 > 0$ so small that $c_1^{(s)}\varepsilon_0^\alpha \leq 1$ and $c_2^{(s)}\varepsilon_0^\alpha \leq 1$ and denote

$$2 + \sigma_1^{(s)} + \sigma_3^{(s)} + c_3^{(s)} + c_4^{(s)}\frac{2}{\gamma-\overline{\gamma}s}K = c_5^{(s)},$$

$$c_5^{(s)} = c_5^{(s)}(d_{0,0}, d_{0,1}, \ldots, d_{s-1,0}).$$

To estimate the last two terms on the right-hand side of inequality (16.34), we represent the corresponding integrals in the form of the infinite sum of integrals over segments of unit length. Then, using inequalities (1.20), (16.2), and (16.23), we establish that the next to the last term on the right-hand side of (16.34) is estimated from above by the value

$$\frac{2}{1-e^{-\gamma}} K c_1 \sigma_3 \Big\{ 1 + c_1 \Big[n + 2 + m d_{1,0} + n d_{0,0} \Big(1 + \frac{1}{2} n d_{0,0} \Big) \Big] \Big\} \varepsilon^{\alpha} \equiv c_6 \varepsilon^{\alpha},$$

and the last term is estimated from above by the value

$$\frac{2}{1-e^{-\gamma+\overline{\gamma}s}} K c_1 \sigma_3 e^{\overline{\gamma}s} \Big[2 + (m+n+2) c_1 + (m+n) c_1 d_{0,0}$$
$$+ n c_1 d_{0,0} \Big(1 + \frac{1}{2} n d_{0,0} \Big) \Big] \varepsilon^{\alpha} \equiv c_6^{(s)} \varepsilon^{\alpha}$$

for $c_0^{(s)} \varepsilon_0^{\alpha} \leq 1$ and $(\sigma_1^{(s)} + \sigma_2^{(s)} + \sigma_3^{(s)}) \varepsilon_0^{\alpha} \leq 1.$

Since $\sigma_0 < 1$ and the constants $c_5^{(s)}$, c_6, and $c_6^{(s)}$ are independent of $d_{s,0}$, $s \geq 3$, we conclude that, for

$$\varepsilon_0^{\alpha} \leq \min \Big\{ \frac{1-\sigma_0}{2n^2 c_1 d_{0,0}}; \frac{1}{d_{s,0}} \Big\}$$

relation (16.34) yields

$$\sup_{\psi,\tau} \| D_{\psi}^{s} Y_{j+1}(\psi,\tau,\varepsilon) \|$$
$$\leq \frac{1+\sigma_0}{2} d_{s,0} \varepsilon^{\alpha} + \Big[c_5^{(s)} (1+m^s) + c_6 + c_6^{(s)} + + \frac{2}{\gamma - s\overline{\gamma}} K n c_1 d_{1,0} \Big] \varepsilon^{\alpha}$$
$$\leq d_{s,0} \varepsilon^{\alpha},$$

$$d_{s,0} = \frac{2}{1-\sigma_0} \Big[c_5^{(s)} (1+m^s) + c_6 + c_6^{(s)} + \frac{2}{\gamma - s\overline{\gamma}} K n c_1 d_{1,0} \Big], \qquad (16.35)$$

for all $(\psi,\tau,\varepsilon) \in G_1$ and $s = \overline{3,l}$.

It follows from the smoothness conditions for the right-hand side of system (16.1) and the functions $Y_j(\psi,\tau,\varepsilon)$ and $\varphi_{\tau,j+1}^{t}(\psi,\varepsilon)$ that estimate (16.35) remains true if we change the order of the differentiation of the function $Y_j(\psi,\tau,\varepsilon)$ with respect to the variables $\psi_1,\ldots,\psi_m$. Also note that inequalities (16.23) and (13.2) guarantee the uniform convergence of the improper integral (16.24) on the set $(\psi,\tau,\varepsilon) \in R^m \times [-T,T] \times (0,\varepsilon_0]$ ($T > 0$ is arbitrary). Therefore, the functions $D_{\psi}^{s} Y_{j+1}(\psi,\tau,\varepsilon)$ are continuous in $(\psi,\tau) \in R^m \times [-T,T]$. Taking into account that T is arbitrary, we conclude that, for every fixed $\varepsilon \in (0, \varepsilon_0]$, the functions $D_\psi^s Y_{j+1}(\psi, \tau, \varepsilon)$, $s = \overline{0, l}$, are continuous for all $(\psi, \tau) \in R^m \times R$. Thus, Lemma 16.1 is proved for $q = 0$ and $s = \overline{0, l}$.

Let us prove the statement of the lemma for $q \geq 1$. Using Lemmas 12.2 and 12.4, we get

$$\left\| \frac{\partial}{\partial \tau} \varphi_{\tau,j+1}^t \right\| \leq c_7^{(0)} \varepsilon^{-1} e^{\overline{\gamma}|t-\tau|}, \quad \left\| \frac{\partial}{\partial \tau} \frac{\partial}{\partial \psi} \varphi_{\tau,j+1}^t \right\| \leq c_7^{(1)} \varepsilon^{\alpha-1} e^{2\overline{\gamma}|t-\tau|}, \quad (16.36)$$

where

$$c_7^{(0)} = c_7(1 + c_1), \quad c_7^{(1)} = \overline{c}_{10}(1 + c_1) \max\left\{1; \frac{1}{\overline{\gamma}}\right\}, \quad \varepsilon_0^\alpha \leq \overline{\gamma} \max\left\{\frac{1}{c_8}, \frac{1}{\overline{c}_{11}}\right\},$$

and c_7, c_8, $\overline{c}_{10}$, and $\overline{c}_{11}$ are the constants defined by Lemmas 12.2 and 12.4. Following the proof of inequality (16.23), one can easily show that

$$\left\| \frac{\partial}{\partial \tau} D_\psi^s \varphi_{\tau,j+1}^t \right\| \leq c_7^{(s)} \varepsilon^{\alpha-1} e^{(s+1)\overline{\gamma}|t-\tau|} \quad (16.37)$$

for all $s = \overline{1, l-1}$, $(\psi, \tau, \varepsilon) \in G_1$, and $t \in R$. Inequalities (16.23), (16.36), and (16.37) yield the uniform convergence of the integral obtained from (16.24) by differentiation with respect to τ under the integral sign on the set

$$\psi \in R^m, \quad \tau \in [-T, T], \quad \varepsilon \in [\underline{\varepsilon}_0, \varepsilon_0],$$

($T > 0$ and $\underline{\varepsilon}_0 \in (0, \varepsilon_0)$ are arbitrary). The smoothness conditions for the right-hand side of system (16.1) and the functions $Y_j(\psi, \tau, \varepsilon)$ and $\varphi_{\tau,j+1}^t(\psi, \varepsilon)$ guarantee the continuity of the functions $\dfrac{\partial}{\partial \tau} D_\psi^s Y_{j+1}(\psi, \tau, \varepsilon)$, $s = \overline{0, l-1}$, in $(\psi, \tau) \in R^m \times [-T, T]$. This yields

$$\frac{\partial}{\partial \tau} D_\psi^s Y_{j+1}(\psi, \tau, \varepsilon) \in C_{\psi,\tau}, \quad (16.38)$$

where $C_{\psi,\tau}$ denotes the set of vector functions $f(\psi, \tau, \varepsilon)$ continuous in $(\psi, \tau) \in R^m \times R$ for every fixed $\varepsilon \in (0, \varepsilon_0]$. Let us write Eq. (13.9) for the function $Y_{j+1} = Y_{j+1}(\psi, \tau, \varepsilon)$. We have

$$\frac{\partial Y_{j+1}}{\partial \tau} = \frac{1}{\varepsilon}\Big[-\frac{\partial Y_{j+1}}{\partial \psi}\omega(\tau) - \varepsilon \frac{\partial Y_{j+1}}{\partial \psi} b(\overline{x}(\tau) + Y_j, \psi, \tau, \varepsilon)\Big] + \varepsilon H(\tau) Y_{j+1} + \varepsilon P(Y_j, \psi, \tau, \varepsilon). \quad (16.39)$$

Here,

$$P(Y_j, \psi, \tau, \varepsilon) = F(Y_j, \tau) + \widetilde{a}(\overline{x}(\tau) + Y_j, \psi, \tau) + \varepsilon A(\overline{x}(\tau) + Y_j, \psi, \tau, \varepsilon);$$

furthermore, this function has $l-1$ continuous partial derivatives with respect to $(\psi, \tau) \in R^m \times R$ for every $\varepsilon \in (0, \varepsilon_0]$. Since

$$D_\psi^s Y_{j+1}(\psi, \tau, \varepsilon) \in C_{\psi,\tau}, \quad s = \overline{0, l}, \tag{16.40}$$

Eq. (16.39) yields

$$D_\psi^\rho \frac{\partial}{\partial \tau} Y_{j+1}(\psi, \tau, \varepsilon) \in C_{\psi,\tau}, \quad \rho = \overline{0, l-1}. \tag{16.41}$$

Conditions (16.38) for $s = 1$ and (16.41) for $\rho = 1$ yield

$$\frac{\partial^2 Y_{j+1}(\psi, \tau, \varepsilon)}{\partial \psi_\nu \partial \tau} = \frac{\partial^2 Y_{j+1}(\psi, \tau, \varepsilon)}{\partial \tau \partial \psi_\nu} \quad \forall (\psi, \tau, \varepsilon) \in G_1, \quad \nu = \overline{1, m}. \tag{16.42}$$

Further, we consider the chain of equalities

$$\frac{\partial^3 Y_{j+1}}{\partial \psi_\nu \partial \psi_\mu \partial \tau} \overset{(16.41)}{=} \frac{\partial^3 Y_{j+1}}{\partial \psi_\mu \partial \psi_\nu \partial \tau} \overset{(16.42)}{=} \frac{\partial^3 Y_{j+1}}{\partial \psi_\mu \partial \tau \partial \psi_\nu}$$

$$\underset{(16.41)}{\overset{(16.38)}{=}} \frac{\partial^3 Y_{j+1}}{\partial \tau \partial \psi_\mu \partial \psi_\nu} \overset{(16.40)}{=} \frac{\partial^3 Y_{j+1}}{\partial \tau \partial \psi_\nu \partial \psi_\mu}, \quad \mu, \nu = \overline{1, m}.$$

Here, the marks above and below the equality signs indicate the relations used. By analogy, one can establish the continuity and, hence, the equality of all partial derivatives of the $(\rho+1)$th order:

$$D_\psi^\rho \frac{\partial}{\partial \tau} Y_{j+1} = D_\psi^{\rho-1} \frac{\partial}{\partial \tau} \frac{\partial}{\partial \psi_\nu} Y_{j+1} = \ldots = \frac{\partial}{\partial \tau} D_\psi^\rho Y_{j+1}.$$

To estimate $\left\| D_\psi^\rho \frac{\partial}{\partial \tau} Y_{j+1} \right\|$, $\rho = \overline{0, l-1}$, we note that the right-hand side of Eq. (16.39) and the right-hand sides of the equations obtained from (16.39) by ρ-fold differentiation with respect to ψ are independent of $D_\psi^\rho \frac{\partial}{\partial \tau} Y_j$. It is also clear that the main contribution to the estimates is made by the term $D_\psi^\rho \Big(\frac{\partial}{\partial \psi} Y_{j+1} \omega(\tau) \Big)$ because the other terms in (16.39) and in the equations differentiated ρ times with respect to ψ have a higher order of smallness with respect to ε as $\varepsilon \to 0$. Consequently,

$$\left\| D_\psi^\rho \frac{\partial}{\partial \tau} Y_{j+1}(\psi, \tau, \varepsilon) \right\| \leq \frac{1}{\varepsilon} \Big[\left\| D_\psi^\rho \frac{\partial}{\partial \psi} Y_{j+1}(\psi, \tau, \varepsilon) \right\| \cdot \|\omega(\tau)\| + \varepsilon \sigma_{\rho,1} \Big],$$

where $\sigma_{\rho,1}$ is a constant that depends on $d_{0,0}, d_{1,0}, \ldots, d_{\rho+1,0}$ but does not depend on $d_{\rho,1}$, $\rho = \overline{0, l-1}$. For $\sigma_{\rho,1}\varepsilon_0^{\alpha} \leq 1$, the last inequality yields

$$\left\| D_{\psi}^{\rho} \frac{\partial}{\partial \tau} Y_{j+1}(\psi, \tau, \varepsilon) \right\| \leq (mc_1 d_{\rho+1,0} + 1)\varepsilon^{\alpha-1} \leq d_{\rho,1}\varepsilon^{\alpha-1}$$

for all $(\psi, \tau, \varepsilon) \in G_1$ and $\rho = \overline{0, l-1}$.

Now assume that all partial derivatives of the $(\rho+\mu)$th order (ρth order with respect to ψ and μth order with respect to τ) of the function $Y_{j+1}(\psi, \tau, \varepsilon)$ are continuous in $(\psi, \tau) \in R^m \times R$ for every fixed $\varepsilon \in (0, \varepsilon_0]$ and such that

$$\left\| D_{\psi}^{\rho} \frac{\partial^{\mu}}{\partial \tau^{\mu}} Y_{j+1}(\psi, \tau, \varepsilon) \right\| \leq d_{\rho,\mu}\varepsilon^{\alpha-\mu} \quad \forall (\psi, \tau, \varepsilon) \in G_1 \tag{16.43}$$

for $0 \leq \mu \leq q < l$ and $0 \leq \rho \leq l - \mu$. Differentiating equality (16.39) q times with respect to τ and ρ times with respect to ψ, one can easily verify that all partial derivatives of the $(\rho + (q+1))$th order of the function $Y_{j+1}(\psi, \tau, \varepsilon)$ are continuous with respect to $\psi, \tau \in R^m \times R$ for every fixed $\varepsilon \in (0, \varepsilon_0]$. Moreover, $D_{\psi}^{\rho} \frac{\partial^{q+1}}{\partial \tau^{q+1}} Y_{j+1}$ depends on $D_{\psi}^{\rho} Y_j, D_{\psi}^{\rho} \frac{\partial}{\partial \tau} Y_j, \ldots, D_{\psi}^{\rho} \frac{\partial^{q}}{\partial \tau^{q}} Y_j$ but does not depend on $D_{\psi}^{\rho} \frac{\partial^{q+1}}{\partial \tau^{q+1}} Y_j$. The smoothness conditions for the right-hand side of system (16.1), inequalities (16.43), and analogous inequalities for Y_j imply that the differentiation of the right-hand side of Eq. (16.39) with respect to ψ does not worsen its order estimates with respect to ε, and each time it is differentiated with respect to τ this order decreases by one. Thus,

$$\begin{aligned}
&\left\| D_{\psi}^{\rho} \frac{\partial^{q+1}}{\partial \tau^{q+1}} Y_{j+1}(\psi, \tau, \varepsilon) \right\| \\
&\qquad \leq \frac{1}{\varepsilon} \left[\left\| D_{\psi}^{\rho} \frac{\partial}{\partial \psi} \frac{\partial^{q}}{\partial \tau^{q}} Y_{j+1}(\psi, \tau, \varepsilon) \right\| \cdot \|\omega(\tau)\| + \varepsilon^{1-q}\sigma_{\rho,q+1} \right] \\
&\qquad \leq [mc_1 d_{\rho+1,q} + 1]\varepsilon^{\alpha-q-1} \leq d_{\rho,q+1}\varepsilon^{\alpha-q-1} \quad \forall (\psi, \tau, \varepsilon) \in G_1
\end{aligned}$$

for $\sigma_{\rho,q+1}\varepsilon_0^{\alpha} \leq 1$ and $0 \leq \rho \leq l - (q+1)$. Here, $\sigma_{\rho,q+1}$ is a constant that depends on $d_{s,\nu}$, $s = \overline{0, \rho+1}$, $\nu = \overline{0, q}$ but does not depend on $d_{\rho,q+1}$.

Thus, according to the principle of mathematical induction, estimate (16.5) holds for all s and q that satisfy the condition $0 \leq s + q \leq l$. Lemma 16.1 is proved.

Proof of Theorem 16.1. We consider iterations (13.7) and fix the constants $d_{s,q}$ and ε_0 for which inequalities (16.5) are satisfied. For $d_{s,q}$ and ε_0 thus chosen, the 2π-periodic (in ψ_ν, $\nu = \overline{1,m}$) function $Y_1(\psi,\tau,\varepsilon)$ defined by formula (13.7) for $j = 1$ satisfies the estimate

$$\left\| D_\psi^s \frac{\partial^q}{\partial \tau^q} Y_1(\psi,\tau,\varepsilon) \right\| \leq d_{s,q}\varepsilon^{\alpha-q} \quad \forall(\psi,\tau,\varepsilon) \in G_1$$

for $0 \leq s+q \leq l$. Using the function $Y_1(\psi,\tau,\varepsilon)$ and formula (13.7) for $j = 2$, we obtain a 2π-periodic (in ψ_ν, $\nu = \overline{1,m}$) function $Y_2(\psi,\tau,\varepsilon)$, which, according to Lemma 16.1, satisfies inequality (16.5), and so on. Thus, relation (13.7) defines a sequence of iterations $Y_j(\psi,\tau,\varepsilon)$, $j \geq 1$, 2π-periodic in ψ_ν, $\nu = \overline{1,m}$, l times continuously differentiable with respect to $(\psi,\tau) \in R^m \times R$ for every fixed $\varepsilon \in (0,\varepsilon_0]$, and satisfying the inequalities

$$\left\| D_\psi^s \frac{\partial^q}{\partial \tau^q} Y_j(\psi,\tau,\varepsilon) \right\| \leq d_{s,q}\varepsilon^{\alpha-q} \quad \forall(\psi,\tau,\varepsilon) \in G_1,$$

$$j \geq 1, \quad 0 \leq s+q \leq l. \tag{16.44}$$

Inequality (16.44) implies that the functions $D_\psi^\rho \frac{\partial^q}{\partial \tau^q} Y_j(\psi,\tau,\varepsilon)$, $j \geq 1$, are uniformly bounded in $(\psi,\tau) \in R^m \times R$. According to the theorem on compactness in the space of continuous functions [KoF], this is sufficient in order that, for every $\varepsilon \in (0,\varepsilon_0]$, the function

$$Y(\psi,\tau,\varepsilon) = \lim_{j\to\infty} Y_j(\psi,\tau,\varepsilon)$$

have continuous derivatives with respect to ψ and τ up to the order $l-1$ that satisfy the Lipschitz condition with respect to ψ and τ from the set $R^m \times [-T,T]$ ($T > 0$ is arbitrary). Moreover, inequalities (16.44) yield

$$\left\| D_\psi^s \frac{\partial^q}{\partial \tau^q} Y(\psi,\tau,\varepsilon) \right\| \leq \overline{c}_2\varepsilon^{\alpha-q}, \quad 0 \leq s+q \leq l-1, \tag{16.45}$$

for all $(\psi,\tau,\varepsilon) \in R^m \times [-T,T] \times (0,\varepsilon_0]$ and $\overline{c}_2 = \max\limits_{0\leq s+q\leq l} d_{s,q}$. Taking into account that T is arbitrary, we conclude that estimates (16.45) hold for any $(\psi,\tau,\varepsilon) \in G_1$. Since $X(\psi,\tau,\varepsilon) = \overline{x}(\tau) + Y(\psi,\tau,\varepsilon)$, inequalities (16.3) follow from (16.45) and condition (a). Theorem 16.1 is proved.

Corollary 3. *If the condition of the boundedness of* $\|\omega(\tau)\|$, $\left\|\frac{d}{d\tau}\omega(\tau)\right\|, \ldots,$ $\left\|\frac{d^l}{d\tau^l}\omega(\tau)\right\|$ *is omitted from the conditions of Theorem 16.1, then the function*

$X(\psi, \tau, \varepsilon)$ *satisfies inequalities (16.3) only for* $q = 0$, *i.e.,*

$$\|D_{\psi}^{s} X(\psi, \tau, \varepsilon)\| \leq c_2 \varepsilon^{\alpha}, \quad (\psi, \tau, \varepsilon) \in G_1, \quad 1 \leq s \leq l-1.$$

The results of the present section remain true for a system of a more general form, namely

$$\begin{aligned} \frac{dx}{d\tau} &= a(x, \tau) + \widetilde{a}(x, \varphi, \tau) + \varepsilon^{\beta} B_1(x, \varphi, \tau, \varepsilon), \\ \frac{d\varphi}{d\tau} &= \frac{\omega(\tau)}{\varepsilon} + b(x, \varphi, \tau, \varepsilon) + \varepsilon^{\delta} B_2(x, \varphi, \tau, \varepsilon). \end{aligned} \tag{16.46}$$

Here, $\min\{\beta; \delta\} \geq \alpha$, a, $\widetilde{a}$, ω, and b satisfy the conditions of Theorem 16.1, and B_1 and B_2 are 2π-periodic (in φ_{ν}, $\nu = \overline{1, m}$) functions that, for every $\varepsilon \in (0, \varepsilon_0]$, have continuous and bounded (by a constant c_1) partial derivatives with respect to x and φ up to an order $l \geq 2$, i.e.,

$$[B_1; B_2] \in C_{x,\varphi}^{l}(\overline{G}, c_1), \tag{16.47}$$

and continuous partial derivatives with respect to x, φ, and τ up to the order $l-1$ that satisfy the inequalities

$$\left\| D_{x,\varphi}^{s} \frac{\partial^q}{\partial \tau^q} [B_1; B_2] \right\| \leq c_1 \varepsilon^{-1-q} \quad \forall (x, \varphi, \tau, \varepsilon) \in \overline{G},$$

$$q \geq 1, \quad s + q \leq l - 1. \tag{16.48}$$

Indeed, in this case, iterations (13.7) for the construction of the integral manifold of system (16.46) are determined by the relations

$$\begin{aligned} Y_j(\psi, \tau, \varepsilon) = \int_{-\infty}^{\infty} & Q(\tau, t)[F(Y_{j-1}, t) + \widetilde{a}(\overline{x}(t) + Y_{j-1}, \varphi_{\tau,j}^{t}, t) \\ & + \varepsilon^{\beta} B_1(\overline{x}(t) + Y_{j-1}, \varphi_{\tau,j}^{t}, t, \varepsilon)]dt, \quad Y_0 \equiv 0, \end{aligned}$$

$$\frac{d\varphi_{\tau,j}^{t}}{dt} = \frac{\omega(\tau)}{\varepsilon} + b(\overline{x}(t) + Y_{j-1}, \varphi_{\tau,j}^{t}, t, \varepsilon) + \varepsilon^{\gamma} B_2(\overline{x}(t) + Y_{j-1}, \varphi_{\tau,j}^{t}, t, \varepsilon),$$

$$\varphi_{\tau,j}^{\tau} = \psi,$$

where $Y_{j-1} = Y_{j-1}(\varphi_{\tau,j}^{t}, t, \varepsilon)$ and $\varphi_{\tau,j}^{t} = \varphi_{\tau,j}^{t}(\psi, \varepsilon)$. Since $\beta \geq \alpha$, $\delta \geq \alpha$, and conditions (16.47) are satisfied, following the scheme of the proof of Lemmas

12.1–12.5 and inequalities (16.23) one can easily verify that these statements and inequalities are true for the functions Y_j and $\varphi^t_{\tau,j}$ constructed above. In this case, only the constants in the corresponding inequalities do change. Therefore, for the functions Y_j and their derivatives with respect to ψ, the following estimate of the form (16.35) is true:

$$\|D^s_\psi Y_j(\psi,\tau,\varepsilon)\| \leq \overline{d}_{s,0}\varepsilon^\alpha \quad \forall(\psi,\tau,\varepsilon)\in G_1,\ \ j\geq 1,\ \ 0\leq s\leq l. \tag{16.49}$$

Using the equality

$$\frac{\partial Y_{j+1}}{\partial\tau} = \frac{1}{\varepsilon}\Big[-\frac{\partial Y_{j+1}}{\partial\psi}(\omega(\tau)+\varepsilon b(\overline{x}(\tau)+Y_j,\psi,\tau,\varepsilon)$$
$$+\varepsilon^{1+\delta}B_2(\overline{x}(t)+Y_j,\psi,\tau,\varepsilon))+\varepsilon H(\tau)Y_{j+1}+\varepsilon F(Y_j,\tau)$$
$$+\varepsilon\widetilde{a}(\overline{x}(\tau)+Y_j,\psi,\tau)+\varepsilon^{1+\beta}B_1(\overline{x}(\tau)+Y_j,\psi,\tau,\varepsilon)\Big], \tag{16.50}$$

where $Y_k = Y_k(\psi,\tau,\varepsilon),\ k=j,j+1,$ we study the character of the estimates for the derivatives of the functions $Y_j(\psi,\tau,\varepsilon),\ j\geq 1,$ with respect to τ. The smoothness conditions for the right-hand side of (16.50) enable one to differentiate this equality $l-1$ times with respect to ψ, and condition (16.47) and inequality (16.49) yield

$$\Big\|D^s_\psi\frac{\partial}{\partial\tau}Y_j(\psi,\tau,\varepsilon)\Big\| \leq \overline{d}_{s,1}\varepsilon^{\alpha-1} \tag{16.51}$$
$$\forall(\psi,\tau,\varepsilon)\in G_1,\quad j\geq 1,\quad 0\leq s\leq l-1$$

where $\overline{d}_{s,1}$ is a certain constant dependent on $d_{\nu,0},\ \nu=\overline{0,s+1}$. Let us differentiate equality (16.50) with respect to τ and use inequalities (16.48) for $q=1$ and (16.51). Then, taking into account that, on the right-hand side of equality (16.50), the coefficients of the functions B_1 and B_2 and their derivatives contain, respectively, the factors $\varepsilon^{1+\beta}$ and $\varepsilon^{1+\delta}$, $\min\{\beta;\delta\}\geq\alpha$, we get

$$\Big\|\frac{\partial^2}{\partial\tau^2}Y_j(\psi,\tau,\varepsilon)\Big\| \leq \overline{d}_{0,2}\varepsilon^{\alpha-2}.$$

According to (16.48) and (16.51), subsequent differentiation with respect to ψ does not worsen the order estimates with respect to ε. Therefore,

$$\Big\|D^s_\psi\frac{\partial^2}{\partial\tau^2}Y_j(\psi,\tau,\varepsilon)\Big\| \leq \overline{d}_{s,2}\varepsilon^{\alpha-2}\quad \forall(\psi,\tau,\varepsilon)\in G_1,$$
$$j\geq 1,\quad 0\leq s\leq l-2.$$

Here, $\overline{d}_{s,2}$ is a constant that depends on $\overline{d}_{\nu,1}$, $\nu = \overline{0, s+1}$ but does not depend on j. By analogy, one can establish the estimates

$$\left\| D_\psi^s \frac{\partial^q}{\partial \tau^q} Y_j(\psi, \tau, \varepsilon) \right\| \leq \overline{d}_{s,q} \varepsilon^{\alpha - q}$$

for all $(\psi, \tau, \varepsilon) \in G_1$, $q = \overline{3, l}$, and $s = \overline{0, l-q}$.

Thus, the following statement is true for $Y(\psi, \tau, \varepsilon) = \lim_{j \to \infty} Y_j(\psi, \tau, \varepsilon)$:

Theorem 16.2. *Suppose that the conditions of Theorem 16.1 for* $A \equiv 0$ *and conditions (16.47) and (16.48) are satisfied. Then, for sufficiently small* $\varepsilon_0 > 0$, *there exists the integral manifold* $x = X(\psi, \tau, \varepsilon) = \overline{x}(\tau) + Y(\psi, \tau, \varepsilon)$ *of system (16.46) for which the function* $Y(\psi, \tau, \varepsilon)$ *is* 2π*-periodic in* ψ_ν, $\nu = \overline{1, m}$, $l-1$ *times continuously differentiable with respect to* $(\psi, \tau) \in R^m \times R$ *for every* $\varepsilon \in (0, \varepsilon_0]$, *and such that*

$$\left\| D_\psi^s \frac{\partial^q}{\partial \tau^q} Y(\psi, \tau, \varepsilon) \right\| \leq \overline{d}_{s,q} \varepsilon^{\alpha - q} \quad \forall (\psi, \tau, \varepsilon) \in G_1, \quad 0 \leq s + q \leq l - 1,$$

and its partial derivatives of the $(l-1)$*th order satisfy the Lipschitz condition with respect to the variables* ψ *and* τ.

Corollary 4. *The function* $Y(\psi, \tau, \varepsilon)$ *constructed in the proof of Theorem 16.2 defines the integral manifold* $y = Y(\psi, \tau, \varepsilon)$ *of the system*

$$\frac{dy}{d\tau} = H(\tau) y + F(y, \tau) + \widetilde{a}(\overline{x}(\tau) + y, \varphi, \tau) + \varepsilon^\beta B_1(\overline{x}(\tau) + y, \varphi, \tau, \varepsilon),$$

$$\frac{d\varphi}{d\tau} = \frac{\omega(\tau)}{\varepsilon} + b(\overline{x}(\tau) + y, \varphi, \tau, \varepsilon) + \varepsilon^\delta B_2(\overline{x}(\tau) + y, \varphi, \tau, \varepsilon). \tag{16.52}$$

The statement below solves the problem of the smoothness of the integral manifold of system (16.1) with respect to the parameter ε.

Theorem 16.3. *If the conditions of Theorem 16.1 are satisfied and the functions* $A(x, \varphi, \tau, \varepsilon)$ *and* $b(x, \varphi, \tau, \varepsilon)$ *have* $l \geq 2$ *continuous and uniformly bounded (by a certain constant) partial derivatives with respect to all variables* $(x, \varphi, \tau, \varepsilon) \in \overline{G}$, *then the integral manifold* $x = X(\psi, \tau, \varepsilon) = \overline{x}(\tau) + Y(\psi, \tau, \varepsilon)$ *of system (16.1) is* $l-1$ *times continuously differentiable with respect to* $(\psi, \tau, \varepsilon) \in G_1$,

$$\left\| D_{\psi}^{s} \frac{\partial^q}{\partial \tau^q} \frac{\partial^r}{\partial \varepsilon^r} Y(\psi, \tau, \varepsilon) \right\| \leq c\varepsilon^{\alpha - q - 2r} \quad \forall (\psi, \tau, \varepsilon) \in G_1, \tag{16.53}$$

$0 \leq s + q + r \leq l - 1$, *and the derivatives of the* $(l-1)$*th order satisfy the Lipschitz condition with respect to the variables* ψ, τ, *and* ε *on the set* $\psi \in R^m$, $\tau \in R$, $\varepsilon \in [\bar{\varepsilon}_0, \varepsilon_0]$, *where* $\bar{\varepsilon}_0$ *is an arbitrary value from the interval* $(0, \varepsilon_0)$.

In view of technical difficulties, we do not prove Theorem 16.3 here. We only note that the iterations $Y_j(\psi, \tau, \varepsilon)$, $j \geq 1$, defined by equality (13.7) are l times continuously differentiable with respect to $(\psi, \tau, \varepsilon) \in G_1$, and their partial derivatives satisfy the inequalities [SPe6]

$$\left\| D_{\psi}^{s} \frac{\partial^q}{\partial \tau^q} \frac{\partial^r}{\partial \varepsilon^r} Y_j(\psi, \tau, \varepsilon) \right\| \leq d_{s,q,r} \varepsilon^{\alpha - q - 2r} \quad \forall (\psi, \tau, \varepsilon) \in G_1, \quad j \geq 1, \tag{16.54}$$

for $0 \leq s + q + r \leq l$. To establish (16.54), one should use the methods proposed in the proof of Lemma 16.1 and the estimates obtained in [Sam2] for oscillation systems with constant frequency vector. In the case of multifrequency systems (16.1) with $\omega = \omega(\tau)$, it is necessary to carefully take into account the measure of the set of points of a time interval of unit length for which the scalar product $(k, \omega(\tau))$ is sufficiently small (k is an integer-valued vector), which substantially affects the character of estimates of oscillation integrals.

17. Asymptotic Expansion of Integral Manifold

Consider a system of ordinary differential equations of the form

$$\begin{aligned} \frac{dx}{d\tau} &= a(x, \tau) + \widetilde{a}(x, \varphi, \tau) + \varepsilon A(x, \varphi, \tau, \varepsilon), \\ \frac{d\varphi}{d\tau} &= \frac{\omega(\tau)}{\varepsilon} + b(x, \varphi, \tau, \varepsilon), \end{aligned} \tag{17.1}$$

where the functions a, $\widetilde{a}$, A, ω, and b are defined on the set $(x, \varphi, \tau, \varepsilon) \in \mathcal{D} \times R^m \times R \times (0, \varepsilon_0] \equiv \overline{G}$, 2π-periodic in φ_ν, $\nu = \overline{1, m}$, and $l \geq 2$ times continuously differentiable with respect to x, φ, and τ for every fixed $\varepsilon \in (0, \varepsilon_0]$, and all their partial derivatives are uniformly bounded in $\overline{G}$ by a constant c_1. We also assume that the function $\widetilde{a}(x, \varphi, \tau)$ averaged with respect to φ over the cube of periods is identically equal to zero and conditions (12.3), (13.2), (13.3), and (16.2) are satisfied. Under these restrictions, in Sections 12–16 we have established the existence of the integral manifold $x = X(\psi, \tau, \varepsilon) = \bar{x}(\tau) + Y(\psi, \tau, \varepsilon)$

$l-1$ times continuously differentiable with respect to $(\psi, \tau) \in R^m \times R$ for every $\varepsilon \in (0, \varepsilon_0]$ and such that the function $Y(\psi, \tau, \varepsilon)$ satisfies inequalities (16.45).

In the present section, we study the problem of the asymptotic expansion of $Y(\psi, \tau, \varepsilon)$ as a function of the parameter ε in the form of a functional sum, namely

$$Y(\psi, \tau, \varepsilon) = \sum_{\nu=0}^{r-1} u_\nu(\psi, \tau, \varepsilon) + v(\psi, \tau, \varepsilon), \tag{17.2}$$

where u_ν and v are defined on the set $G_1 = R^m \times R \times (0, \varepsilon_0]$ and satisfy the estimates

$$\|u_\nu(\psi, \tau, \varepsilon)\| \le \sigma_\nu \varepsilon^{\frac{\nu}{p}}, \quad \nu = \overline{0, r-1}, \quad \|v(\psi, \tau, \varepsilon)\| \le \sigma_r \varepsilon^{\frac{r}{p}} \tag{17.3}$$

for all $(\psi, \tau, \varepsilon) \in G_1$ and $2 \le r \le l-2$. Here, the integer $p = \dfrac{1}{\alpha}$ is determined by condition (12.3), and $\sigma_\mu = \text{const}, \mu = \overline{0, r}$.

Lemma 17.1. *Suppose that the conditions formulated above are satisfied and a function $f(\varphi, \tau, \varepsilon)$ is 2π-periodic in φ_ν, $\nu = \overline{1, m}$, r times continuously differentiable with respect to $(\varphi, \tau) \in R^m \times R$ for every $\varepsilon \in (0, \varepsilon_0]$, $1 \le r \le l$, and such that*

$$\left\| D_\varphi^s \frac{\partial^q}{\partial \tau^q} f(\varphi, \tau, \varepsilon) \right\| \le \sigma \varepsilon^{\alpha - q} \quad \forall (\varphi, \tau, \varepsilon) \in G_1, \quad 0 \le s + q \le r. \tag{17.4}$$

Then, for sufficiently small $\varepsilon_0 > 0$, there exists the integral manifold $y = Y(\psi, \tau, \varepsilon)$ of the system

$$\begin{aligned} \frac{dy}{d\tau} &= H(\tau) y + \frac{\partial \tilde{a}(\overline{x}(\tau), \varphi, \tau)}{\partial x} y + f(\varphi, \tau, \varepsilon), \\ \frac{d\varphi}{d\tau} &= \frac{\omega(\tau)}{\varepsilon} + b(\overline{x}(\tau), \varphi, \tau, \varepsilon) \end{aligned} \tag{17.5}$$

for which the function $Y(\psi, \tau, \varepsilon)$ is 2π-periodic in ψ_ν, $\nu = \overline{1, m}$, r times continuously differentiable with respect to ψ and τ for every fixed $\varepsilon \in (0, \varepsilon_0]$, and such that

$$\left\| D_\psi^s \frac{\partial^q}{\partial \tau^q} Y(\psi, \tau, \varepsilon) \right\| \le \overline{\sigma} \varepsilon^{\alpha - q} \quad \forall (\psi, \tau, \varepsilon) \in G_1, \quad 0 \le s + q \le r. \tag{17.6}$$

Proof. To construct the function $Y(\psi, \tau, \varepsilon)$, we consider the iterations

$$Y_{j+1}(\psi, \tau, \varepsilon) = \int_{-\infty}^{\infty} Q(\tau, t)\Big[\frac{\partial}{\partial x}\tilde{a}(\overline{x}(t), \varphi_\tau^t, t) Y_j(\varphi_\tau^t, t, \varepsilon) + f(\varphi_\tau^t, t, \varepsilon)\Big] dt, \quad (17.7)$$

where $\varphi_t^\tau = \varphi_t^\tau(\psi, \varepsilon)$ is a solution of the second equation of system (17.5) that takes the value ψ for $\tau = t$, and $Y_0 \equiv 0$. It follows from Theorem 16.2 that each function $Y_j(\psi, \tau, \varepsilon)$, $j \geq 1$, is r times continuously differentiable with respect to $(\psi, \tau) \in R^m \times R$ for every fixed value of $\varepsilon \in (0, \varepsilon_0]$, and

$$\left\| D_\psi^s \frac{\partial^q}{\partial \tau^q} Y_j(\psi, \tau, \varepsilon) \right\| \leq \overline{d}_{s,q} \varepsilon^{\alpha - q} \quad \forall (\psi, \tau, \varepsilon) \in G_1, \quad (17.8)$$

$$j \geq 0, \quad 0 \leq s + q \leq r.$$

Note that condition (17.8) is also satisfied for $r = l$ because the function $\frac{\partial}{\partial x}\tilde{a}(\overline{x}(\tau), \varphi, \tau) y$ is $l - 1$ times continuously differentiable with respect to y, φ, and τ and, according to (16.2), has l continuous derivatives with respect to y and φ. Denote

$$Z_{j+1}(\psi, \tau, \varepsilon) = Y_{j+1}(\psi, \tau, \varepsilon) - Y_j(\psi, \tau, \varepsilon).$$

Then it follows from (17.7) that

$$\begin{aligned} \sup_{G_1} \|Z_{j+1}(\psi, \tau, \varepsilon)\| &\leq \int_{-\infty}^{\infty} K e^{-\gamma|t-\tau|} \sup_{\varphi,\tau} \left\| \frac{\partial}{\partial x}\tilde{a}(\overline{x}(\tau), \varphi, \tau) \right\| dt \sup_{G_1} \|Z_j(\psi, \tau, \varepsilon)\| \\ &= \sigma_0 \sup_{G_1} \|Z_j(\psi, \tau, \varepsilon)\|. \end{aligned}$$

According to condition (13.3), the constant σ_0 is less than 1; therefore, the last relation guarantees the convergence of the numerical series

$$\sum_{j=1}^{\infty} \sup_{G_1} \|Z_j(\psi, \tau, \varepsilon)\|,$$

and, hence, the uniform convergence of the sequence $\{Y_j(\psi,\tau,\varepsilon)\}$ on the set G_1. Further, we assume that each numerical series

$$\sum_{j=1}^{\infty} \sup_{G_1} \|D_\psi^\nu Z_j(\psi,\tau,\varepsilon)\|, \quad \nu=\overline{1,s-1}, \quad s\le r, \tag{17.9}$$

is also convergent. Consider the equality

$$D_\psi^s Z_{j+1}(\psi,\tau,\varepsilon) = \int_{-\infty}^{\infty} Q(\tau,t) D_\psi^s \Big[\frac{\partial}{\partial x}\widetilde{a}(\overline{x}(t),\varphi_\tau^t,t) Z_j(\varphi_\tau^t,t,\varepsilon)\Big] dt.$$

To estimate the last integral, we differentiate the product in the square brackets and use the following inequalities of the form (16.6) and (16.23):

$$\Big\|\frac{\partial}{\partial\psi}(\varphi_\tau^t(\psi,\varepsilon)-\psi)\Big\| \le \underline{c}_0^{(1)} \varepsilon^\alpha e^{\overline{\gamma}|t-\tau|}, \qquad \Big\|\frac{\partial}{\partial\psi}\varphi_\tau^t(\psi,\varepsilon)\Big\| \le c_0^{(1)} e^{\overline{\gamma}|t-\tau|},$$

$$\|D_\psi^s \varphi_\tau^t(\psi,\varepsilon)\| \le \underline{c}_0^{(s)} \varepsilon^\alpha e^{\overline{\gamma}s|t-\tau|}, \quad s\ge 2, \ \overline{\gamma}=\frac{\gamma}{2l}. \tag{17.10}$$

Taking into account that $\dfrac{\partial}{\partial x}\widetilde{a}(\overline{x}(\tau),\varphi,t)$ has r bounded derivatives with respect to φ, we get

$$\|D_\psi^s Z_{j+1}(\psi,\tau,\varepsilon)\| \le K \int_{-\infty}^{\infty} e^{-\gamma|t-\tau|} \Big[\sigma^{(s)} \sum_{\nu=0}^{s-1} \sup_{G_1} \|D_\psi^\nu Z_j(\psi,\tau,\varepsilon)\| e^{\overline{\gamma}s|t-\tau|}$$

$$+ \sup_{\varphi,\tau}\Big\|\frac{\partial}{\partial x}\widetilde{a}(\overline{x}(\tau),\varphi,\tau)\Big\| \cdot \|D_\psi^s Z_j(\varphi_\tau^t,t,\varepsilon)\|\Big] dt. \tag{17.11}$$

Here, $\sigma^{(s)}$ is a constant independent of ε and j. Applying the scheme of the proof of Lemma 16.1, we obtain

$$\|D_\psi^s Z_j(\varphi_\tau^t,t,\varepsilon)\| \le \sum_{\nu=0}^{s-1} \sup_{G_1} \|D_\psi^\nu Z_j(\psi,\tau,\varepsilon)\| \sum_\beta c_{\nu\beta} \|D_\psi \varphi_\tau^t\|^{\beta_1} \|D_\psi^s \varphi_\tau^t\|^{\beta_s}$$

$$+ \Big\| \sum_{p_1+\ldots+p_m=s} \frac{\partial^s Z_j(\varphi_\tau^t,t,\varepsilon)}{\partial\varphi_1^{p_1}\ldots\partial\varphi_m^{p_m}}$$

$$\times \sum \prod_{\mu=1}^m \prod_{\nu=1}^m \big(\delta_{\nu\mu} + \frac{\partial}{\partial\psi_\nu}(\varphi_\tau^{t,\mu}-\psi_\mu)\big)^{\beta_\nu^{(\mu)}} \Big\|,$$

where $\varphi_\tau^t = (\varphi_\tau^{t,1}, \dots, \varphi_\tau^{t,m})$, $\delta_{\nu,\mu}$ is the Kronecker symbol, the symbol $\sum$ means summation over all $\beta_\nu^{(\mu)}$ satisfying the conditions

$$\sum_{\mu=1}^{m} \beta_\nu^{(\mu)} = s_\nu, \quad \sum_{\nu=1}^{m} \beta_\nu^{(\mu)} = p_\nu, \quad \nu, \mu = \overline{1, m},$$

and

$$D_\psi^s = \frac{\partial^s}{\partial \psi_1^{s_1} \dots \partial \psi_m^{s_m}}.$$

Taking inequalities (17.10) into account, we get

$$\begin{aligned}
&\|D_\psi^s Z_j(\varphi_\tau^t, t, \varepsilon)\| \\
&\qquad \le \sup_{G_1} \|D_\psi^s Z_j(\psi, \tau, \varepsilon)\| \\
&\qquad\qquad + \overline{\sigma}^{(s)} \Big[\sum_{\nu=1}^{s-1} \sup_{G_1} \|D_\psi^\nu Z_j(\psi, \tau, \varepsilon)\| + \varepsilon^\alpha L_j(s) \Big] e^{\overline{\gamma} s |t-\tau|}, \qquad (17.12)
\end{aligned}$$

$$L_j(s) = \sum_{p_1 + \dots + p_m = s} \sup_{G_1} \Big\| \frac{\partial^s Z_j(\psi, \tau, \varepsilon)}{\partial \psi_1^{p_1} \dots \partial \psi_m^{p_m}} \Big\|.$$

Inequalities (17.11) and (17.12) yield

$$\begin{aligned}
L_{j+1}(s) \le \Big[\sigma_0 + \frac{2}{\gamma - s\overline{\gamma}} K \overline{\sigma}^{(s)} s^m \varepsilon_0^\alpha \Big] L_j(s) \\
+ \frac{2}{\gamma - s\overline{\gamma}} K s^m \Big(\sigma^{(s)} + \overline{\sigma}^{(s)} \sup_{\varphi, \tau} \Big\| \frac{\partial}{\partial x} \widetilde{a}(\overline{x}(\tau), \varphi, \tau) \Big\| \Big) \\
\times \sum_{\nu=1}^{s-1} \sup_{G_1} \|D_\psi^\nu Z_j(\psi, \tau, \varepsilon)\|. \qquad (17.13)
\end{aligned}$$

Since, for sufficiently small $\varepsilon_0 > 0$, the constant in the square brackets on the right-hand side of (17.13) is less than 1 and series (17.9) are convergent, it follows from (17.13) that each series

$$\sum_{j=1}^{\infty} \sup_{G_1} \|D_\psi^s Z_j(\psi, \tau, \varepsilon)\| \qquad (17.14)$$

is convergent. Thus, according to the principle of mathematical induction, each numerical series (17.14) is convergent for $0 \leq s \leq r$. We now write a partial differential equation for the function $Y_{j+1} = Y_{j+1}(\psi, \tau, \varepsilon)$:

$$\frac{\partial Y_{j+1}}{\partial \tau} = -\frac{\partial Y_{j+1}}{\partial \psi}\left[\frac{\omega(\tau)}{\varepsilon} + b(\overline{x}(\tau), \psi, \tau, \varepsilon)\right] + H(\tau) Y_{j+1} + \frac{\partial}{\partial x}\widetilde{a}(\overline{x}(\tau), \psi, \tau) Y_j + f(\psi, \tau, \varepsilon).$$

This yields

$$\frac{\partial Z_{j+1}}{\partial \tau} = -\frac{\partial Z_{j+1}}{\partial \psi}\left[\frac{\omega(\tau)}{\varepsilon} + b(\overline{x}(\tau), \psi, \tau, \varepsilon)\right] + H(\tau) Z_{j+1} + \frac{\partial}{\partial x}\widetilde{a}(\overline{x}(\tau), \psi, \tau) Z_j, \tag{17.15}$$

where $Z_\nu = Z_\nu(\psi, \tau, \varepsilon)$ for $\nu = j, j+1$. Let us fix an arbitrary $\varepsilon \in (0, \varepsilon_0]$. It follows from Eq. (17.15) and the condition of the boundedness of the functions ω, b, and $\dfrac{\partial \widetilde{a}}{\partial x}$ and their derivatives that the series

$$\sum_{j=1}^{\infty} \sup_{\psi, \tau} \left\| \frac{\partial}{\partial \tau} Z_j(\psi, \tau, \varepsilon) \right\|$$

is convergent. Differentiating equality (17.15) ν times, $1 \leq \nu \leq r-1$, with respect to ψ and using the convergence of series (17.14) for $s = \overline{0, r}$, we establish the convergence of each series

$$\sum_{j=1}^{\infty} \sup_{\psi, \tau} \left\| D_\psi^s \frac{\partial}{\partial \tau} Z_j(\psi, \tau, \varepsilon) \right\|, \quad 0 \leq s \leq r-1.$$

Further, differentiating equality (17.15) with respect to τ and ν times with respect to ψ, $0 \leq \nu \leq r-2$, we establish the convergence of the series

$$\sum_{j=1}^{\infty} \sup_{\psi, \tau} \left\| D_\psi^s \frac{\partial^2}{\partial \tau^2} Z_j(\psi, \tau, \varepsilon) \right\|, \quad 0 \leq s \leq r-2,$$

and so on. Thus, all numerical series

$$\sum_{j=1}^{\infty} \sup_{\psi, \tau} \left\| D_\psi^s \frac{\partial^q}{\partial \tau^q} Z_j(\psi, \tau, \varepsilon) \right\|, \quad 0 \leq s + q \leq r,$$

are convergent for any value of the small parameter $\varepsilon \in (0, \varepsilon_0]$. This is sufficient for the limit function

$$Y(\psi, \tau, \varepsilon) = \lim_{j \to \infty} Y_j(\psi, \tau, \varepsilon)$$

to have r continuous derivatives with respect to $(\psi, \tau) \in R^m \times R$ for every fixed $\varepsilon \in (0, \varepsilon_0]$. Passing to the limit as $j \to \infty$ in inequalities (17.8), we obtain estimates (17.6). Lemma 17.1 is proved.

Remark 4. System (17.5) satisfies all conditions of Theorem 16.2, which guarantees that the function $Y(\psi, \tau, \varepsilon)$ is smooth with respect to ψ and τ up to the order $r - 1$, $r \geq 2$, and the derivatives of the $(r - 1)$th order satisfy the Lipschitz condition. Since system (17.5) is linear with respect to y and the equations for φ are independent of y, Lemma 17.1 establishes the smoothness of the function $Y(\psi, \tau, \varepsilon)$ with respect to ψ and τ up to the order r, which can be equal to 1.

By analogy, using estimate (1.20) for oscillation integrals, one can prove the following statement:

Lemma 17.2. *Under the conditions imposed on system (17.1), there exists the integral manifold* $y = Y(\psi, \tau, \varepsilon)$ *of the equations*

$$\frac{dy}{d\tau} = H(\tau)y + \widetilde{a}(\overline{x}(\tau), \varphi, \tau) + \frac{\partial}{\partial x}\widetilde{a}(\overline{x}(\tau), \varphi, \tau)y,$$

$$\frac{d\varphi}{d\tau} = \frac{\omega(\tau)}{\varepsilon} + b(\overline{x}(\tau), \varphi, \tau, \varepsilon),$$

where $Y(\psi, \tau, \varepsilon)$ *is* 2π*-periodic in* ψ_ν, $\nu = \overline{1, m}$, *and* l *times continuously differentiable with respect to* ψ *and* τ *for every value of* $\varepsilon \in (0, \varepsilon_0]$, $\varepsilon_0 > 0$ *is sufficiently small, and*

$$\left\| D_\psi^s \frac{\partial^q}{\partial \tau^q} Y(\psi, \tau, \varepsilon) \right\| \leq c\varepsilon^{\alpha - q} \qquad \forall (\psi, \tau, \varepsilon) \in G_1, \quad 0 \leq s + q \leq l.$$

To establish relations (17.2) and (17.3), we rewrite Eq. (14.13) for the function $X(\psi,\tau,\varepsilon)=\overline{x}(\tau)+Y(\psi,\tau,\varepsilon)$ in the form

$$\frac{\partial Y}{\partial \tau}+\frac{\partial Y}{\partial \psi}\Big[\frac{\omega(\tau)}{\varepsilon}+b(\overline{x}(\tau),\psi,\tau,\varepsilon)\Big]$$
$$=H(\tau)Y+F(Y,\tau)+\widetilde{a}(\overline{x}(\tau)+Y,\psi,\tau)+\varepsilon A(\overline{x}(\tau)+Y,\psi,\tau,\varepsilon)$$
$$+\frac{\partial Y}{\partial \psi}\Big[b(\overline{x}(\tau),\psi,\tau,\varepsilon)-b(\overline{x}(\tau)+Y,\psi,\tau,\varepsilon)\Big], \qquad (17.16)$$

where

$$H(\tau)=\frac{\partial}{\partial x}a(\overline{x}(\tau),\tau), \quad F(Y,\tau)=a(\overline{x}(\tau)+Y,\tau)-a(\overline{x}(\tau),\tau)-H(\tau)Y.$$

We now substitute the value of Y from (17.2) into (17.16) and then expand the right-hand side into the sum over values of the same order $\varepsilon^{\frac{\nu}{p}}$, assuming that u_ν and its derivatives $\dfrac{\partial u_\nu}{\partial \psi_\mu}$, $\mu=\overline{1,m}$, are values of order $\varepsilon^{\frac{\nu}{p}}$. Equating the expression on the left-hand side of (17.16) for $Y=u_\nu$ to the term of order $\varepsilon^{\frac{\nu}{p}}$ of the indicated expansion of the right-hand side of (17.16), we obtain a partial differential equation for the determination of the function $u_\nu=u_\nu(\psi,\tau,\varepsilon)$. It follows from estimate (16.45) for $s=q=0$ that $u_0(\psi,\tau,\varepsilon)\equiv 0$ for any $(\psi,\tau,\varepsilon)\in G_1$.

Now consider the following equation for the determination of u_1:

$$\frac{\partial u_1}{\partial \tau}+\frac{\partial u_1}{\partial \psi}\Big[\frac{\omega(\tau)}{\varepsilon}+b(\overline{x}(\tau),\psi,\tau,\varepsilon)\Big]$$
$$=H(\tau)u_1+\widetilde{a}(\overline{x}(\tau),\psi,\tau)+\frac{\partial \widetilde{a}(\overline{x}(\tau),\psi,\tau)}{\partial x}u_1.$$

It is clear that the solution of this equation is the integral manifold $y=u_1(\psi,\tau,\varepsilon)$ of the system

$$\frac{dy}{d\tau}=H(\tau)y+\widetilde{a}(\overline{x}(\tau),\varphi,\tau)+\frac{\partial \widetilde{a}(\overline{x}(\tau),\varphi,\tau)}{\partial x}y,$$
$$\frac{d\varphi}{d\tau}=\frac{\omega(\tau)}{\varepsilon}+b(\overline{x}(\tau),\varphi,\tau,\varepsilon).$$

According to Lemma 17.2, the integral manifold $y=u_1(\psi,\tau,\varepsilon)$ of this system exists; moreover, the function $u_1(\psi,\tau,\varepsilon)$ is 2π-periodic in ψ_ν, $\nu=\overline{1,m}$, and

l times continuously differentiable with respect to $(\psi, \tau) \in R^m \times R$ for every fixed ε, and its derivatives satisfy the estimates

$$\left\| D_\psi^s \frac{\partial^q}{\partial \tau^q} u_1(\psi, \tau, \varepsilon) \right\| \leq d_{s,q,1} \varepsilon^{\frac{1}{p} - q} \tag{17.17}$$

$$\forall (\psi, \tau, \varepsilon) \in G_1, \quad 0 \leq s + q \leq l.$$

Further, we write an equation for the determination of $u_\nu(\psi, \tau, \varepsilon)$ for $\nu \geq 2$:

$$\begin{aligned} \frac{\partial u_\nu}{\partial \tau} &+ \frac{\partial u_\nu}{\partial \psi} \Big[\frac{\omega(\tau)}{\varepsilon} + b(\overline{x}(\tau), \psi, \tau, \varepsilon) \Big] \\ &= H(\tau) u_\nu + \frac{\partial \widetilde{a}(\overline{x}(\tau), \psi, \tau)}{\partial x} u_\nu \\ &\qquad + f_\nu \Big(\psi, \tau, \varepsilon, u_1(\psi, \tau, \varepsilon), \ldots, u_{\nu-1}(\psi, \tau, \varepsilon), \\ &\qquad\qquad \frac{\partial}{\partial \psi} u_1(\psi, \tau, \varepsilon), \ldots, \frac{\partial}{\partial \psi} u_{\nu-1}(\psi, \tau, \varepsilon) \Big). \end{aligned}$$

It follows from the smoothness conditions for the right-hand side of system (17.1) that f_ν is a polynomial of at most νth degree with respect to $u_1, \ldots, u_{\nu-1}$, $\frac{\partial}{\partial \psi} u_1, \ldots, \frac{\partial}{\partial \psi} u_{\nu-1}$ whose coefficients are $l - \nu$ times continuously differentiable with respect to ψ and τ for fixed $\varepsilon \in (0, \varepsilon_0]$, and all their partial derivatives are uniformly bounded in G_1. Moreover, if the functions $u_\mu(\psi, \tau, \varepsilon)$, $\mu = \overline{1, \nu - 1}$, are 2π-periodic in ψ_k, $k = \overline{1, m}$, and $l - \mu$ times continuously differentiable with respect to ψ and τ, and their derivatives satisfy the inequalities

$$\left\| D_\psi^s \frac{\partial^q}{\partial \tau^q} u_\mu(\psi, \tau, \varepsilon) \right\| \leq d_{s,q,\mu} \varepsilon^{\frac{\mu}{p} - q} \quad \forall (\psi, \tau, \varepsilon) \in G_1 \tag{17.19}$$

for $0 \leq s + q \leq l - \mu$, then f_ν, as a function of ψ, τ, and ε, is obviously $l - \nu$ times continuously differentiable with respect to ψ and τ for every $\varepsilon \in (0, \varepsilon_0]$, 2π-periodic in ψ_k, $k = \overline{1, m}$, and such that

$$\left\| D_\psi^s \frac{\partial^q}{\partial \tau^q} f_\nu \right\| \leq \overline{d}_{s,q,\nu} \varepsilon^{\frac{\nu}{p} - q} \quad \forall (\psi, \tau, \varepsilon) \in G_1, \ \ 0 \leq s + q \leq l - \nu,$$

where $\overline{d}_{s,q,\nu}$ is a certain constant independent of ε. We set

$$u_\nu = \varepsilon^{\frac{\nu - 1}{p}} \overline{u}_\nu(\psi, \tau, \varepsilon).$$

Then it follows from (17.18) that $y = \overline{u}_\nu(\psi, \tau, \varepsilon)$ is the integral manifold of the system

$$\frac{dy}{d\tau} = H(\tau)y + \frac{\partial \widetilde{a}(\overline{x}(\tau), \varphi, \tau)}{\partial x} y + \overline{f}_\nu(\varphi, \tau, \varepsilon),$$

$$\frac{d\varphi}{d\tau} = \frac{\omega(\tau)}{\varepsilon} + b(\overline{x}(\tau), \varphi, \tau, \varepsilon),$$

where

$$\overline{f}_\nu(\varphi, \tau, \varepsilon) = \varepsilon^{\frac{1-\nu}{p}} f_\nu\Big(\varphi, \tau, \varepsilon, u_1(\varphi, \tau, \varepsilon), \ldots, \frac{\partial}{\partial \varphi} u_{\nu-1}(\varphi, \tau, \varepsilon)\Big).$$

Taking into account that

$$\left\| D_\varphi^s \frac{\partial^q}{\partial \tau^q} \overline{f}_\nu(\varphi, \tau, \varepsilon) \right\| \leq \overline{d}_{s,q,\nu} \varepsilon^{\alpha - q}, \quad \alpha = \frac{1}{p},$$

$$\forall (\varphi, \tau, \varepsilon) \in G_1, \quad 0 \leq s + q \leq l - \nu,$$

we conclude that, according to Lemma 17.1, the function $\overline{u}_\nu(\psi, \tau, \varepsilon)$ is 2π-periodic in ψ_k, $k = \overline{1, m}$, has continuous partial derivatives with respect to ψ and τ for every $\varepsilon \in (0, \varepsilon_0]$ up to the order $l - \nu$, and satisfies the estimates

$$\left\| D_\psi^s \frac{\partial^q}{\partial \tau^q} \overline{u}_\nu(\psi, \tau, \varepsilon) \right\| \leq d_{s,q,\nu} \varepsilon^{\alpha - q} \quad \forall (\psi, \tau, \varepsilon) \in G_1, \ 0 \leq s + q \leq l - \nu.$$

This yields estimate (17.19) with $\mu = \nu$ for the function $u_\nu = \overline{u}_\nu \varepsilon^{\frac{\nu-1}{p}}$. Thus, according to the principle of mathematical induction, every function $u_\mu(\psi, \tau, \varepsilon)$, $\mu = \overline{1, r-1}$, is 2π-periodic in ψ_ν, $\nu = \overline{1, m}$, has continuous (in $(\psi, \tau) \in R^m \times R$ for every $\varepsilon \in (0, \varepsilon_0]$) partial derivatives to within the order $l - \mu$, and satisfies estimates (17.19).

Let us determine the asymptotic character of expansion (17.2). For this purpose, we denote

$$u(\psi, \tau, \varepsilon) = \sum_{\nu=1}^{r-1} u_\nu(\psi, \tau, \varepsilon) \tag{17.20}$$

and change the variables in Eq. (17.16) as follows:

$$Y = u(\psi, \tau, \varepsilon) + \varepsilon^{\frac{r-1}{p}} z(\psi, \tau, \varepsilon). \tag{17.21}$$

For z, we obtain the following equation:

$$\frac{\partial z}{\partial \tau} + \frac{\partial z}{\partial \psi}\Big[\frac{\omega(\tau)}{\varepsilon} + b(\overline{x}(\tau), \psi, \tau, \varepsilon) + \varepsilon^{\alpha} B_2(z, \psi, \tau, \varepsilon)\Big]$$

$$= H(\tau) z + \frac{\partial \widetilde{a}(\overline{x}(\tau), \psi, \tau)}{\partial x} z + \varepsilon^{\alpha} B_1(z, \psi, \tau, \varepsilon), \quad (17.22)$$

where

$$B_1(z, \psi, \tau, \varepsilon) = \varepsilon^{-\frac{r}{p}} \Big\{ \widetilde{a}(\overline{x}(\tau) + u + \varepsilon^{\frac{r-1}{p}} z, \psi, \tau)$$

$$- \widetilde{a}(\overline{x}(\tau), \psi, \tau) - \frac{\partial \widetilde{a}(\overline{x}(\tau), \psi, \tau)}{\partial x} (u + \varepsilon^{\frac{r-1}{p}} z)$$

$$+ \frac{\partial u}{\partial \psi}\Big[b(\overline{x}(\tau), \psi, \tau, \varepsilon) - b(\overline{x}(\tau) + u + \varepsilon^{\frac{r-1}{p}} z, \psi, \tau, \varepsilon)\Big]$$

$$+ F(u + \varepsilon^{\frac{r-1}{p}} z, \tau) + \varepsilon A(\overline{x}(\tau) + u + \varepsilon^{\frac{r-1}{p}} z, \psi, \tau, \varepsilon)$$

$$- \sum_{\nu=1}^{r-1} f_{\nu}(\psi, \tau, \varepsilon, u_1(\psi, \tau, \varepsilon), \ldots, \frac{\partial}{\partial \psi} u_{\nu-1}(\psi, \tau, \varepsilon) \Big\}, \quad f_1 \equiv 0,$$

$$B_2(z, \psi, \tau, \varepsilon) = \varepsilon^{-\frac{1}{p}} \big[b(\overline{x}(\tau) + u + \varepsilon^{\frac{r-1}{p}} z, \psi, \tau, \varepsilon) - b(\overline{x}(\tau), \psi, \tau, \varepsilon)\big].$$

Using properties of the functions $u_{\nu}(\psi, \tau, \varepsilon)$, $\nu = \overline{1, r-1}$, and the smoothness conditions for the right-hand side of system (17.1), we establish that, for sufficiently small $\varepsilon_0 > 0$, the functions $B_j(z, \psi, \tau, \varepsilon)$, $j = 1, 2$, are defined on the set

$$\|z\| \le \Delta, \quad \psi \in R^m, \quad \tau \in R, \quad \varepsilon \in (0, \varepsilon_0]$$

($\Delta > 0$ is fixed) and 2π-periodic in ψ_k, $k = \overline{1, m}$, and, for every fixed $\varepsilon \in (0, \varepsilon_0]$, they have continuous partial derivatives with respect to z, ψ, and τ up to the order $l - r$ inclusive that satisfy an inequality of the form

$$\left\| D^s_{z,\psi} \frac{\partial^q}{\partial \tau^q} B_j(z, \psi, \tau, \varepsilon) \right\| \le c \varepsilon^{-q}, \quad 0 \le s + q \le l - r, \quad j = 1, 2. \quad (17.23)$$

A function $z(\psi, \tau, \varepsilon)$ that is a solution of Eq. (17.22) defines the integral manifold $y = z(\psi, \tau, \varepsilon)$ of the system

$$\frac{dy}{d\tau} = H(\tau)y + \frac{\partial \widetilde{a}(\overline{x}(\tau), \varphi, \tau)}{\partial x} y + \varepsilon^{\alpha} B_1(y, \varphi, \tau, \varepsilon),$$
$$\frac{d\varphi}{d\tau} = \frac{\omega(\tau)}{\varepsilon} + b(\overline{x}(\tau), \varphi, \tau, \varepsilon) + \varepsilon^{\alpha} B_2(y, \varphi, \tau, \varepsilon), \tag{17.24}$$

which has the same form as system (16.52) for $F \equiv 0$, $\widetilde{a}(\overline{x}(\tau) + y, \varphi, \tau) = \frac{\partial}{\partial x}\widetilde{a}(\overline{x}(\tau), \varphi, \tau)y$, and $\beta = \delta = \alpha$. Therefore, according to Theorem 16.2, for $l - r \geq 2$ there exists the integral manifold $y = z(\psi, \tau, \varepsilon)$ of system (17.24) that satisfies the estimates

$$\left\| D_{\psi}^{s} \frac{\partial^{q}}{\partial \tau^{q}} z(\psi, \tau, \varepsilon) \right\| \leq d_{s,q} \varepsilon^{\alpha - q} \quad \forall (\psi, \tau, \varepsilon) \in G_1, \;\; 0 \leq s + q \leq l - r - 1.$$

It follows from the change of variables (17.21) that the function

$$X(\psi, \tau, \varepsilon) = \overline{x}(\tau) + u(\psi, \tau, \varepsilon) + v(\psi, \tau, \varepsilon), \tag{17.25}$$

where

$$v = \varepsilon^{\frac{r-1}{p}} z(\psi, \tau, \varepsilon), \quad \left\| D_{\psi}^{s} \frac{\partial^{q}}{\partial \tau^{q}} v(\psi, \tau, \varepsilon) \right\| \leq d_{s,q} \varepsilon^{\frac{r}{p} - q} \tag{17.26}$$

for all $(\psi, \tau, \varepsilon) \in G_1$, $0 \leq s + q \leq l - r - 1$, determines the integral manifold of system (17.1). Thus, the following statement is true:

Theorem 17.1. *Suppose that the conditions imposed above on system (17.1) are satisfied for* $l \geq r + 2$, $r \geq 2$. *Then, for sufficiently small* $\varepsilon_0 > 0$, *the function* $X(\psi, \tau, \varepsilon)$ *that defines the integral manifold of system (17.1) for* $(\psi, \tau, \varepsilon) \in G_1$ *admits the asymptotic decomposition (17.25) in which the functions* $u(\psi, \tau, \varepsilon)$ *and* $v(\psi, \tau, \varepsilon)$ *satisfy conditions (17.19), (17.20), and (17.26).*

Corollary 5. *If*

$$\frac{\partial \widetilde{a}(\overline{x}(\tau), \varphi, \tau)}{\partial x} \equiv 0 \quad \forall (\varphi, \tau) \in R^m \times R,$$

then, according to Lemmas 17.1 and 17.2, the functions u_ν are determined in explicit form by the following formulas:

$$u_1(\psi, \tau, \varepsilon) = \int_{-\infty}^{\infty} Q(\tau, t)\widetilde{a}(\overline{x}(t), \varphi_\tau^t(\psi, \varepsilon), t)dt,$$

$$u_\nu(\psi, \tau, \varepsilon)$$

$$= \int_{-\infty}^{\infty} Q(\tau, t) f_\nu(\varphi_\tau^t(\psi, \varepsilon), t, \varepsilon, u_1(\varphi_\tau^t(\psi, \varepsilon), t, \varepsilon), \ldots, u_{\nu-1}(\varphi_\tau^t(\psi, \varepsilon), t, \varepsilon),$$

$$\frac{\partial}{\partial \varphi} u_1(\varphi_\tau^t(\psi, \varepsilon), t, \varepsilon), \ldots, \frac{\partial}{\partial \varphi} u_{\nu-1}(\varphi_\tau^t(\psi, \varepsilon), t, \varepsilon))dt, \quad 2 \le \nu \le r-1.$$

18. Decomposition of Equations in a Neighborhood of Asymptotically Stable Integral Manifold

Consider the system of ordinary differential equations

$$\frac{dx}{d\tau} = a(x, \varphi, \tau, \varepsilon), \quad \frac{d\varphi}{d\tau} = \frac{\omega(\tau)}{\varepsilon} + b(x, \varphi, \tau, \varepsilon), \tag{18.1}$$

where $x \in D \subset R^n$, $\varphi \in R^m$, $m \ge 2$, $\tau \in R$, $\varepsilon \in (0, \varepsilon_0]$ is a small parameter, the vector functions a and b are defined on the set $\overline{G} = D \times R^m \times R \times (0, \varepsilon_0]$, 2π-periodic in φ_ν, $\nu = \overline{1, m}$, and thrice continuously differentiable with respect to x, φ, and τ for every fixed $\varepsilon \in (0, \varepsilon_0]$, and all their partial derivatives are uniformly bounded in $\overline{G}$ by a constant c_1 independent of ε. Assume that

$$a(x, \varphi, \tau, \varepsilon) = a(x, \tau) + \widetilde{a}(x, \varphi, \tau) + \varepsilon A(x, \varphi, \tau, \varepsilon),$$

where the function $\widetilde{a}(x, \varphi, \tau)$ averaged with respect to φ over the cube of periods is identically equal to zero, and the Fourier coefficients $c_k(x, \tau, \varepsilon)$ of the function $c(x, \varphi, \tau, \varepsilon) = [\widetilde{a}(x, \varphi, \tau); b(x, \varphi, \tau, \varepsilon)]$ satisfy the inequality

$$\sum_{k \ne 0} \left[\|k\|^3 \sup_{\overline{G}} \|c_k\| + \|k\|^2 \left(\sup_{\overline{G}} \left\| \frac{\partial c_k}{\partial \tau} \right\| + \sup_{\overline{G}} \left\| \frac{\partial c_k}{\partial x} \right\| \right) \right] \le c_1. \tag{18.2}$$

Consider the system of equations of the first approximation for slow variables averaged with respect to all angular variables φ:

$$\frac{\partial \overline{x}}{\partial \tau} = a(\overline{x}, \tau).$$

Assume that there exists a solution $\overline{x} = \overline{x}(\tau)$ of this system defined on the entire axis that lies, together with a certain ρ-neighborhood of it, in the domain D and for which the normal fundamental matrix $Q(\tau, t)$ of solutions of the variational equation

$$\frac{dz}{d\tau} = \frac{\partial a(\overline{x}(\tau), \tau)}{\partial x} z$$

satisfies the estimate

$$\|Q(\tau, t)\| \leq K e^{-\gamma(\tau - t)} \quad \forall \tau \geq t \in R, \tag{18.3}$$

where $\gamma > 0$ and $K \geq 1$ are certain constants. Let

$$\sigma_0 = \frac{2}{\gamma} K \sup_{\varphi, \tau} \left\| \frac{\partial}{\partial x} \widetilde{a}(\overline{x}(\tau), \varphi, \tau) \right\| < 1. \tag{18.4}$$

We also impose certain restrictions on the components $\omega_\nu(\tau)$, $\nu = \overline{1, m}$, of the frequency vector $\omega(\tau)$. Assume that the functions

$$\omega_\nu^{(\mu)}(\tau) \equiv \frac{d^\mu}{d\tau^\mu} \omega_\nu(\tau), \quad \nu = \overline{1, m}, \ \ \mu = \overline{0, p-1}, \ \ p \geq m,$$

are uniformly continuous on the entire axis, and

$$\|(W_p^T(\tau) W_p(\tau))^{-1} W_p^T(\tau)\| \leq c_1 \quad \forall \tau \in R, \tag{18.5}$$

where

$$W_p(\tau) = (\omega_\nu^{(\mu-1)}(\tau))_{\nu, \mu = 1}^{m, p}$$

and $W_p^T(\tau)$ is the transposed matrix. As proved above, under these conditions there exists an asymptotically stable integral manifold $x = X(\psi, \tau, \varepsilon) = \overline{x}(\tau) + Y(\psi, \tau, \varepsilon)$ of system (18.1) for which the function $Y(\psi, \tau, \varepsilon)$ is 2π-periodic in ψ_ν, $\nu = \overline{1, m}$, twice continuously differentiable with respect to ψ and τ for every value of ε, and such that its second derivatives satisfy the Lipschitz condition and

$$\|Y(\psi, \tau, \varepsilon)\| + \left\| \frac{\partial}{\partial \psi} Y(\psi, \tau, \varepsilon) \right\| + \sum_{\nu=1}^{m} \left\| \frac{\partial^2}{\partial \psi \partial \psi_\nu} Y(\psi, \tau, \varepsilon) \right\| \leq \overline{c}_1 \varepsilon^\alpha,$$

$$\sum_{\nu=1}^{m} \left\| \frac{\partial^2}{\partial \psi \partial \psi_\nu} Y(\psi, \tau, \varepsilon) - \frac{\partial^2}{\partial \psi \partial \psi_\nu} Y(\overline{\psi}, \tau, \varepsilon) \right\| \leq \overline{c}_1 \varepsilon^\alpha \|\psi - \overline{\psi}\| \tag{18.6}$$

$$\forall (\psi, \tau, \varepsilon) \in G_1, \quad \overline{\psi} \in R^m, \quad G_1 = R^m \times R \times (0, \varepsilon_0], \quad \alpha = \frac{1}{p}.$$

Performing the change of variables $y = x - X(\varphi, \tau, \varepsilon)$ in (18.1), we obtain the system

$$\frac{dy}{d\tau} = a(y + X(\varphi, \tau, \varepsilon), \varphi, \tau, \varepsilon) - a(X(\varphi, \tau, \varepsilon), \varphi, \tau, \varepsilon)$$

$$- \frac{\partial X(\varphi, \tau, \varepsilon)}{\partial \varphi}[b(y + X(\varphi, \tau, \varepsilon), \varphi, \tau, \varepsilon) - b(X(\varphi, \tau, \varepsilon), \varphi, \tau, \varepsilon)],$$

$$\frac{d\varphi}{d\tau} = \frac{\omega(\tau)}{\varepsilon} + b(y + X(\varphi, \tau, \varepsilon), \varphi, \tau, \varepsilon). \tag{18.7}$$

For every value of the small parameter ε, the right-hand side of this system has continuous partial derivatives of the first order with respect to y, φ, and τ, and the derivatives with respect to y and φ satisfy the Lipschitz condition with respect to y and φ with a Lipschitz constant independent of ε.

In the present section, we decompose Eqs. (18.7) in a neighborhood of the integral manifold $y \equiv 0$ by introducing new variables according to the formula

$$\varphi = \psi + \Phi(y, \psi, \tau, \varepsilon) \quad (\Phi(0, \psi, \tau, \varepsilon) \equiv 0), \tag{18.8}$$

which reduces system (18.7) to the form

$$\frac{dy}{d\tau} = a(y + X(\psi + \Phi, \tau, \varepsilon), \psi + \Phi, \tau, \varepsilon) - a(X(\psi + \Phi, \tau, \varepsilon), \psi + \Phi, \tau, \varepsilon)$$

$$- \frac{\partial X(\psi + \Phi, \tau, \varepsilon)}{\partial \varphi}[b(y + X(\psi + \Phi, \tau, \varepsilon), \psi + \Phi, \tau, \varepsilon)$$

$$- b(X(\psi + \Phi, \tau, \varepsilon), \psi + \Phi, \tau, \varepsilon)],$$

$$\frac{d\psi}{d\tau} = \frac{\omega(\tau)}{\varepsilon} + b(X(\psi, \tau, \varepsilon), \psi, \tau, \varepsilon), \tag{18.9}$$

For $\omega = \text{const}$, results concerning the canonical form of a dynamical system in a neighborhood of an invariant torus are presented in Chapter 4. One should also note the work [SaS], where the decomposition of equations was carried out for systems with slowly varying phase, and the monograph [StS], where the decomposition of singularly perturbed equations was carried out. We write the following partial differential equation for the determination of the function $\Phi = \Phi(y, \psi, \tau, \varepsilon)$:

$$\begin{aligned}
\frac{\partial \Phi}{\partial \tau} + \frac{\partial \Phi}{\partial \psi}\Big[\frac{\omega(\tau)}{\varepsilon} &+ b(X(\psi,\tau,\varepsilon),\psi,\tau,\varepsilon)\Big] \\
&+ \frac{\partial \Phi}{\partial y}\Big[a(y + X(\psi+\Phi,\tau,\varepsilon),\psi+\Phi,\tau,\varepsilon) \\
&- a(X(\psi+\Phi,\tau,\varepsilon),\psi+\Phi,\tau,\varepsilon) \\
&- \frac{\partial X(\psi+\Phi,\tau,\varepsilon)}{\partial \varphi}(b(yX(\psi+\Phi,\tau,\varepsilon),\psi+\Phi,\tau,\varepsilon) \\
&- b(X(\psi+\Phi,\tau,\varepsilon),\psi+\Phi,\tau,\varepsilon)\Big] \\
= b(y + X(\psi+\Phi,\tau,\varepsilon),&\psi+\Phi,\tau,\varepsilon) - b(X(\psi,\tau,\varepsilon),\psi,\tau,\varepsilon). \qquad (18.10)
\end{aligned}$$

We construct a solution of Eq. (18.10) by the method of successive approximations, defining these approximations by the formula

$$\Phi^{j+1}(y,\psi,\tau,\varepsilon) = -\int_{\tau}^{\infty}[b^j - \widetilde{b}]d\xi, \quad \Phi^0 \equiv 0, \qquad (18.11)$$

where

$$b^j = b(y_\tau^{\xi,j} + X(\psi_\tau^\xi + \Phi^j(y_\tau^{\xi,j},\psi_\tau^\xi,\xi,\varepsilon),\xi,\varepsilon),\psi_\tau^\xi + \Phi^j(y_\tau^{\xi,j},\psi_\tau^\xi,\xi,\varepsilon),\xi,\varepsilon),$$
$$\widetilde{b} = b(X(\psi_\tau^\xi,\xi,\varepsilon),\psi_\tau^\xi,\xi,\varepsilon).$$

Here, $(y_\tau^{\xi,j},\psi_\tau^\xi) = (y_\tau^{\xi,j}(y,\psi,\varepsilon),\psi_\tau^\xi(\psi,\varepsilon))$ is a solution of the Cauchy problem

$$\frac{d}{d\xi}y_\tau^{\xi,j} = a^j - \widetilde{a}^j - \frac{\partial X^j}{\partial \varphi}(b^j - \widetilde{b}^j), \quad y_\tau^{\tau,j} = y, \qquad (18.12)$$

$$\frac{d}{d\xi}\psi_\tau^\xi = \frac{\omega(\xi)}{\varepsilon} + \widetilde{b}, \quad \psi_\tau^\tau = \psi, \qquad (18.13)$$

where $a^j = a(y_\tau^{\xi,j} + X^j,\psi_\tau^\xi + \Phi^j,\xi,\varepsilon)$, $\widetilde{a}^j = a(X^j,\psi_\tau^\xi + \Phi^j,\xi,\varepsilon)$, $X^j = X(\psi_\tau^\xi + \Phi^j,\xi,\varepsilon)$, $\widetilde{b}^j = b(X^j,\psi_\tau^\xi + \Phi^j,\xi,\varepsilon)$, and $\Phi^j = \Phi^j(y_\tau^{\xi,j},\psi_\tau^\xi,\xi,\varepsilon)$.

If the order of differentiation and integration can be changed, then one can easily verify that the function $\Phi^{j+1} = \Phi^{j+1}(y,\psi,\tau,\varepsilon)$ defined by equality (18.11) satisfies the partial differential equation

$$\frac{\partial\Phi^{j+1}}{\partial\tau}+\frac{\partial\Phi^{j+1}}{\partial\psi}\Big[\frac{\omega(\tau)}{\varepsilon}+b(X(\psi,\tau,\varepsilon),\psi,\tau,\varepsilon)\Big]$$

$$+\frac{\partial\Phi^{j+1}}{\partial y}\Big[a(y+X(\psi+\Phi^j,\tau,\varepsilon),\psi+\Phi^j,\tau,\varepsilon)$$

$$-a(X(\psi+\Phi^j,\tau,\varepsilon),\psi+\Phi^j,\tau,\varepsilon)$$

$$-\frac{\partial X(\psi+\Phi^j,\tau,\varepsilon)}{\partial\varphi}(b(y+X(\psi+\Phi^j,\tau,\varepsilon),\psi+\Phi^j,\tau,\varepsilon)$$

$$-b(X(\psi+\Phi^j,\tau,\varepsilon),\psi+\Phi^j,\tau,\varepsilon)\Big]$$

$$=b(y+X(\psi+\Phi^j,\tau,\varepsilon),\psi+\Phi^j,\tau,\varepsilon)-b(X(\psi,\tau,\varepsilon),\psi,\tau,\varepsilon). \quad (18.14)$$

Assuming that the norm of the matrix $\frac{\partial b}{\partial\varphi}$ is sufficiently small, i.e.,

$$\sup_{\overline{G}}\left\|\frac{\partial b(x,\varphi,\tau,\varepsilon)}{\partial\varphi}\right\|<\frac{\gamma}{K}\min\left\{\frac{1}{2};\frac{1}{K}\right\}, \quad (18.15)$$

we prove that the sequence $\{\Phi^j(y,\psi,\tau,\varepsilon)\}$ converges uniformly to a certain function $\Phi(y,\psi,\tau,\varepsilon)$ on the set $y\in P_h\equiv\{y\colon y\in R^n, \|y\|\le h\}$, $\psi\in R^m$, $\tau\in R$, $\varepsilon\in(0,\varepsilon_0]$, provided that $\varepsilon_0>0$ and $h>0$ are sufficiently small. Moreover, we establish the convergence of the sequences $\left\{\frac{\partial}{\partial y}\Phi^j\right\}$, $\left\{\frac{\partial}{\partial\psi}\Phi^j\right\}$, and $\left\{\frac{\partial}{\partial\tau}\Phi^j\right\}$ to $\frac{\partial}{\partial y}\Phi$, $\frac{\partial}{\partial\psi}\Phi$, and $\frac{\partial}{\partial\tau}\Phi$, respectively. Then, passing to the limit as $j\to\infty$ in Eq. (18.14), we obtain equality (18.10) for the function $\Phi(y,\psi,\tau,\varepsilon)$ constructed above.

Theorem 18.1. *Suppose that the conditions imposed on system (18.1) and conditions (18.2)–(18.5) and (18.15) are satisfied. Then, for sufficiently small* $h>0$ *and* $\varepsilon_0>0$, *there exists a change of variables (18.8) that reduces system (18.1) to the decomposed form (18.9), where the function* $\Phi(y,\psi,\tau,\varepsilon)$ *is* 2π-*periodic in* ψ_ν, $\nu=\overline{1,m}$, *continuously differentiable with respect to* y, ψ, *and* τ *for every fixed* $\varepsilon\in(0,\varepsilon_0]$, *and such that*

$$\|\Phi\|\le d_1\|y\|, \quad \left\|\frac{\partial\Phi}{\partial y}\right\|\le d_2, \quad \left\|\frac{\partial\Phi}{\partial\psi}\right\|\le d_3\|y\| \quad (18.16)$$

$\forall (y,\psi,\tau,\varepsilon) \in P_h \times R^m \times R \times (0,\varepsilon_0] \equiv \underline{G}$, and its partial derivatives with respect to y and ψ satisfy the Lipschitz condition:

$$\left\| \frac{\partial}{\partial y}\Phi(\overline{y},\overline{\psi},\tau,\varepsilon) - \frac{\partial}{\partial y}\Phi(y,\psi,\tau,\varepsilon)\right\| \le \mu(\|\overline{y}-y\| + \|\overline{\psi}-\psi\|),$$

$$\left\| \frac{\partial}{\partial \psi}\Phi(\overline{y},\psi,\tau,\varepsilon) - \frac{\partial}{\partial \psi}\Phi(y,\psi,\tau,\varepsilon)\right\| \le \nu\|\overline{y}-y\|, \tag{18.17}$$

$$\left\| \frac{\partial}{\partial \psi}\Phi(y,\overline{\psi},\tau,\varepsilon) - \frac{\partial}{\partial \psi}\Phi(y,\psi,\tau,\varepsilon)\right\| \le \nu\|y\|\cdot\|\overline{\psi}-\psi\|.$$

Here, $d_1 - d_3$, μ, and ν are constants independent of ε and h.

We prove Theorem 18.1 in the next section. Here, we establish certain properties of a solution of the Cauchy problem (18.12).

Lemma 18.1. *Suppose that the conditions of Theorem 18.1 are satisfied and, for every $\varepsilon \in (0,\varepsilon_0]$ and certain $j \ge 0$, the function $\Phi^j(y,\psi,\tau,\varepsilon)$ is continuously differentiable with respect to y, ψ, and τ. Then one can find constants $\overline{\varepsilon}_0 > 0$, $h_1 > 0$, and $\gamma_1 \in \left(\frac{\gamma}{2},\gamma\right)$ such that, for all $(y,\psi,\tau,\varepsilon) \in \underline{G}$, $h \le h_1$, $\varepsilon_0 \le \overline{\varepsilon}_0$, and $\xi \ge \tau$, the solution $y_\tau^{\xi,j} = y_\tau^{\xi,j}(y,\psi,\varepsilon)$ of the Cauchy problem (18.12) satisfies the inequality*

$$\|y_\tau^{\xi,j}\| \le K\|y\|e^{-\gamma_1(\xi-\tau)}.$$

Proof. It follows from (18.12) and the smoothness conditions for the right-hand side of system (18.1) that

$$\frac{dy_\tau^{\xi,j}}{d\xi} = H(\xi)y_\tau^{\xi,j} + \frac{\partial\widetilde{a}(\overline{x}(\xi),\psi_\tau^\xi + \Phi^j,\xi)}{\partial x}y_\tau^{\xi,j} + F_1(y_\tau^{\xi,j},\psi_\tau^\xi,\Phi^j,\xi,\varepsilon), \tag{18.18}$$

where

$$\|F_1\| \le \sum_{\nu=1}^{n}\sup_{\overline{G}}\left\|\frac{\partial^2}{\partial x\partial x_\nu}a(x,\varphi,\tau,\varepsilon)\right\|(\|y_\tau^{\xi,j}\| + \overline{c}_1\varepsilon^\alpha)\|y_\tau^{\xi,j}\|$$

$$+\,\varepsilon\sup_{\overline{G}}\left\|\frac{\partial}{\partial x}A(x,\varphi,\tau,\varepsilon)\right\|\,\|y_\tau^{\xi,j}\| + \overline{c}_1\varepsilon^\alpha\sup_{\overline{G}}\left\|\frac{\partial}{\partial x}b(x,\varphi,\tau,\varepsilon)\right\|\,\|y_\tau^{\xi,j}\|$$

$$\le c_2(\|y_\tau^{\xi,j}\| + \varepsilon_0^\alpha), \quad c_2 = nc_1(1+n+\overline{c}_1), \quad H(\xi) = \frac{\partial}{\partial x}a(\overline{x}(\xi),\xi).$$

For the function $y_\tau^{\xi,j}$, we can write the representation

$$y_\tau^{\xi,j} = Q(\xi,\tau)y + \int_\tau^\xi Q(\xi,l)\Big[\frac{\partial}{\partial x}\widetilde{a}(\overline{x}(l),\psi_\tau^l$$

$$+ \Phi^j(y_\tau^{l,j},\psi_\tau^l,l,\varepsilon),l)y_\tau^{l,j} + F_1(y_\tau^{l,j},\psi_\tau^l,\Phi^j(y_\tau^{l,j},\psi_\tau^l,l,\varepsilon),l,\varepsilon)\Big]dl,$$

whence

$$\|y_\tau^{\xi,j}\| \le K\|y\|e^{-\gamma(\xi-\tau)} + \int_\tau^\xi e^{-\gamma(\xi-l)}\Big[K\sup_{\varphi,\tau}\Big\|\frac{\partial}{\partial x}\widetilde{a}(\overline{x}(\tau),\varphi,\tau)\Big\|$$

$$+ Kc_2(\|y_\tau^{l,j}\| + \varepsilon_0^\alpha)\Big]\|y_\tau^{l,j}\|dl.$$

Assume that $\|y_\tau^{\xi,j}\| < 2K\|y\|$ on the maximum half-interval $\xi \in [\tau,T)$. Then, denoting

$$z_\tau^\xi = \|y_\tau^{\xi,j}\|e^{\gamma(\xi-\tau)},$$

we obtain the inequality

$$z_\tau^\xi \le K\|y\| + \int_\tau^\xi \Big[K\sup_{\varphi,\tau}\Big\|\frac{\partial}{\partial x}\widetilde{a}(\overline{x}(t),\varphi,\tau)\Big\| + Kc_2(\varepsilon_0^\alpha + 2K\|y\|)\Big]z_\tau^l dl,$$

whose solution (according to the Gronwall–Bellman lemma) satisfies the estimate

$$z_\tau^\xi \le K\|y\|\exp\Big\{\Big[K\sup_{\varphi,\tau}\Big\|\frac{\partial}{\partial x}\widetilde{a}(\overline{x}(t),\varphi,\tau)\Big\| + Kc_2(\varepsilon_0^\alpha + 2K\|y\|)\Big](\xi-\tau)\Big\}.$$

Taking into account that, according to (18.4), we have $\sigma_0 < 1.$ and assuming that

$$2K^2c_2h \le \frac{1}{8}\gamma(1-\sigma_0), \quad Kc_2\varepsilon_0^\alpha \le \frac{1}{8}\gamma(1-\sigma_0),$$

we deduce from the last estimate that

$$\|y_\tau^{\xi,j}\| \le K\|y\|e^{-\gamma_1(\xi-\tau)}, \quad \gamma_1 = \frac{3-\sigma_0}{4}, \tag{18.19}$$

for all $y \in P_h$, $\psi \in R^m$, $\xi \in [\tau,T)$, and $\varepsilon \in (0,\varepsilon_0]$. It is obvious that

$$\|y_\tau^{\xi,j}\| \le K\|y\| < 2K\|y\| \quad \forall\xi \in [\tau,T).$$

Therefore, relation (18.19) holds for all $\xi \in [\tau,\infty)$. Lemma 18.1 is proved.

Lemma 18.2. *If the conditions of Lemma 18.1 are satisfied and the function* $\Phi^j(y,\psi,\tau,\varepsilon)$ *satisfies the inequality* $\left\|\frac{\partial}{\partial y}\Phi^j\right\| \leq d_2,$ *then, for sufficiently small* $h > 0$ *and* $\varepsilon_0 > 0,$ *the following estimate is true:*

$$\left\|\frac{\partial}{\partial y}y_\tau^{\xi,j}(y,\psi,\varepsilon)\right\| \leq Ke^{-\gamma_1(\xi-\tau)} \qquad \forall(y,\psi,\tau,\varepsilon) \in \underline{G}, \quad \xi \geq \tau. \tag{18.20}$$

Proof. Differentiating Eq. (18.12) with respect to y and using the equality $\frac{\partial}{\partial y}y_\tau^{\tau,j}(y,\psi,\varepsilon) = E_n,$ where E_n is the n-dimensional identity matrix, we get

$$\begin{aligned}\frac{\partial}{\partial y}y_\tau^{\xi,j} = Q(\xi,\tau) + \int_\tau^\xi Q(\xi,l)\Big[\frac{\partial}{\partial x}\widetilde{a}(\overline{x}(l),\psi_\tau^l + \Phi^j(y_\tau^{l,j},\psi_\tau^l,l,\varepsilon),l)\\ + F_2\Big(y_\tau^{l,j},\psi_\tau^l,\Phi^j(y_\tau^{l,j},\psi_\tau^l,l,\varepsilon),\\ \frac{\partial}{\partial y}\Phi^j(y_\tau^{l,j},\psi_\tau^l,l,\varepsilon),l,\varepsilon\Big)\Big]\frac{\partial}{\partial y}y_\tau^{l,j}\,dl,\end{aligned} \tag{18.21}$$

where

$$\|F_2\| \leq c_3(\varepsilon_0^\alpha + \|y\|),$$

$$c_3 = nc_1K[n + n\overline{c}_1 + c_1 + d_2(n\overline{c}_1 + n\overline{c}_1^2 + m + 2m\overline{c}_1)].$$

Equality (18.21) yields

$$\left\|\frac{\partial}{\partial y}y_\tau^{\xi,j}\right\| e^{\gamma(\xi-\tau)} \leq K + \int_\tau^\xi e^{\gamma(l-\tau)}\Big[\frac{\gamma}{2}\sigma_0 + Kc_3(\varepsilon_0^\alpha + h)\Big]\left\|\frac{\partial}{\partial y}y_\tau^{l,j}\right\| dl,$$

Solving this inequality, we obtain estimate (18.20) under the conditions $Kc_3\varepsilon_0^\alpha \leq \frac{1}{8}\gamma(1-\sigma_0)$ and $Kc_3h \leq \frac{1}{8}\gamma(1-\sigma_0).$ Lemma 18.2 is proved.

Lemma 12.1–12.3 yield the following estimates for a solution $\psi_\tau^\xi = \psi_\tau^\xi(\psi,\varepsilon)$ of the Cauchy problem (18.13):

$$\left\|\frac{\partial}{\partial\psi}(\varphi_\tau^\xi - \psi)\right\| \leq c_4\varepsilon^\alpha(1 + |\xi-\tau|)e^{c_4\varepsilon^\alpha|\xi-\tau|},$$

$$\Big\|\frac{\partial}{\partial\tau}\psi_\tau^\xi\Big\| \le c_4\Big(\frac{1}{\varepsilon}\|\omega(\tau)\|+1\Big)\,e^{c_4\varepsilon^\alpha|\xi-\tau|}, \tag{18.22}$$

$$\sum_{\nu=1}^{m}\Big\|\frac{\partial^2}{\partial\psi\partial\psi_\nu}\varphi_\tau^\xi\Big\| \le c_4\varepsilon^\alpha(1+|\xi-\tau|^2)e^{c_4\varepsilon^\alpha|\xi-\tau|}.$$

Here, $\psi\in R^m,\ \varepsilon\in(0,\varepsilon_0],\ \tau\in R,\ \xi\in R,$, and c_4 is the constant equal to the greatest constant in the corresponding estimates in Lemmas 12.1–12.3. Using (18.22) and repeating the scheme of the proof of Lemmas 18.1 and 18.2, we establish the following statements:

Lemma 18.3. *Let*

$$\Big\|\frac{\partial}{\partial y}\Phi^j(y,\psi,\tau,\varepsilon)\Big\| \le d_2, \quad \Big\|\frac{\partial}{\partial\psi}\Phi^j(y,\psi,\tau,\varepsilon)\Big\| \le d_3\|y\|$$

$$\forall(y,\psi,\tau,\varepsilon)\in\underline{G}.$$

Then, for sufficiently small $h>0$ *and* $\varepsilon_0>0$ *and all* $(y,\psi,\tau,\varepsilon)\in\underline{G}$ *and* $\xi\ge\tau,$ *the following inequalities are true:*

$$\Big\|\frac{\partial}{\partial\psi}y_\tau^{\xi,j}\Big\| \le c_5\|y\|(1+\|y\|d_3)e^{-\gamma_2(\xi-\tau)},\quad \gamma_2=\frac{5-\sigma_0}{8}\gamma, \tag{18.23}$$

$$\Big\|\frac{\partial}{\partial\tau}y_\tau^{\xi,j}\Big\| \le c_5\|y\|\Big(\frac{1}{\varepsilon}\|\omega(\tau)\|+d_3\|y\|\Big)\,e^{-\gamma_2(\xi-\tau)}, \tag{18.24}$$

where the constant c_5 *depends on* d_2 *and does not depend on* d_3 *and* ε.

Lemma 18.4. *If the conditions of Lemma 18.3 are satisfied and the function* $\dfrac{\partial}{\partial y}\Phi^j(y,\psi,\tau,\varepsilon)$ *satisfies the Lipschitz condition*

$$\Big\|\frac{\partial}{\partial y}\Phi^j(\overline{y},\overline{\psi},\tau,\varepsilon)-\frac{\partial}{\partial y}\Phi^j(y,\psi,\tau,\varepsilon)\Big\| \le \mu(\|y-\overline{y}\|+\|\psi-\overline{\psi}\|) \tag{18.25}$$

$$\forall y,\overline{y}\in P_h,\quad \psi,\overline{\psi}\in R^m,\quad \tau\in R,\quad \varepsilon\in(0,\varepsilon_0],$$

then the following estimate is true:

$$\Big\|\frac{\partial}{\partial y}y_\tau^{\xi,j}(\overline{y},\overline{\psi},\varepsilon)-\frac{\partial}{\partial y}y_\tau^{\xi,j}(y,\psi,\varepsilon)\Big\| \le c_6(1+h\mu)e^{-\gamma_3(\xi-\tau)}(\|y-\overline{y}\|+\|\psi-\overline{\psi}\|),$$

where $\xi\ge\tau,\ \gamma_3=\frac{1}{16}(9-\sigma_0)\gamma,\ \varepsilon_0>0$ *and* $h>0$ *are sufficiently small, and* c_6 *is a constant independent of* $\mu,\ h,$ *and* ε.

Lemma 18.5. *Suppose that the conditions of Lemma 18.4 are satisfied and the following estimates are true:*

$$\left\| \frac{\partial}{\partial \psi} \Phi^j(\overline{y}, \psi, \tau, \varepsilon) - \frac{\partial}{\partial \psi} \Phi^j(y, \psi, \tau, \varepsilon) \right\| \leq \nu \|y - \overline{y}\|, \quad \nu = \text{const},$$

$$\left\| \frac{\partial}{\partial \psi} \Phi^j(y, \overline{\psi}, \tau, \varepsilon) - \frac{\partial}{\partial \psi} \Phi^j(y, \psi, \tau, \varepsilon) \right\| \leq \nu \|y\| \|\psi - \overline{\psi}\|. \tag{18.26}$$

Then one can find a constant c_7 *independent of* ν, h, *and* ε *and such that the following inequalities hold for all* $\xi \geq \tau$:

$$\left\| \frac{\partial}{\partial \psi} y_\tau^{\xi,j}(\overline{y}, \psi, \varepsilon) - \frac{\partial}{\partial \psi} y_\tau^{\xi,j}(y, \psi, \varepsilon) \right\| \leq c_7(1 + h\nu) e^{-\gamma_4(\xi - \tau)} \|y - \overline{y}\|,$$

$$\left\| \frac{\partial}{\partial \psi} y_\tau^{\xi,j}(y, \overline{\psi}, \varepsilon) - \frac{\partial}{\partial \psi} y_\tau^{\xi,j}(y, \psi, \varepsilon) \right\| \leq c_7(1 + h\nu) \|y\| e^{-\gamma_4(\xi - \tau)} \|\psi - \overline{\psi}\|,$$

where $\gamma_4 = \dfrac{1}{32}(17 - \sigma_0)\gamma$.

Further, we denote

$$\|\Phi^{j+1}(y, \psi, \tau, \varepsilon) - \Phi^j(y, \psi, \tau, \varepsilon)\| = \|y\| v^j(y, \psi, \tau, \varepsilon),$$

$$\left\| \frac{\partial}{\partial \psi} (\Phi^{j+1}(y, \psi, \tau, \varepsilon) - \Phi^j(y, \psi, \tau, \varepsilon)) \right\| = \|y\| v_j(y, \psi, \tau, \varepsilon),$$

$$v^j(0, \psi, \tau, \varepsilon) \equiv 0, \quad v_j(0, \psi, \tau, \varepsilon) \equiv 0.$$

We use the following lemma for the investigation of the convergence of the iterations defined by (18.11):

Lemma 18.6. *Suppose that, for all* $(y, \psi, \tau, \varepsilon) \in \underline{G}$ *and* $j \geq 0$, *the inequalities*

$$\left\| \frac{\partial}{\partial y} \Phi^j(y, \psi, \tau, \varepsilon) \right\| \leq d_2, \quad \left\| \frac{\partial}{\partial \psi} \Phi^j(y, \psi, \tau, \varepsilon) \right\| \leq d_3 \|y\|$$

are true and the Lipschitz conditions (18.25) and (18.26) are satisfied. Then there exist constants $\underline{\varepsilon}_0 > 0$, $h_2 > 0$, $\gamma_5 > \dfrac{\gamma}{2}$, *and* c_8 *such that, for* $j \geq 0$, $(y, \psi, \tau, \varepsilon) \in \underline{G}$, $\varepsilon_0 \leq \underline{\varepsilon}_0$, $h \leq h_2$, *and* $\xi \geq \tau$, *the following estimates are true:*

$$\|y_\tau^{\xi,j+1} - y_\tau^{\xi,j}\| \le c_8 \|y\|^2 \sup_{\underline{G}} v^j e^{-\gamma_5(\xi-\tau)}, \tag{18.27}$$

$$\begin{aligned} &\Big\| \frac{\partial}{\partial y}(y_\tau^{\xi,j+1} - y_\tau^{\xi,j}) \Big\| \\ &\quad \le c_8 \|y\| \Big[\sup_{\underline{G}} v^j \\ &\qquad + \sup_{\underline{G}} \Big\| \frac{\partial}{\partial y}(\Phi^{j+1}(y,\psi,\tau,\varepsilon) - \Phi^j(y,\psi,\tau,\varepsilon)) \Big\| \Big] e^{-\gamma_5(\xi-\tau)}, \end{aligned} \tag{18.28}$$

$$\begin{aligned} \Big\| \frac{\partial}{\partial \psi}(y_\tau^{\xi,j+1} - y_\tau^{\xi,j}) \Big\| &\le c_8 \|y\| \Big[\sup_{\underline{G}} \Big\| \frac{\partial}{\partial y}(\Phi^{j+1}(y,\psi,\tau,\varepsilon) - \Phi^j(y,\psi,\tau,\varepsilon)) \Big\| \\ &\quad + \|y\| (\sup_{\underline{G}} v^j + \sup_{\underline{G}} v_j) \Big] e^{-\gamma_5(\xi-\tau)}, \end{aligned} \tag{18.29}$$

where $y_\tau^{\xi,s} = y_\tau^{\xi,s}(y,\psi,\varepsilon)$ *for* $s = j, j+1$.

Proof. Rewriting Eq. (18.12) in the form

$$\begin{aligned} \frac{d}{d\xi} y_\tau^{\xi,j} = H(\xi) y_\tau^{\xi,j} &+ \int_0^1 \Big[\frac{\partial}{\partial x} a(t y_\tau^{\xi,j} + X^j, \psi_\tau^\xi + \Phi^j, \xi, \varepsilon) - H(\xi) \Big] dt\, y_\tau^{\xi,j} \\ &- \frac{\partial X^j}{\partial \varphi} \int_0^1 \frac{\partial}{\partial x} b(t y_\tau^{\xi,j} + X^j, \psi_\tau^\xi + \Phi^j, \xi, \varepsilon) dt\, y_\tau^{\xi,j}, \end{aligned}$$

one can easily obtain the estimate

$$\begin{aligned} \|y_\tau^{\xi,j+1} - y_\tau^{\xi,j}\| \le \int_\tau^\xi K e^{-\gamma(\xi-l)} \Big\{ &\Big\| \int_0^1 \Big[\frac{\partial a^{j+1}}{\partial x} - H(l) \Big] dt \Big\| \|y_\tau^{l,j+1} - y_\tau^{l,j}\| \\ &+ \Big\| \int_0^1 \Big[\frac{\partial a^{j+1}}{\partial x} - \frac{\partial a^j}{\partial x} \Big] dt \Big\| \, \|y_\tau^{l,j}\| \\ &+ \Big\| \frac{\partial X^{j+1}}{\partial \varphi} - \frac{\partial X^j}{\partial \varphi} \Big\| n c_1 \|y_\tau^{l,j+1}\| \\ &+ \Big\| \frac{\partial X^j}{\partial \varphi} \Big\| \Big\| \int_0^1 \Big[\frac{\partial b^{j+1}}{\partial x} y_\tau^{l,j+1} - \frac{\partial b^j}{\partial x} y_\tau^{l,j} \Big] dt \Big\| \Big\} dl. \end{aligned} \tag{18.30}$$

We now use Lemma 18.1 and inequalities (18.6). Since

$$\left\|\frac{\partial a^{j+1}}{\partial x} - H(l)\right\| \leq \sup_{\varphi,\tau}\left\|\frac{\partial}{\partial x}\widetilde{a}(\overline{x}(\tau),\varphi,\tau)\right\| + nc_1(1+n\overline{c}_1+Kn)(\varepsilon^\alpha + \|y\|),$$

$$\left\|\frac{\partial a^{j+1}}{\partial x} - \frac{\partial a^j}{\partial x}\right\| \|y_\tau^{l,j}\| \leq Knc_1(n+m)(1+\overline{c}_1)\|y\|e^{-\gamma_1(l-\tau)}(\|y_\tau^{l,j+1} - y_\tau^{l,j}\|$$
$$+ \|\Phi^{j+1}(y_\tau^{l,j+1},\psi_\tau^l,l,\varepsilon) - \Phi^j(y_\tau^{l,j},\psi_\tau^l,l,\varepsilon)\|,$$

$$\left\|\frac{\partial X^{j+1}}{\partial \varphi} - \frac{\partial X^j}{\partial \varphi}\right\| nc_1\|y_\tau^{l,j+1}\| + \left\|\frac{\partial X^j}{\partial \varphi}\right\| \left\|\int_0^1\left[\frac{\partial b^{j+1}}{\partial x}y_\tau^{l,j+1} - \frac{\partial b^j}{\partial x}y_\tau^{l,j}\right]dt\right\|$$
$$\leq Knc_1\overline{c}_1(1+n\overline{c}_1+n+m)[(\varepsilon^\alpha + \|y\|)\|y_\tau^{l,j+1} - y_\tau^{l,j}\|$$
$$+ \varepsilon^\alpha\|y\|e^{-\gamma_1(l-\tau)}\|\Phi^{j+1}(y_\tau^{l,j+1},\psi_\tau^l,l,\varepsilon) - \Phi^j(y_\tau^{l,j},\psi_\tau^l,l,\varepsilon)\|],$$

estimate (18.30) can be rewritten in the form

$$\|y_\tau^{\xi,j+1} - y_\tau^{\xi,j}\| \leq \int_\tau^\xi e^{-\gamma(\xi-l)}\left[K\sup_{\varphi,\tau}\left\|\frac{\partial}{\partial x}\widetilde{a}(\overline{x}(\tau),\varphi,\tau)\right\|\right.$$
$$\left. + (\|y\| + \varepsilon^\alpha)3c_9K\right]\|y_\tau^{l,j+1} - y_\tau^{l,j}\|dl$$
$$+ 2c_9K\|y\|\int_\tau^\xi e^{-\gamma(\xi-l)-\gamma_1(l-\tau)}\|\Phi^{j+1} - \Phi^j\|dl,$$

where $c_9 = Knc_1(1+\overline{c}_1)(1+n\overline{c}_1+n+m)$. Taking into account that the derivative of each function $\Phi^j(y,\psi,\tau,\varepsilon)$, $j \geq 0$, with respect to y is bounded from above by the constant d_2, we get

$$\|\Phi^{j+1}(y_\tau^{l,j+1},\psi_\tau^l,l,\varepsilon) - \Phi^j(y_\tau^{l,j},\psi_\tau^l,l,\varepsilon)\|$$
$$\leq \|y_\tau^{l,j+1}\|\sup_{\underline{G}} v^j + d_2\|y_\tau^{l,j+1} - y_\tau^{l,j}\|$$
$$\leq K\|y\|e^{-\gamma_1(l-\tau)}\sup_{\underline{G}} v^j + d_2\|y_\tau^{l,j+1} - y_\tau^{l,j}\|, \tag{18.31}$$

whence

$$\|y_\tau^{\xi,j+1} - y_\tau^{\xi,j}\| \le \int\limits_\tau^\xi e^{-\gamma(\xi-l)}\Big[\frac{\gamma}{2}\sigma_0 + Kc_9(3+2d_2)(\varepsilon_0^\alpha + h)\Big]\|y_\tau^{l,j+1} - y_\tau^{l,j}\|dl$$

$$+\frac{2c_9}{2\gamma_1-\gamma}K^2 e^{-\gamma(\xi-l)}\|y\|^2 \sup_{\underline{G}} v^j. \tag{18.32}$$

We choose $h > 0$ and $\varepsilon_0 > 0$ so small that

$$Kc_9(3+2d_2)\varepsilon_0^\alpha \le \frac{1}{8}\gamma(1-\sigma_0), \quad Kc_9(3+2d_2)h \le \frac{1}{8}\gamma(1-\sigma_0).$$

Then, solving inequality (18.32), we obtain the estimate

$$\|y_\tau^{\xi,j+1} - y_\tau^{\xi,j}\| \le \frac{2c_9}{2\gamma_1-\gamma}K^2 e^{-\frac{\gamma}{4}(3-\sigma_0)(\xi-\tau)}\|y\|^2 \sup_{\underline{G}} v^j,$$

the form of which coincides with the form of estimate (18.27). Inequalities (18.28) and (18.29) can be established by analogy.

19. Proof of Theorem 18.1

Consider iterations (18.11). Since $\Phi^0 \equiv 0$ satisfies all conditions of Lemmas 18.1–18.3, it follows from (18.11) for $j = 0$ that

$$\|\Phi^1(y,\psi,\tau,\varepsilon)\| \le \sup_{\underline{G}}\Big\|\frac{\partial b}{\partial x}\Big\|\int\limits_\tau^\infty \|y_\tau^{\xi,0}\|d\xi \le \frac{n}{\gamma_1}c_1K\|y\| \le d_1\|y\|,$$

$$\Big\|\frac{\partial}{\partial y}\Phi^1(y,\psi,\tau,\varepsilon)\Big\| \le \sup_{\overline{G}}\Big\|\frac{\partial b}{\partial x}\Big\|\int\limits_\tau^\infty \Big\|\frac{\partial}{\partial y}y_\tau^{\xi,0}\Big\|d\xi \le \frac{n}{\gamma_1}c_1K \le d_2.$$

To estimate $\Big\|\dfrac{\partial}{\partial\psi}\Phi^1(y,\psi,\tau,\varepsilon)\Big\|$, we use inequalities (18.22), the first of which yields

$$\Big\|\frac{\partial}{\partial\psi}\psi_\tau^\xi(\psi,\varepsilon)\Big\| \le m + \Big\|\frac{\partial}{\partial\psi}(\psi_\tau^\xi(\psi,\varepsilon)-\psi)\Big\|$$

$$\le m + c_4\varepsilon^\alpha(1+\xi-\tau)e^{c_4\varepsilon^\alpha(\xi-\tau)}$$

$$\le m + e^{2c_4\varepsilon^\alpha(\xi-\tau)} \le (m+1)e^{\frac{1}{4}\gamma(\xi-\tau)}$$

for $\xi \geq \tau$ and $2c_4\varepsilon_0^\alpha \leq \frac{1}{4}\gamma_1$. Then

$$\Big\|\frac{\partial}{\partial\psi}\Phi^1(y,\psi,\tau,\varepsilon)\Big\| \leq \Big[2nc_1c_5\frac{1}{\gamma_2} + n^2c_1\bar{c}_1K\frac{1}{\gamma_1} + mnc_1(m+1)K\frac{4}{3\gamma_1}\Big]\|y\|$$
$$\leq d_3\|y\|,$$

provided that $hd_3 \leq 1$. The constants d_1, d_2, and d_3 in the inequalities presented above will be fixed in what follows. Consider the integral obtained from (18.11) for $j = 0$ by differentiation with respect to τ under the integral sign. Estimates (18.22) and (18.24) yield

$$\int_\tau^\infty \Big\|\frac{\partial}{\partial\tau}(b^0 - \widetilde{b})\Big\| d\xi$$
$$\leq c_5\|y\|\Big(\frac{\|\omega(\tau)\|}{\varepsilon} + 1\Big)nc_1\int_\tau^\infty e^{-\gamma_2(\xi-\tau)}d\xi$$
$$+ (n\bar{c}_1\varepsilon_0^\alpha + m)nc_1c_4K\|y\|\Big(\frac{\|\omega(\tau)\|}{\varepsilon} + 1\Big)\int_\tau^\infty e^{-\frac{3}{4}\gamma_1(\xi-\tau)}d\xi.$$

The estimates presented above guarantee the uniform convergence of the improper integrals

$$\int_\tau^\infty (b^0 - \widetilde{b})d\xi, \quad \int_\tau^\infty \frac{\partial}{\partial y}(b^0 - \widetilde{b})d\xi, \quad \int_\tau^\infty \frac{\partial}{\partial\psi}(b^0 - \widetilde{b})d\xi, \quad \int_\tau^\infty \frac{\partial}{\partial\tau}(b^0 - \widetilde{b})d\xi$$

on the set

$$y \in P_h, \quad \psi \in R^m, \quad \tau \in [-T, T], \quad \varepsilon \in [\Delta, \varepsilon_0], \tag{19.1}$$

where $T > 0$ and $\Delta > 0$ $(\Delta < \varepsilon_0)$ are arbitrary. Then, for every $\varepsilon \in [\Delta, \varepsilon_0]$, the function $\Phi^1(y,\psi,\tau,\varepsilon)$ is continuously differentiable with respect to y, ψ, and τ from set (19.1) and satisfies Eq. (18.14) with $j = 0$ for these values of y, ψ, τ, and ε. Since Δ and T are arbitrary, we obtain relation (18.14) for all $(y,\psi,\tau,\varepsilon) \in \underline{G}$.

Now assume that, for all $j = \overline{1,s}$, $s > 1$, the functions $\Phi^j(y,\psi,\tau,\varepsilon)$ are continuously differentiable with respect to $(y,\psi,\tau) \in P_h \times R^m \times R$ for every value of the parameter ε and satisfy the inequalities

$$\|\Phi^j\| \le d_1\|y\|, \qquad \left\|\frac{\partial}{\partial y}\Phi^j\right\| \le d_2, \qquad \left\|\frac{\partial}{\partial \psi}\Phi^j\right\| \le d_3\|y\| \tag{19.2}$$

and Eq. (18.14) $\forall(y,\psi,\tau,\varepsilon) \in \underline{G}$. Using Lemmas 18.1–18.3 for $j = s$ and estimates (18.15) and (18.22), we deduce from (18.11) that

$$\|\Phi^{s+1}(y,\psi,\tau,\varepsilon)\| \le \left[nc_1\frac{1}{\gamma_1}K(1+\overline{c}_1\varepsilon_0^\alpha d_1) + \frac{1}{\gamma_1}K\sup_{\overline{G}}\left\|\frac{\partial b}{\partial \varphi}\right\|d_1\right]\|y\|.$$

Since $\gamma_1 > \dfrac{\gamma}{2}$ and

$$\frac{K}{\gamma_1}\sup_{\overline{G}}\left\|\frac{\partial b}{\partial \varphi}\right\| < \frac{2}{\gamma}K\sup_{\overline{G}}\left\|\frac{\partial b}{\partial \varphi}\right\| = \overline{\sigma}_0 < 1,$$

for $d_1\varepsilon_0^\alpha \le 1$ we get

$$\|\Phi^{s+1}(y,\psi,\tau,\varepsilon)\| \le \left[nc_1\frac{1}{\gamma_1}K(1+\overline{c}_1) + \overline{\sigma}_0 d_1\right]\|y\| \le d_1\|y\|,$$

where

$$d_1 = \frac{nc_1(1+\overline{c}_1)K}{\gamma_1(1-\sigma_0)}.$$

By analogy, for $d_2\varepsilon_0^\alpha \le 1$ we obtain

$$\left\|\frac{\partial}{\partial y}\Phi^{s+1}(y,\psi,\tau,\varepsilon)\right\| \le \frac{2nc_1(1+\overline{c}_1)}{\gamma_1(1-\sigma_0)}K \equiv d_2.$$

We now represent $\dfrac{\partial}{\partial \psi}\Phi^{s+1}(y,\psi,\tau,\varepsilon)$ in the form

$$\frac{\partial}{\partial \psi}\Phi^{s+1}(y,\psi,\tau,\varepsilon)$$

$$= -\int_\tau^\infty \Big[\frac{\partial b^s}{\partial y} + \Big(\frac{\partial b^s}{\partial y}\frac{\partial X^s}{\partial \varphi} + \frac{\partial b^s}{\partial \varphi}\Big)\frac{\partial \Phi^s}{\partial y}\Big]\frac{\partial y_\tau^{\xi,s}}{\partial \psi}d\xi$$

$$-\int_\tau^\infty \Big[\frac{\partial(b^s-\widetilde{b})}{\partial y}\frac{\partial X^s}{\partial \varphi} + \frac{\partial \widetilde{b}}{\partial y}\Big(\frac{\partial X^s}{\partial \varphi} - \frac{\partial X}{\partial \varphi}\Big) + \frac{\partial(b^s-\widetilde{b})}{\partial \varphi}\Big]\frac{\partial \psi_\tau^\xi}{\partial \psi}d\xi$$

$$-\int_\tau^\infty \Big[\frac{\partial b^s}{\partial y}\frac{\partial X^s}{\partial \varphi}\frac{\partial \Phi^s}{\partial \varphi}\frac{\partial \psi_\tau^\xi}{\partial \psi} + \frac{\partial b^s}{\partial \varphi}\frac{\partial \Phi^s}{\partial \varphi}\frac{\partial(\psi_\tau^\xi - \psi)}{\partial \psi} + \frac{\partial b^s}{\partial \varphi}\frac{\partial \Phi^s}{\partial \varphi}\Big]d\xi$$

and use the inequalities

$$\|\Phi^s(y_\tau^{\xi,s},\psi_\tau^\xi,\xi,\varepsilon)\| \le d_1K\|y\|e^{-\gamma_1(\xi-\tau)},$$

$$\left\|\frac{\partial}{\partial\varphi}\Phi^s(y_\tau^{\xi,s},\psi_\tau^\xi,\xi,\varepsilon)\right\| \le d_3K\|y\|e^{-\gamma_1(\xi-\tau)}, \quad \xi\ge\tau.$$

Then, for $\overline{c}_1\varepsilon_0^\alpha \le 1$, $d_3\varepsilon_0^\alpha\le 1$, and $hd_3\le 1$, we get

$$\left\|\frac{\partial}{\partial\psi}\Phi^{s+1}(y,\psi,\tau,\varepsilon)\right\| \le (c_{10}+\overline{\sigma}_0 d_3)\|y\| \le \frac{c_{10}}{1-\overline{\sigma}_0}\|y\| \equiv d_3\|y\|,$$

where

$$c_{10} = \frac{2}{\gamma_2}(n+(n+m)d_2)c_1c_5 + \frac{4}{3\gamma_1}c_1K(m+1)[(n+m)^2(1+2d_1)+nd_1]$$
$$+\frac{4}{3\gamma_1}n(m+1)c_1\overline{c}_1K + \frac{4}{9\gamma_1^2}(4+3\gamma_1)mc_1c_4K.$$

Thus, by the method of mathematical induction, we establish that, for sufficiently small $h>0$ and $\varepsilon_0>0$, inequalities (19.2) with the constants d_1, d_2, and d_3 defined above hold for all $j\ge 0$ and $(y,\psi,\tau,\varepsilon)\in \underline{G}$. Moreover, the fact that the norms of the matrices

$$y_\tau^{\xi,j}, \quad \frac{\partial}{\partial y}y_\tau^{\xi,j}, \quad \frac{\partial}{\partial\psi}y_\tau^{\xi,j}, \quad \frac{\partial}{\partial\tau}y_\tau^{\xi,j}$$

tend exponentially to zero as $\xi\to\infty$ guarantees the uniform convergence [on set (19.1)] of the improper integral (18.11) and the integrals obtained from it by differentiation with respect to y, ψ, and τ under the integral sign. Therefore, the functions $\Phi^j(y,\psi,\tau,\varepsilon)$, $j\ge 0$, are continuously differentiable with respect to y, ψ, and τ for every fixed ε from set (19.1) and satisfy Eq. (18.14). Since Δ and T are arbitrary, equality (18.14) holds for all $(y,\psi,\tau,\varepsilon)\in\underline{G}$, and the functions $\Phi^j(y,\psi,\tau,\varepsilon)$ have continuous partial derivatives of the first order with respect to $(y,\psi,\tau)\in P_h\times R^m\times R$ for every $\varepsilon\in(0,\varepsilon_0]$.

We now prove that the matrices $\frac{\partial}{\partial y}\Phi^j(y,\psi,\tau,\varepsilon)$ and $\frac{\partial}{\partial\psi}\Phi^j(y,\psi,\tau,\varepsilon)$ satisfy the Lipschitz condition with respect to the variables y and ψ. Using Lemma 18.4 for $j=0$ and the inequalities

$$\|\psi_\tau^\xi(\psi,\varepsilon)-\psi_\tau^\xi(\overline{\psi},\varepsilon)\| \le [1+c_4\varepsilon^\alpha(1+\xi-\tau)e^{c_4\varepsilon^\alpha(\xi-\tau)}]\|\psi-\overline{\psi}\|,$$

$$\|y_\tau^{\xi,j}(y,\psi,\varepsilon)-y_\tau^{\xi,j}(\overline{y},\overline{\psi},\varepsilon)\|$$
$$\le Ke^{-\gamma_1(\xi-\tau)}]\|y-\overline{y}\| + c_5h(1+hd_3)e^{-\gamma_2(\xi-\tau)}\|\psi-\overline{\psi}\|, \quad \xi\ge\tau, \tag{19.3}$$

which follow from relations (18.22) and (18.23) and Lemma 18.2, we deduce the following estimate from (18.11):

$$\Big\|\frac{\partial}{\partial y}\Phi^1(y,\psi,\tau,\varepsilon)-\frac{\partial}{\partial y}\Phi^1(\overline{y},\overline{\psi},\tau,\varepsilon)\Big\|$$

$$\leq Knc_1\Big[\frac{c_6}{\gamma_3}+\frac{n}{2\gamma_1}K+nc_5h\frac{1}{\gamma_1+\gamma_2}(1+hd_3)$$

$$+(m+n\overline{c}_1\varepsilon_0^\alpha)\Big(\frac{1}{\gamma_2}+\frac{4}{4\gamma_2-\gamma_1}\Big)\Big](\|y-\overline{y}\|+\|\psi-\overline{\psi}\|)$$

$$\leq \mu(\|y-\overline{y}\|+\|\psi-\overline{\psi}\|).$$

Here, $\mu>0$ is a constant, which will be fixed in what follows. Assume that inequalities (18.25) are satisfied for all $j=\overline{1,s},\ \ s>1$. Then

$$\Big\|\frac{\partial}{\partial y}\Phi^{s+1}(y,\psi,\tau,\varepsilon)-\frac{\partial}{\partial y}\Phi^{s+1}(\overline{y},\overline{\psi},\tau,\varepsilon)\Big\|$$

$$\leq c_1(n+d_2(n+m))\int_\tau^\infty\Big\|\frac{\partial}{\partial y}y_\tau^{\xi,s}(y,\psi,\varepsilon)-\frac{\partial}{\partial y}y_\tau^{\xi,s}(\overline{y},\overline{\psi},\varepsilon)\Big\|d\xi$$

$$+\Big\{c_1(1+d_2+d_3h)K[(1+d_2)(n+m)^2+nd_2]$$

$$+\Big(nc_1\overline{c}_1\varepsilon_0^\alpha+\sup_{\overline{G}}\Big\|\frac{\partial b}{\partial\varphi}\Big\|\Big)K\mu\Big\}\int_\tau^\infty(\|y_\tau^{\xi,s}(y,\psi,\varepsilon)-y_\tau^{\xi,s}(\overline{y},\overline{\psi},\varepsilon)\|$$

$$+\|\psi_\tau^\xi(\psi,\varepsilon)-\psi_\tau^\xi(\overline{\psi},\varepsilon)\|)e^{-\gamma_1(\xi-\tau)}d\xi.$$

Choosing $h>0$ and $\varepsilon_0>0$ so small that $h\mu\leq 1$ and $\varepsilon_0^\alpha\mu\leq 1$ and using Lemma 18.4 and estimates (19.3), we get

$$\Big\|\frac{\partial}{\partial y}\Phi^{s+1}(y,\psi,\tau,\varepsilon)-\frac{\partial}{\partial y}\Phi^{s+1}(\overline{y},\overline{\psi},\tau,\varepsilon)\Big\|$$

$$\leq c_{11}(\|y-\overline{y}\|+\|\psi-\overline{\psi}\|)+\sup_{\overline{G}}\Big\|\frac{\partial b}{\partial\varphi}\Big\|\frac{1}{2\gamma_1}K^2\mu\|y-\overline{y}\|$$

$$+\sup_{\overline{G}}\Big\|\frac{\partial b}{\partial\varphi}\Big\|\frac{1}{\gamma_1}K\mu\|\psi-\overline{\psi}\|, \tag{19.4}$$

where

$$c_{11} = c_1[n + d_2(n+m)]\frac{2}{\gamma_3}c_6$$
$$+ c_1K[(2+d_2)(nd_2 + (1+d_2)(n+m)^2 + n\bar{c}_1)]\Big[\frac{2c_5}{(\gamma_1+\gamma_2)d_3} + \frac{K}{2\gamma_1}\Big]$$
$$+ mc_1K\Big[\frac{2c_5}{\gamma_1+\gamma_2} + \frac{4}{9\gamma_1^2}c_4(4+3\gamma_1)\Big].$$

Since $\gamma_1 > \frac{1}{2}\gamma$ and, according to condition (18.15), the constant

$$\underline{\sigma}_0 = \frac{K}{\gamma}\sup_{\underline{G}}\Big\|\frac{\partial b}{\partial \varphi}\Big\| \max\{2; K\}$$

is less than $1,$ it follows from inequality (19.4) that

$$\Big\|\frac{\partial}{\partial y}\Phi^{s+1}(y,\psi,\tau,\varepsilon) - \frac{\partial}{\partial y}\Phi^{s+1}(\overline{y},\overline{\psi},\tau,\varepsilon)\Big\|$$
$$\le (c_{11} + \mu\underline{\sigma}_0)(\|y-\overline{y}\| + \|\psi - \overline{\psi}\|) \le \mu(\|y-\overline{y}\| + \|\psi-\overline{\psi}\|),$$
$$\mu = \frac{c_{11}}{1-\underline{\sigma}_0}.$$

Thus, we have established inequalities (18.25) for all $j \ge 0.$

By analogy, using Lemma 18.5, for sufficiently small $h > 0$ and $\varepsilon_0 > 0$ one can prove inequalities (18.26) for all $j \ge 0$ with the constant ν independent of ε and $h.$

Consider the problem of the convergence of the sequence $\{\Phi^j(y,\psi,\tau,\varepsilon)\}$ on the set $\underline{G}$. In view of inequalities (18.27) and (18.31), relation (18.11) yields

$$\|\Phi^{j+1}(y,\psi,\tau,\varepsilon) - \Phi^j(y,\psi,\tau,\varepsilon)\|$$
$$\le \Big[(n + (m + nc_1\varepsilon_0^{\alpha})d_2)\frac{c_1}{\gamma_5}\|y\|^2 + \sup_{\underline{G}}\Big\|\frac{\partial b}{\partial\varphi}\Big\|\frac{1}{\gamma_1}K\|y\|\Big] \sup_{\underline{G}} v^{j-1}$$

or

$$\sup_{\underline{G}} v^j \le \Big[\overline{\sigma}_0 + c_1(n+m)d_2\frac{1}{\gamma_5}h\Big] \sup_{\underline{G}} v^{j-1}. \tag{19.5}$$

For small $h > 0,$ the expression in the square brackets on the right-hand side of inequality (19.5) can be estimated by a constant c_{12} less than $1.$ Consequently, the inequalities

$$\sup_{\underline{G}} v^j \leq c_{12} \sup_{\underline{G}} v^{j-1}, \quad c_{12} < 1, \quad \sup_{\underline{G}} v^0 \leq d_1$$

imply that the numerical series $\sum\limits_{j=0}^{\infty} \sup\limits_{\underline{G}} v^j$ is convergent, and, hence, the functional sequence $\{\Phi^j(y, \psi, \tau, \varepsilon)\}$ converges uniformly on the set $\underline{G}$ to the limit function

$$\Phi(y, \psi, \tau, \varepsilon) = \lim_{j \to \infty} \Phi^j(y, \psi, \tau, \varepsilon)$$

continuous in y, ψ, and τ.

By analogy, using estimates (18.28) and (18.29) we prove the uniform convergence of the sequences $\left\{\frac{\partial}{\partial \psi} \Phi^j(y, \psi, \tau, \varepsilon)\right\}$ and $\left\{\frac{\partial}{\partial y} \Phi^j(y, \psi, \tau, \varepsilon)\right\}$ on the set $\underline{G}$.

Consider the sequence $\left\{\frac{\partial}{\partial \tau} \Phi^j(y, \psi, \tau, \varepsilon)\right\}$ each element of which is determined by equality (18.14). The smoothness conditions for the right-hand side of system (18.1) and the uniform convergence of the sequences $\{\Phi^j\}$, $\left\{\frac{\partial}{\partial \psi} \Phi^j\right\}$, and $\left\{\frac{\partial}{\partial y} \Phi^j\right\}$ on the set $\underline{G}$ yield the uniform convergence of the sequence $\left\{\frac{\partial}{\partial \tau} \Phi^j\right\}$ on set (19.1). Therefore, the limit function $\Phi(y, \psi, \tau, \varepsilon)$ is continuous in y, ψ, and τ for every fixed ε from set (19.1). Passing to the limit as $j \to \infty$ in Eq. (18.4), we obtain Eq. (18.10) for all y, ψ, τ, and ε from set (19.1). Since T and Δ are arbitrary, we get Eq. (18.10) for any $(y, \psi, \tau, \varepsilon) \in \underline{G}$. The Lipschitz conditions (18.17) follow from inequalities (18.25) and (18.26). Theorem 18.1 is proved.

20. Investigation of Second-Order Oscillation Systems

Consider a system of weakly connected oscillators with slowly varying parameters of the form

$$\frac{d^2 x_\nu}{dt^2} + \omega_\nu^2(\tau) x_\nu = \varepsilon f_\nu\Big(x, \frac{dx}{dt}, \tau\Big), \tag{20.1}$$

where $\nu = \overline{1, m}$, $m \geq 2$, $\tau = \varepsilon t$ is "slow" time, f_ν are polynomials in $x = (x_1, \ldots, x_m)$ and $\frac{dx}{dt} = \Big(\frac{dx_1}{dt}, \ldots, \frac{dx_m}{dt}\Big)$ of degree not higher than $N \geq 0$ with coefficients $l \geq 1$ times continuously differentiable with respect to $\tau \in [0, L]$, $\omega_\nu(\tau) > 0$ for any $\tau \in [0, L]$ and $\nu = \overline{1, m}$, and ε is a small positive parameter. Systems of the form (20.1) are encountered in numerous problems of nonlinear mechanics [Mit3, Mit4].

In the present section, we apply the results obtained above to the investigation of properties of solutions of Eqs. (20.1). For this purpose, we rewrite these equations in the form of a system:

$$\begin{aligned} \frac{dx_\nu}{dt} &= y_\nu, \\ \frac{dy_\nu}{dt} &= -\omega_\nu^2(\tau)x_\nu + \varepsilon f_\nu(x, y, \tau), \quad \nu = \overline{1, m}. \end{aligned} \tag{20.2}$$

Using the general scheme of the investigation of oscillations [BoM2, Mit2, MiS1–MiS3], we pass to amplitude–phase variables r_ν and φ_ν in Eqs. (20.2) according to the formulas

$$x_\nu = r_\nu \sin\varphi_\nu, \quad y_\nu = r_\nu\omega_\nu(\tau)\cos\varphi_\nu. \quad \nu = \overline{1, m}, \tag{20.3}$$

As a result, we obtain the equations

$$\begin{aligned} \frac{dr_\nu}{d\tau} &= -r_\nu \cos^2\varphi_\nu \frac{d}{d\tau}\ln\omega_\nu(\tau) + \frac{f_\nu \cos\varphi_\nu}{\omega_\nu(\tau)}, \quad \nu = \overline{1, m}, \\ \frac{d\varphi_\nu}{d\tau} &= \frac{\omega_\nu(\tau)}{\varepsilon} + \sin\varphi_\nu\cos\varphi_\nu \frac{d}{d\tau}\ln\omega_\nu(\tau) - \frac{f_\nu \sin\varphi_\nu}{r_\nu\omega_\nu(\tau)}, \end{aligned} \tag{20.4}$$

where

$$\begin{aligned} f_\nu &= f_\nu(r\sin\varphi, r\omega(\tau)\cos\varphi, \tau) \\ &\equiv f_\nu(r_1\sin\varphi_1, \ldots, r_m\sin\varphi_m, r_1\omega_1(\tau)\cos\varphi_1, \ldots, r_m\omega_m(\tau)\cos\varphi_m, \tau). \end{aligned}$$

We construct the following system averaged over all angular variables $\varphi = (\varphi_1, \ldots, \varphi_m)$:

$$\frac{d\overline{r}_\nu}{d\tau} = -\frac{\overline{r}_\nu}{2}\frac{d}{d\tau}\ln\omega_\nu(\tau) + g_\nu(\overline{r}, \tau), \quad \frac{d\overline{\varphi}_\nu}{d\tau} = \frac{\omega_\nu(\tau)}{\varepsilon} + \widetilde{g}_\nu(\overline{r}, \tau)\frac{1}{\overline{r}_\nu}, \tag{20.5}$$

where

$$\begin{aligned} g_\nu &= \frac{(2\pi)^{-m}}{\omega_\nu(\tau)} \int_0^{2\pi} \cdots \int_0^{2\pi} f_\nu(\overline{r}\sin\varphi, \overline{r}\omega(\tau)\cos\varphi, \tau)\cos\varphi_\nu d\varphi_1 \ldots d\varphi_m, \\ \widetilde{g}_\nu &= -\frac{(2\pi)^{-m}}{\omega_\nu(\tau)} \int_0^{2\pi} \cdots \int_0^{2\pi} f_\nu(\overline{r}\sin\varphi, \overline{r}\omega(\tau)\cos\varphi, \tau)\sin\varphi_\nu d\varphi_1 \ldots d\varphi_m. \end{aligned} \tag{20.6}$$

Analyzing relations (20.6), we establish that g_ν and $\widetilde{g}_\nu$ are not identically equal to zero if and only if $f_\nu\Big(x, \dfrac{dx}{dt}, \tau\Big)$ have terms of the form

$$f_\nu^I\Big(x^2, \Big(\frac{dx}{dt}\Big)^2, \tau\Big)\frac{dx_\nu}{dt} + f_\nu^{II}\Big(x^2, \Big(\frac{dx}{dt}\Big)^2, \tau\Big)\, x_\nu,$$

where

$$x^2 = (x_1^2, \dots, x_m^2), \quad \Big(\frac{dx}{dt}\Big)^2 = \Big(\Big(\frac{dx_1}{dt}\Big)^2, \dots, \Big(\frac{dx_m}{dt}\Big)^2\Big).$$

In this case, the function f_ν^I introduces nonzero terms in g_ν, and the function f_ν^{II} introduces nonzero terms in $\widetilde{g}_\nu$. Therefore,

$$g_\nu(\overline{r}, \tau) = \overline{r}_\nu a_\nu(\overline{r}^2, \tau), \quad \widetilde{g}_\nu(\overline{r}, \tau) = \overline{r}_\nu b_\nu(\overline{r}^2, \tau),$$

where $a_\nu(z, \tau)$ and $b_\nu(z, \tau)$ are polynomials in z of degree not higher than $E\Big\{\dfrac{1}{2}N\Big\}$, and $E\{k\}$ is the integer part of the number k. Thus, setting $\overline{r}_\nu^2 = z_\nu$, $\nu = \overline{1, m}$, we can rewrite the averaged system (20.5) in the form

$$\begin{aligned} \frac{dz_\nu}{d\tau} &= \Big[-\frac{d}{d\tau}\ln\omega_\nu(\tau) + 2a_\nu(z, \tau)\Big]\, z_\nu, \\ \frac{d\overline{\varphi}_\nu}{d\tau} &= \frac{\omega_\nu(\tau)}{\varepsilon} + b_\nu(z, \tau), \quad \nu = \overline{1, m}. \end{aligned} \tag{20.7}$$

For Eqs. (20.1), we introduce the initial conditions

$$x_\nu|_{t=0} = x_\nu^0, \quad \frac{dx_\nu}{dt}|_{t=0} = \dot{x}_\nu^0, \quad (x_\nu^0)^2 + (\dot{x}_\nu^0)^2 > 0, \quad \nu = \overline{1, m}. \tag{20.8}$$

Then the corresponding initial conditions for the amplitude–phase variables take the form

$$z_\nu|_{\tau=0} = z_\nu^0, \quad \overline{\varphi}_\nu|_{\tau=0} = \varphi_\nu^0, \quad \nu = \overline{1, m}, \tag{20.9}$$

where $z_\nu^0 = (x_\nu^0)^2 + \Big(\dfrac{1}{\omega_\nu(0)}\dot{x}_\nu^0\Big)^2 > 0$ and φ_ν^0 is one of solutions of the system of equations

$$x_\nu^0 = \sqrt{z_\nu^0}\sin\varphi_\nu^0, \quad \dot{x}_\nu^0 = \sqrt{z_\nu^0}\,\omega_\nu(0)\cos\varphi_\nu^0. \tag{20.10}$$

Note that the averaged Cauchy problem (20.7), (20.9) decomposes into the following two problems:

$$\frac{dz_\nu}{d\tau} = \Big[-\frac{d}{d\tau}\ln\omega_\nu(\tau) + 2a_\nu(z, \tau)\Big] z_\nu, \quad z_\nu|_{\tau=0} = z_\nu^0, \quad \nu = \overline{1, m}, \tag{20.11}$$

$$\frac{d\overline{\varphi}_\nu}{d\tau} = \frac{\omega_\nu(\tau)}{\varepsilon} + b_\nu(z, \tau), \quad \varphi_\nu|_{\tau=0} = \varphi_\nu^0, \quad \nu = \overline{1, m}. \tag{20.12}$$

If

$$z(\tau, z^0) = (z_1(\tau, z^0), \dots, z_m(\tau, z^0)), \quad z^0 = (z_1^0, \dots, z_m^0),$$

is a solution of the Cauchy problem (20.11), then a solution of the Cauchy problem (20.12) is given by the formula

$$\overline{\varphi}_\nu(\tau, z^0, \varphi_\nu^0, \varepsilon) = \varphi_\nu^0 + \frac{1}{\varepsilon}\int\limits_0^\tau [\omega_\nu(\tau) + \varepsilon b(z(\tau, z^0), \tau)]d\tau, \quad \nu = \overline{1, m}.$$

Theorem 20.1. *Suppose that the following conditions are satisfied:*

(i) $\omega(\tau) \in C_{[0,L]}^{p-1}, \ p \geq m;$

(ii) $\det(W_p^T(\tau)W_p(\tau)) \neq 0 \ \ \forall\, \tau \in [0, L];$

(iii) there exists a solution $z = z(\tau, z^0)$ *of problem (20.11) defined on* $[0, L]$.

Then one can find constants c_1 *and* $\varepsilon_0 > 0$ *such that a solution* $x = x(t, \varepsilon)$ *of the Cauchy problem (20.1), (20.8) is defined for all* $t \in [0, L\varepsilon^{-1}]$ *and* $\varepsilon \in (0, \varepsilon_0]$ *and*

$$\begin{aligned} &|x_\nu(t, \varepsilon) - \sqrt{z_\nu(\varepsilon t, z^0)} \sin \overline{\varphi}_\nu(\varepsilon t, z^0, \varphi_\nu^0, \varepsilon)| \\ &\qquad + \left|\frac{dx_\nu(t, \varepsilon)}{dt} - \sqrt{z_\nu(\varepsilon t, z^0)}\omega_\nu(\varepsilon t) \cos \overline{\varphi}_\nu(\varepsilon t, z^0, \varphi_\nu^0, \varepsilon)\right| \\ &\quad \leq c_1 \varepsilon^{\frac{1}{p}}, \qquad \nu = \overline{1, m}. \end{aligned} \tag{20.13}$$

Proof. It is obvious that each component $z_\nu(\tau, z_0)$ of the solution $z = z(\tau, z^0)$ of the Cauchy problem (20.11) does not vanish on the segment $[0, L]$. Denote

$$\min_{1 \leq \nu \leq m} \ \min_{\tau \in [0, L]} z_\nu(\tau, z^0) = 2\rho > 0,$$

$$\max_{1 \leq \nu \leq m} \ \max_{\tau \in [0, L]} z_\nu(\tau, z^0) = \Delta.$$

This implies that the curve $z = z(\tau, z^0)$ lies in the cube

$$\Pi = \{z \colon z \in R^m, \rho \leq z_\nu \leq \Delta + \rho, \nu = \overline{1, m}\}$$

together with its ρ-neighborhood for any $\tau \in [0, L]$. Moreover, the right-hand side of system (20.4) satisfies all conditions of Theorem 2.1. Therefore, the solution $(r(\tau, \sqrt{z^0}, \varphi^0, \varepsilon); \varphi(\tau, \sqrt{z^0}, \varphi^0, \varepsilon))$ of the Cauchy problem (20.4), (20.9) is defined for all $\tau \in [0, L]$ and $\varepsilon \in (0, \varepsilon_0]$ ($\varepsilon_0 > 0$ is sufficiently small) and satisfies the estimates

$$|r_\nu(\tau, \sqrt{z^0}, \varphi^0, \varepsilon) - \sqrt{z^\nu(\tau, z^0)}| + |\varphi_\nu(\tau, \sqrt{z^0}, \varphi^0, \varepsilon) - \overline{\varphi}_\nu(\tau, z^0, \varphi_\nu^0, \varepsilon)|$$
$$\leq c\varepsilon^{\frac{1}{p}}, \qquad \nu = \overline{1, m}, \tag{20.14}$$

where c is a certain constant independent of ε. Relations (20.3) and (20.14) yield inequality (20.13) with

$$c_1 = c(1 + \Delta)(1 + \max_{1 \leq \nu \leq m} \max_{\tau \in [0, L]} \omega_\nu(\tau)).$$

Theorem 20.1 is proved.

For Eqs. (20.1), we now introduce boundary conditions of the form

$$x_\nu|_{t=t_\nu} = x_\nu^0, \quad \frac{dx_\nu}{dt}|_{t=t_\nu} = \dot{x}_\nu^0, \quad \nu = \overline{1, m}, \tag{20.15}$$

where $0 \leq t_1 < t_2 < \ldots < t_m \leq L\varepsilon^{-1}$ and $(x_\nu^0)^2 + (\dot{x}_\nu^0)^2 > 0$. Problem (20.1), (20.15) may arise in the case where, in the course of the investigation of properties of a system of oscillators, one can determine the values of x and $\dfrac{dx}{dt}$ at given time for only one oscillator (e.g., the number of measuring devices is insufficient).

In the variables r and φ, conditions (20.15) can be rewritten as follows:

$$r_\nu|_{\tau=\tau_\nu} = \sqrt{z_\nu^0}, \quad \varphi_\nu|_{\tau=\tau_\nu} = \varphi_\nu^0, \quad \nu = \overline{1, m}, \tag{20.16}$$

where $\tau_\nu = \varepsilon t_\nu$, and z_ν^0 and φ_ν^0 have the same meaning as in (20.9).

The multipoint problem (20.1), (20.15) generates the averaged problem

$$\frac{dz_\nu}{d\tau} = \Big[- \frac{d}{d\tau} \ln \omega_\nu(\tau) + 2a_\nu(z, \tau) \Big] z_\nu, \quad z_\nu|_{\tau=\tau_\nu} = z_\nu^0, \quad \nu = \overline{1, m}, \tag{20.17}$$

$$\frac{d\overline{\varphi}_\nu}{d\tau} = \frac{\omega_\nu(\tau)}{\varepsilon} + b_\nu(z, \tau), \quad \varphi_\nu|_{\tau=\tau_\nu} = \varphi_\nu^0, \quad \nu = \overline{1, m}, \tag{20.18}$$

which is obviously much simpler than problem (20.4), (20.16). Indeed, if a solution $z = z(\tau, y^0)$, $z(0, y^0) = y^0$ of problem (20.17) is obtained, then problem

(20.18) decomposes into m mutually independent Cauchy problems whose solutions $\overline{\varphi}_\nu = \overline{\varphi}_\nu(\tau, y^0, \psi^0_\nu, \varepsilon)$, $\overline{\varphi}_\nu(0, y^0, \psi^0_\nu, \varepsilon) = \psi^0_\nu$ are determined as follows:

$$\overline{\varphi}_\nu(\tau, y^0, \psi^0_\nu, \varepsilon) = \psi^0_\nu + \frac{1}{\varepsilon}\int\limits_0^\tau [\omega_\nu(\tau) + \varepsilon b_\nu(z(\tau, y^0), \tau)]d\tau,$$

where

$$\psi^0_\nu = \varphi^0_\nu - \frac{1}{\varepsilon}\int\limits_0^{\tau_\nu} [\omega_\nu(\tau) + \varepsilon b_\nu(z(\tau, y^0), \tau)]d\tau.$$

Theorem 20.2. *Suppose that conditions (i) and (ii) of Theorem 20.1 are satisfied and there exists a solution* $z = z(\tau, y^0)$, $y^0 = (y^0_1, \ldots, y^0_m)$, *of problem (20.17) defined on* $[0, L]$ *and such that the matrix*

$$A = \left(\frac{\partial z_\nu(\tau_\nu, y^0)}{\partial y^0_\mu}\right)^m_{\nu,\mu=1}$$

is nondegenerate. Then, for every $\varepsilon \in (0, \varepsilon_1]$, *where* $\varepsilon_1 > 0$ *is sufficiently small, there exists a unique solution* $x = x(t, \varepsilon)$ *of problem (20.1), (20.15) defined for all* $t \in [0, L\varepsilon^{-1}]$ *and such that*

$$\begin{aligned} &|x_\nu(t,\varepsilon) - \sqrt{z_\nu(\varepsilon t, y^0)}\sin\overline{\varphi}_\nu(\varepsilon t, y^0, \psi^0_\nu, \varepsilon)| \\ &\quad + \left|\frac{dx_\nu(t,\varepsilon)}{dt} - \sqrt{z_\nu(\varepsilon t, y^0)}\omega_\nu(\varepsilon t)\cos\overline{\varphi}_\nu(\varepsilon t, y^0, \psi^0_\nu, \varepsilon)\right| \\ &\qquad \le \overline{c}_1 \varepsilon^{\frac{1}{p}}, \qquad \nu = \overline{1, m}, \end{aligned} \tag{20.19}$$

where $\overline{c}_1$ *is a certain constant independent of* ε.

Proof. By analogy with the proof of Theorem 20.1, we establish that the curve $z = z(\tau, y^0)$ lies in the cube Π together with its ρ-neighborhood. Further, we use Theorem 8.2. The matrix S defined by (8.20) for problem (20.17), (20.18) has the form

$$S = \begin{pmatrix} A & 0 \\ B & E_m \end{pmatrix},$$

where E_m is the m-dimensional identity matrix, 0 is the zero matrix, and

$$B = \left(\int\limits_0^{\tau_\nu} \frac{\partial b_\nu(z(\tau, y^0), \tau)}{\partial z}\frac{\partial z(\tau, y^0)}{\partial y^0_\mu}d\tau\right)^m_{\nu,\mu=1}.$$

Since $\det S = \det A \neq 0$, i.e., $\|S^{-1}\| \leq c_2 = \text{const}$, we conclude that, according to Theorem 8.2, for every $\varepsilon \in (0, \varepsilon_1]$, where ε_1 is sufficiently small, there exists a unique solution $(r(\tau, \varepsilon); \varphi(\tau, \varepsilon))$ of problem (20.4), (20.16) that satisfies the inequality

$$\sum_{\nu=1}^{m} [|r_\nu(\tau, \varepsilon) - \sqrt{z_\nu(\tau, y^0)}| + |\varphi_\nu(\tau, \varepsilon) - \overline{\varphi}_\nu(\tau, y^0, \psi_\nu^0, \varepsilon)|] \leq \overline{c}_2 \varepsilon^{\frac{1}{p}}$$

for all $\tau \in [0, L]$ and $\varepsilon \in (0, \varepsilon_1]$. Taking into account formulas (20.3) and the last inequality, we get estimate (20.19). Theorem 20.2 is proved.

Corollary 6. *If the polynomials* $f_\nu\Big(x, \dfrac{dx}{dt}, \tau\Big)$ *do not contain terms of the form* $f_\nu'\Big(x^2, \Big(\dfrac{dx}{dt}\Big)^2, \tau\Big)\dfrac{dx_\nu}{dt}$, *i.e.,* $a_\nu(z, \tau) \equiv 0 \ \ \forall \nu = \overline{1, m}$, *then problem (20.17) decomposes into* m *Cauchy problems whose solutions are given by the formulas*

$$z_\nu(\tau, y^0) = \frac{y_\nu^0 \omega_\nu(0)}{\omega_\nu(\tau)}, \quad y_\nu^0 = \frac{z_\nu^0 \omega_\nu(\tau_\nu)}{\omega_\nu(0)}, \quad \nu = \overline{1, m}.$$

In this case, we have

$$\det A = \det \operatorname{diag}\Big(\frac{\omega_1(0)}{\omega_1(\tau_1)}, \ldots, \frac{\omega_m(0)}{\omega_m(\tau_m)}\Big) = \prod_{\nu=1}^{m} \frac{\omega_\nu(0)}{\omega_\nu(\tau_\nu)} \neq 0.$$

We have considered above one of the simplest versions of boundary conditions. Note that Theorem 8.2 can also be used in the case where conditions (20.15) are replaced by the more general conditions

$$\Phi\Big(x|_{t=t_1}, \frac{dx}{dt}|_{t=t_1}, \ldots, x|_{t=t_k}, \frac{dx}{dt}|_{t=t_k}, \varepsilon\Big) = 0,$$

where $0 \leq t_1 < t_2 < \ldots < t_k \leq L\varepsilon^{-1}$, $k \geq 2$, and Φ is a $2m$-dimensional vector function.

Now assume that the following conditions are satisfied:

(1°) the coefficients of the polynomials $f_\nu\Big(x, \dfrac{dx}{dt}, \tau\Big)$ are defined and bounded together with their derivatives with respect to τ up to an order $l \geq 2$ for all $\tau \in R$;

(2°) the frequencies $\omega_\nu(\tau)$, $\nu = \overline{1,m}$, and their derivatives up to an order $p \geq m$ are uniformly bounded for all $\tau \in R$ and

$$\det\left(W_p^T(\tau)W_p(\tau)\right) \geq c_3 = \text{const} > 0, \quad \omega_\nu(\tau) \geq c_4 = \text{const} > 0;$$

(3°) the averaged equations (20.5) for the slow variables $\overline{r}$ have a bounded solution $\overline{r} = \xi(\tau) = (\xi_1(\tau), \dots, \xi_m(\tau))$ defined on the entire axis and such that $\xi_\nu(\tau) \geq 2\rho = \text{const} > 0$, $\nu = \overline{1,m}$;

(4°) the normal fundamental matrix $Q(\tau, \overline{\tau})$ of the variational system corresponding to the solution $\overline{r} = \xi(\tau)$ satisfies the estimate

$$\|Q(\tau, \overline{\tau})\| \leq K e^{-\gamma(\tau - \overline{\tau})} \quad \forall \tau \geq \overline{\tau} \in R,$$

where $K \geq 1$ and $\gamma > 0$ are certain constants;

(5°) the following inequality is true:

$$\frac{2}{\gamma} K \sup_{\varphi,\tau} \left\| \frac{\partial}{\partial r} \widetilde{a}(\xi(\tau), \varphi, \tau) \right\| < 1,$$

where

$$\widetilde{a}(r, \varphi, \tau) = (\widetilde{a}_1(r, \varphi, \tau), \dots, \widetilde{a}_m(r, \varphi, \tau)),$$

$$\widetilde{a}_\nu(r, \varphi, \tau) = -r_\nu \left(\cos^2 \varphi_\nu - \frac{1}{2} \right) \frac{d}{d\tau} \ln \omega_\nu(\tau) + \frac{1}{\omega_\nu(\tau)} f_\nu(r \sin \varphi, r\omega(\tau) \cos \varphi, \tau) - g_\nu(r, \tau).$$

Under these conditions, in Sections 12–17 we have proved the existence and studied properties of the asymptotically stable integral manifold $r = R(\psi, \tau, \varepsilon) = (R_1(\psi, \tau, \varepsilon), \dots, R_m(\psi, \tau, \varepsilon))$ of system (20.4), on which the equations for the fast variables φ_ν, $\nu = \overline{1,m}$, have the form

$$\frac{d\varphi_\nu}{d\tau} = \frac{1}{\varepsilon} \omega_\nu(\tau) + \sin \varphi_\nu \cos \varphi_\nu \frac{d}{d\tau} \ln \omega_\nu(\tau) - \frac{\sin \varphi_\nu}{R_\nu(\varphi, \tau, \varepsilon)\omega_\nu(\tau)} f_\nu(R(\varphi, \tau, \varepsilon) \sin \varphi, R(\varphi, \tau, \varepsilon)\omega(\tau) \cos \varphi, \tau). \quad (20.20)$$

It follows from the definition of integral manifold [MiLy] that if $\varphi = \varphi_{\tau_0}^{\tau}(\psi, \varepsilon)$, $\varphi_{\tau_0}^{\tau_0}(\psi, \varepsilon) = \psi \in R^m$, is a solution of Eqs. (20.20), then

$$r = R(\varphi_{\tau_0}^{\tau}(\psi, \varepsilon), \tau, \varepsilon), \quad \varphi = \varphi_{\tau_0}^{\tau}(\psi, \varepsilon) \tag{20.21}$$

is a solution of system (20.4) for all $\tau \in R$. In this case, in view of formula (20.3), we have the bounded solution

$$x = R(\varphi_{\tau_0}^{\tau}(\psi, \varepsilon), \tau, \varepsilon) \sin \varphi_{\tau_0}^{\tau}(\psi, \varepsilon),$$

$$y = R(\varphi_{\tau_0}^{\tau}(\psi, \varepsilon), \tau, \varepsilon)\omega(\tau) \cos \varphi_{\tau_0}^{\tau}(\psi, \varepsilon)$$

of system (20.2), which is defined for $\tau \in R$, $\psi \in R^m$, and $\varepsilon \in (0, \varepsilon_2]$, where $\varepsilon_2 > 0$ is sufficiently small. Thus, in the $(2m+2)$-dimensional space of variables x, y, τ, and ε, the relation

$$(x; y) = \Gamma(\psi, \tau, \varepsilon) \equiv (R(\psi, \tau, \varepsilon) \sin \psi; R(\psi, \tau, \varepsilon)\omega(\tau) \cos \psi) \tag{20.22}$$

is the equation of a surface that possesses the following property: if $(x^0; \dot{x}^0) \in \Gamma(\psi, \tau_0, \varepsilon)$, then the solution $(x_{t_0}^t(x^0, \dot{x}^0, \varepsilon); y_{t_0}^t(x^0, \dot{x}^0, \varepsilon))$ of the Cauchy problem

$$\frac{dx_\nu}{dt} = y_\nu, \quad \frac{dy_\nu}{dt} = -\omega_\nu^2(\tau)x_\nu + \varepsilon f_\nu(x, y, \tau), \quad \nu = \overline{1, m},$$

$$x_\nu|_{t=t_0} = x^0, \quad y_\nu|_{t=t_0} = \dot{x}^0, \quad t_0 = \tau_0\varepsilon^{-1}, \tag{20.23}$$

is defined for all $t \in R$ and $\varepsilon \in (0, \varepsilon_2]$, bounded, and lying on the surface Γ.

The asymptotic stability of the integral manifold $r = R(\psi, \tau, \varepsilon)$ of system (20.4) means (see Theorem 15.1) that if $r|_{\tau=\tau_0} = r^0$ lies in a certain small neighborhood of the point $R(\psi, \tau_0, \varepsilon)$, then, as $\tau \to \infty$, the slow variables $r_{\tau_0}^{\tau}(r^0, \psi^0, \varepsilon)$ of every solution

$$(r_{\tau_0}^{\tau}(r^0, \psi^0, \varepsilon); \varphi_{\tau_0}^{\tau}(r^0, \psi^0, \varepsilon)),$$

$$r_{\tau_0}^{\tau_0}(r^0, \psi^0, \varepsilon) = r^0; \varphi_{\tau_0}^{\tau_0}(r^0, \psi^0, \varepsilon) = \psi^0, \quad \psi^0 \in R^m,$$

of system (20.4) tend exponentially to the curve $r = R(\varphi_{\tau_0}^{\tau}(r^0, \psi^0, \varepsilon), \tau, \varepsilon)$, which lies on the surface $r = R(\psi, \tau, \varepsilon)$. Taking into account formula (20.3), we establish that, as $t \to \infty$, every solution $(x_{t_0}^t(x^0, \dot{x}^0, \varepsilon); y_{t_0}^t(x^0, \dot{x}^0, \varepsilon))$ of the Cauchy problem (20.23) tends exponentially to the curve

$$(x; y) = (R(\varphi_{\tau_0}^{\tau}(r^0, \psi^0, \varepsilon), \tau, \varepsilon) \sin \varphi_{\tau_0}^{\tau}(r^0, \psi^0, \varepsilon),$$

$$R(\varphi_{\tau_0}^{\tau}(r^0, \psi^0, \varepsilon), \tau, \varepsilon)\omega(\tau) \cos \varphi_{\tau_0}^{\tau}(r^0, \psi^0, \varepsilon), \quad \tau = \varepsilon t,$$

which lies on the surface Γ under the condition that the point $(x^0, \dot{x}^0)$ lies in a small neighborhood of the point $\Gamma(\psi^0, \tau_0, \varepsilon)$. Here,

$$r^0 = (r_1^0, \ldots, r_m^0), \quad \psi^0 = (\psi_1^0, \ldots, \psi_m^0),$$

$$r_\nu^0 = \sqrt{(x_\nu^0)^2 + (\dot{x}_\nu^0 \omega_\nu^{-1}(\tau_0))^2},$$

and ψ_ν^0 is one of solutions of the system of equations

$$x_\nu^0 = r_\nu^0 \sin \psi_\nu^0, \quad \dot{x}_\nu^0 = r_\nu^0 \omega_\nu(\tau_0) \cos \psi_\nu^0.$$

This reasoning enables us to apply Theorems 15.1 and 16.3 to system (20.2) and obtain the following corollary of these statements:

Theorem 20.3. *If conditions (1°)–(5°) are satisfied, then one can find a sufficiently small* $\varepsilon_2 > 0$ *and a sufficiently large constant* c_5 *such that, for all* $(\tau, \varepsilon) \in R \times (0, \varepsilon_2]$, *there exists the asymptotically stable integral manifold* $(x; y) = \Gamma(\psi, \tau, \varepsilon)$ *of Eqs. (20.2) for which the function* Γ *is* 2π*-periodic in* ψ, $k = \min\{p; l\} - 1$ *times continuously differentiable with respect to* $(\psi, \tau, \varepsilon) \in R^m \times R \times (0, \varepsilon_2]$, *and such that*

$$\left\| D_\psi^s \frac{\partial^q}{\partial \tau^q} \frac{\partial^{\bar{q}}}{\partial \varepsilon^{\bar{q}}} (\Gamma(\psi, \tau, \varepsilon) - \widetilde{\Gamma}(\psi, \tau)) \right\| \le c_5 \varepsilon^{\frac{1}{p} - q - 2\bar{q}},$$

where $0 \le s + q + \bar{q} \le k$ *and* $\widetilde{\Gamma}(\psi, \tau) = (\xi(\tau) \sin \psi; \xi(\tau)\omega(\tau) \cos \psi)$.

21. Weakening of Conditions in the Theorem on Integral Manifold

In this section, we return to the problem on the integral manifold of the oscillation system

$$\begin{aligned} \frac{dx}{d\tau} &= a(x, \tau) + \widetilde{a}(x, \varphi, \tau) + \varepsilon A(x, \varphi, \tau, \varepsilon), \\ \frac{d\varphi}{d\tau} &= \frac{\omega(\tau)}{\varepsilon} + b(x, \varphi, \tau, \varepsilon). \end{aligned} \tag{21.1}$$

Note that, in Sections 12–14, we have proved the existence and established properties of the integral manifold $x = X(\varphi, \tau, \varepsilon)$ in the case where the right-hand sides of Eqs. (21.1) are twice continuously differentiable with respect to x, φ, and τ and the norm of the matrix $P = \dfrac{\partial}{\partial x}\widetilde{a}(x,\varphi,\tau)$ is sufficiently small [inequality (13.3)]. In what follows, we omit the condition that $\|P\|$ is small and study analogous problems. Note that the method proposed here requires an increase in the smoothness order by one and certain additional restrictions on the Fourier coefficients of the function $\widetilde{a}(x,\varphi,\tau)$.

Assume that

$$[a,b,A] \in C^1_\tau(\overline{G},\sigma_1) \cap C^2_{x,\tau}(\overline{G},\sigma_1),$$

$$\frac{\partial a}{\partial x} \in C^1_\tau(\overline{G},\sigma_1), \widetilde{a} \in C^3_{x,\tau}(\overline{G},\sigma_1),$$

$$\sum_{k\neq 0}\Big[\|k\|^2 \sup\|b_k\| + \|k\|\Big(\sup\Big\|\frac{\partial b_k}{\partial \tau}\Big\| + \sup\Big\|\frac{\partial b_k}{\partial x}\Big\|\Big)\Big] \leq \sigma_1,$$

$$\sum_{k\neq 0}\Big[\|k\|^2 \sup\|a_k\| + \|k\|\Big(\sup\Big\|\frac{\partial a_k}{\partial \tau}\Big\| + \sup\Big\|\frac{\partial a_k}{\partial x}\Big\|\Big)$$
$$+ \sup\Big\|\frac{\partial^2 a_k}{\partial x \partial \tau}\Big\| + \sum_{j=1}^{n}\sup\Big\|\frac{\partial^2 a_k}{\partial x \partial x_j}\Big\|$$
$$+ \frac{1}{\|k\|}\Big(\sup\Big\|\frac{\partial^2 a_k}{\partial \tau^2}\Big\| + \sum_{j=1}^{n}\sup\Big\|\frac{\partial^3 a_k}{\partial x \partial x_j \partial \tau}\Big\|$$
$$+ \sum_{j,s=1}^{n}\sup\Big\|\frac{\partial^3 a_k}{\partial x \partial x_j \partial x_s}\Big\|\Big)\Big] \leq \sigma_1. \tag{21.2}$$

Here, the supremum is taken over all $(x,\varphi,\tau,\varepsilon) \in \overline{G}$. Note that the notation used in this section is the same as in Sections 12–14.

Assume that the components $\omega_\nu(\tau)$, $\nu = \overline{1,m}$, of the frequency vector $\omega(\tau)$ and their derivatives with respect to τ up to an order $p-1$ $(p \geq m)$ are uniformly continuous on the entire axis and

$$\|(W_p^T(\tau)W_p(\tau))^{-1}W_p^T(\tau)\| \leq \sigma_2, \quad \|\omega(\tau)\| + \Big\|\frac{d}{d\tau}\omega(\tau)\Big\| \leq \sigma_2, \tag{21.3}$$

where, as above, $W_p(\tau)$ and $W_p^T(\tau)$ denote the matrix $\Big(\dfrac{d^{s-1}}{d\tau^{s-1}}\omega_\nu(\tau)\Big)_{\nu,s=1}^{m,p}$ and its transpose, respectively, and σ_2 is a constant.

Consider the system of equations of the first approximation for slow variables averaged over all angular variables φ, namely

$$\frac{d\overline{x}}{d\tau} = a(\overline{x}, \tau),$$

and assume that there exists its solution $\overline{x} = \overline{x}(\tau)$ defined for all $\tau \in R$ and such that $\overline{x}(\tau) \in \mathcal{D}_\rho$ for certain $\rho > 0$ ($\mathcal{D}_\rho$ is the set of points that belong to the bounded domain $\mathcal{D}$ together with their ρ-neighborhoods). Assume that the variational system

$$\frac{dz}{d\tau} = H(\tau)z, \quad H(\tau) = \frac{\partial}{\partial x} a(\overline{x}(\tau), \tau),$$

is hyperbolic and the Green matrix $Q(\tau, t)$ satisfies the inequality

$$\|Q(\tau, t)\| \leq K e^{-\gamma|\tau - t|} \quad \forall \tau, t \in R, \tag{21.4}$$

where $\gamma > 0$ and $K \geq 1$ are certain constants.

In system (21.1), we set

$$x = z + \varepsilon u(z, \varphi, \tau, \mu), \quad u = \sum_{k \neq 0} \frac{1 - h_\mu((\overline{k}, \omega(\tau)))}{i(\overline{k}, \omega(\tau))} a_k(z, \tau) e^{i(k, \varphi)}, \tag{21.5}$$

where i is the imaginary unit, $\overline{k} = \dfrac{k}{\|k\|}$,

$$h_\mu(t) = \int_{-\infty}^{\infty} \nu_{2\mu}(l) \omega_\mu(t - l) dl,$$

$\nu_{2\mu}(l) \equiv 1$ for $|l| \leq 2\mu$, $\nu_{2\mu}(l) \equiv 0$ for $|l| > 2\mu$, and $\omega_\mu(l)$ is the averaging kernel [Mik], namely

$$\omega_\mu(l) = \begin{cases} 0, & |l| \geq \mu, \\ \dfrac{1}{\mu} \sigma_3 e^{-\frac{\mu^2}{\mu^2 - l^2}}, & |l| < \mu, \end{cases} \qquad \sigma_3 = \int_{-1}^{1} e^{-\frac{1}{1 - l^2}} dl.$$

We fix the averaging radius $\mu < 1$ in what follows. The function $h_\mu(t)$ thus constructed is infinitely differentiable for all $t \in R$ and finite, $0 \leq h_\mu(t) \leq 1$

for any $t \in R$, $h_\mu(t) \equiv 1$ for $|t| \le \mu$, $h_\mu(t) \equiv 0$ for $|t| > 3\mu$, and, for all integer $q \ge 0$, the following estimates are satisfied:

$$\left| \frac{d^q}{dt^q} h_\mu(t) \right| \le c_q t^{-q} \overline{h}_\mu(t), \tag{21.6}$$

where c_q are constants, $\overline{h}_\mu(t) \equiv 0$ for $|t| \le \mu$ and $|t| \ge 3\mu$, and $\overline{h}_\mu(t) \equiv 1$ for $\mu < |t| < 3\mu$.

If a positive ε_0 is sufficiently small, then, for all $\varepsilon \in (0, \varepsilon_0]$, the change of variables (21.5) reduces system (21.1) to the form

$$\frac{dz}{d\tau} = a(z, \tau) + \delta(z, \varphi, \tau, \mu) + \varepsilon v(z, \varphi, \tau, \mu) + \varepsilon A_1(z, \varphi, \tau, \varepsilon, \mu),$$

$$\frac{d\varphi}{d\tau} = \frac{\omega(\tau)}{\varepsilon} + \widetilde{b}(z, \varphi, \tau, \varepsilon, \mu), \tag{21.7}$$

where

$$\widetilde{b} = b(z + \varepsilon u, \varphi, \tau, \varepsilon), \quad v = -\frac{\partial u}{\partial \tau}, \quad u = u(z, \varphi, \tau, \mu),$$

$$\delta = \sum_{k \neq 0} a_k(z, \tau) h_\mu((\overline{k}, \omega(\tau))) e^{i(k,\varphi)},$$

$$A_1 = B - \frac{\partial u}{\partial z}\left(E_n + \varepsilon \frac{\partial u}{\partial z}\right)^{-1} (a(z, \tau) + \delta + \varepsilon v + \varepsilon B),$$

$$B = A(z + \varepsilon u, \varphi, \tau, \varepsilon) - \frac{\partial u}{\partial \varphi}\widetilde{b} + \frac{1}{\varepsilon}[a(z + \varepsilon u, \tau)$$

$$- a(z, \tau) + \widetilde{a}(z + \varepsilon u, \varphi, \tau) - \widetilde{a}(z, \varphi, \tau)].$$

Taking conditions (21.2), (21.3), and (21.6) into account, one can easily establish the existence of a constant σ_4 such that

$$\|u\| + \left\|\frac{\partial u}{\partial z}\right\| + \left\|\frac{\partial u}{\partial \varphi}\right\| \le \frac{\sigma_4}{\mu}, \quad \|v\| \le \frac{\sigma_4}{\mu^2}, \quad \|A_1\| \le \frac{\sigma_4}{\mu}(1 + \varepsilon\|v\|) \tag{21.8}$$

for all $(z, \varphi, \tau, \varepsilon) \in \mathcal{D}_{\frac{1}{2}\rho} \times R^m \times R \times (0, \varepsilon_0]$. Note that the restriction $z \in \mathcal{D}_{\frac{1}{2}\rho}$ and the inequality $\sigma_4 \varepsilon \mu^{-1} < \frac{1}{2}\rho$ guarantee that the point $z + \varepsilon u$ belongs to the domain $\mathcal{D}$.

Lemma 21.1. *Suppose that* $f(t) = (f_1(t), \ldots, f_m(t)) \in \mathcal{D}$ *and* $\theta(t) = (\theta_1(t), \ldots, \theta_m(t)) \in R^m$ *are arbitrary continuous (for* $t \in R$*) functions and conditions (21.2)–(21.4) are satisfied. Then there exist constants* σ_5 *and* σ_6 *independent of* μ *and such that the following estimates hold for all* $\tau \in R$:

$$\Big\| \int_{-\infty}^{\infty} Q(\tau, t)\delta(f(t), \theta(t), t, \mu)dt \Big\| \le \sigma_5 \mu^{\frac{1}{p-1}}, \tag{21.9}$$

$$\Big\| \int_{-\infty}^{\infty} Q(\tau, t)v(f(t), \theta(t), t, \mu)\, dt \Big\| \le \frac{\sigma_6}{\mu}. \tag{21.10}$$

Proof. According to condition (21.3) and the results of Section 1, for arbitrary $\tau \in R$ there exist $\Delta > 0$ independent of τ and $\overline{k}$ and an integer $r = r(\tau, \overline{k}) \in [0, p-1]$ such that the following inequality holds for all $t \in [\tau - \Delta, \tau + \Delta]$:

$$\Big| \frac{d^r}{dt^r}(\overline{k}, \omega(t)) \Big| \ge \frac{1}{2p\sigma_2}. \tag{21.11}$$

If $r \ge 1$, then the last inequality implies that the functions $(\overline{k}, \omega(t)) + c$ and $\dfrac{d}{dt}(\overline{k}, \omega(t))$ $(c = \text{const})$ can take the zero value on the segment $[\tau - \Delta, \tau + \Delta]$ at at most 2^{p-1} points. Moreover, $[\tau - \Delta, \tau + \Delta]$ can be divided into two sets $M(\tau)$ and $N(\tau)$ of segments such that $M(\tau)$ consists of $l_1 \le 2^{p-1} - 1$ segments whose lengths do not exceed $2\overline{\mu} = \text{const}$, and $N(\tau)$ consists of $l_2 \le 2^{p-1}$ segments on each of which the following inequality is satisfied:

$$|(\overline{k}, \omega(t))| \ge \frac{1}{2p\sigma_2}\, \overline{\mu}^{p-1}. \tag{21.12}$$

First, we prove estimate (21.9). We have

$$\Big\| \int_{-\infty}^{\infty} Q\delta dt \Big\| \le \sum_{s=-\infty}^{\infty} \sum_{k \ne 0} \Big\| \int_{\tau+2s\Delta}^{\tau+2(s+1)\Delta} Q a_k h_\mu((\overline{k}, \omega(t)))\, dt \Big\|$$

$$\le K \sum_{s=-\infty}^{\infty} e^{-2|s|\gamma\Delta} \sum_{k \ne 0} \sup_{\overline{G}} \|a_k\| \Bigg(\int_{M(\tau+(2s+1)\Delta)} h_\mu((\overline{k}, \omega(t)))dt$$

$$+ \int_{N(\tau+(2s+1)\Delta)} h_\mu((\overline{k}, \omega(t)))dt \Bigg). \tag{21.13}$$

We set $\overline{\mu}^{p-1} = 7p\sigma_2\mu$. It follows from inequality (21.12) and the definition of the function $h_\mu(t)$ that $h_\mu((\overline{k},\omega(t))) \equiv 0$ on the set $N(\tau+(2s+1)\Delta)$; therefore,

$$\Big\| \int_{-\infty}^{\infty} Q\delta dt \Big\| \leq 2K \sum_{s=0}^{\infty} e^{-2s\gamma\Delta} \frac{2^p}{\overline{\mu}} \sum_{k\neq 0} \sup_{\overline{G}} \|a_k\| \leq \frac{2^{p+1}\sigma_1 K}{1-e^{-2\gamma\Delta}}\overline{\mu}.$$

This yields estimate (21.9) with the constant

$$\sigma_5 = 2^{p+1} K\sigma_1 \frac{1}{1-e^{-2\gamma\Delta}} (7p\sigma_2)^{\frac{1}{p-1}}.$$

Let us prove estimate (21.10). Taking into account the definition of the function v and relation (21.6) for $q=1$, we get

$$\Big\| \int_{-\infty}^{\infty} Qv dt \Big\| \leq K \sum_{s=-\infty}^{\infty} e^{-2|s|\gamma\Delta} \sum_{k\neq 0} \frac{1}{\|k\|}$$

$$\times \Big(\sup_{\overline{G}} \Big\| \frac{\partial a_k}{\partial \tau} \Big\| + \sup_{\overline{G}} \|a_k\| \Big) \int_{\tau+2s\Delta}^{\tau+2(s+1)\Delta} g_\mu(\overline{k},t) dt,$$

where

$$g_\mu(\overline{k},t) = [1-h_\mu((\overline{k},\omega(t)))] \Big[\frac{1}{|(\overline{k},\omega(t))|} + \Big| \frac{d}{dt} \frac{1}{(\overline{k},\omega(t))} \Big| \Big]$$

$$+ c_1 \overline{h}_\mu((\overline{k},\omega(t))) \Big| \frac{d}{dt} \frac{1}{(\overline{k},\omega(t))} \Big|.$$

If inequality (21.11) holds for $r=0$, then, obviously,

$$\int_{\tau+2s\Delta}^{\tau+2(s+1)\Delta} g_\mu(\overline{k},t) dt \leq 4p\sigma_2\Delta(1+2p\sigma_2(1+c_1)). \tag{21.14}$$

Assume that inequality (21.11) holds for $r \geq 1$. According to the arguments presented above, the segment $[\tau+2s\Delta, \tau+2(s+1)\Delta]$ can be decomposed into finitely many segments $[\alpha_j, \beta_j]$ on each of which the functions $\mu - |(\overline{k},\omega(t))|$ and $\dfrac{d}{dt}(k,\omega(t))$ do not change their signs. If $\mu - |(\overline{k},\omega(t))| \geq 0 \ \ \forall t \in [\alpha_j, \beta_j]$, then

$$\int_{\alpha_j}^{\beta_j} g_\mu(\overline{k}, t)dt = 0, \tag{21.15}$$

and if $\mu - |(k, \omega(t))| \leq 0$ for $t \in [\alpha_j, \beta_j]$, then

$$\begin{aligned} \int_{\alpha_j}^{\beta_j} g_\mu(\overline{k}, t)dt &\leq \frac{\beta_j - \alpha_j}{\mu} + (1 + c_1) \int_{\alpha_j}^{\beta_j} \Big| \frac{d}{dt} \frac{1}{(\overline{k}, \omega(t))} \Big| dt \\ &= \frac{\beta_j - \alpha_j}{\mu} + (1 + c_1) \Big| \int_{\alpha_j}^{\beta_j} \frac{d}{dt} \frac{1}{(\overline{k}, \omega(t))} dt \Big| \\ &\leq \Big[\beta_j - \alpha_j + 2(1 + c_1) \Big] \frac{1}{\mu}. \end{aligned} \tag{21.16}$$

Combining inequalities (21.13), (21.14), and (21.16) and equality (21.15), we obtain estimate (21.10). The lemma is proved.

We now transform system (21.7) using the change of variables $z = \overline{x}(\tau) + y$, $\|y\| \leq \frac{1}{2}\rho$ as follows:

$$\begin{aligned} \frac{dy}{d\tau} &= H(\tau)y + F(y, \tau) + \delta(\overline{x}(\tau) + y, \varphi, \tau, \mu) \\ &\quad + \varepsilon v(\overline{x}(\tau) + y, \varphi, \tau, \mu) + \varepsilon A_1(\overline{x}(\tau) + y, \varphi, \tau, \varepsilon, \mu), \\ \frac{d\varphi}{d\tau} &= \frac{\omega(\tau)}{\varepsilon} + \widetilde{b}(\overline{x}(\tau) + y, \varphi, \tau, \varepsilon, \mu), \end{aligned} \tag{21.17}$$

where $F(y, \tau) = a(\overline{x}(\tau) + y, \tau) - a(\overline{x}(\tau), \tau) - H(\tau)y$ and $\|F\| \leq \frac{1}{2} n^2 \sigma_1 \|y\|^2$.

We define the integral manifold of Eqs. (21.17) as the limit of the following iterations as $j \to \infty$:

$$\begin{aligned} Y_j(\psi, \tau, \varepsilon, \mu) = \int_{-\infty}^{\infty} Q(\tau, t)[&F(Y_{j-1}, t) + \delta(\overline{x}(t) + Y_{j-1}, \varphi^t_{\tau,j}, t, \mu) \\ &+ \varepsilon v(\overline{x}(t) + Y_{j-1}, \varphi^t_{\tau,j}, t, \mu) \\ &+ \varepsilon A_1(\overline{x}(t) + Y_{j-1}, \varphi^t_{\tau,j}, t, \varepsilon, \mu)]dt, \quad j \geq 0, \end{aligned} \tag{21.18}$$

where $Y_0 \equiv 0$, $Y_{j-1} = Y_{j-1}(\varphi^t_{\tau,j}, t, \varepsilon, \mu)$, and $\varphi^t_{\tau,j} = \varphi^t_{\tau,j}(\psi, \varepsilon, \mu)$ is a solution of the Cauchy problem

$$\frac{d}{dt}\varphi^t_{\tau,j} = \frac{\omega(t)}{\varepsilon} + \widetilde{b}(\overline{x}(t) + Y_{j-1}, \varphi^t_{\tau,j}, t, \varepsilon, \mu), \quad \varphi^\tau_{\tau,j} = \psi \in R^m.$$

Theorem 21.1. *If conditions (21.2)–(21.4) are satisfied, then one can find constants* d_s, $s = \overline{1,6}$, *independent of* ε *and* $\mu = \mu(\varepsilon)$ *and such that, for sufficiently small* ε_0, *the functions* $Y_j = Y_j(\psi, \tau, \varepsilon, \mu(\varepsilon))$ *are* 2π*-periodic in each component* ψ_ν, $\nu = \overline{1,m}$, *of the vector* ψ, *twice continuously differentiable with respect to* ψ *and* τ, *and such that the following inequalities hold for all* $(\psi, \tau, \varepsilon) \in R^m \times R \times (0, \varepsilon_0] = G_1$:

$$\|Y_j\| \le d_1 \varepsilon^{\frac{1}{p}}, \quad \left\|\frac{\partial Y_j}{\partial \psi}\right\| \le d_2 \varepsilon^{\frac{1}{p}}, \quad \sum_{\nu=1}^{m} \left\|\frac{\partial^2 Y_j}{\partial \psi \partial \psi_\nu}\right\| \le d_3 \varepsilon^{\frac{1}{p}}, \tag{21.19}$$

$$\left\|\frac{\partial Y_j}{\partial \tau}\right\| \le d_3 \varepsilon^{\frac{1}{p}-1}, \quad \left\|\frac{\partial^2 Y_j}{\partial \psi \partial \tau}\right\| \le d_4 \varepsilon^{\frac{1}{p}-1}, \quad \left\|\frac{\partial^2 Y_j}{\partial \tau^2}\right\| \le d_6 \varepsilon^{\frac{1}{p}-2}. \tag{21.20}$$

Proof. Consider iterations (21.18). The fact that the functions Y_j are smooth with respect to ψ and τ and periodic in ψ_ν, $\nu = \overline{1,m}$, can be established by analogy with Section 13. Let us prove the first inequality in (21.19). Using Lemma 21.1 and estimates (21.8), we get

$$\sup_{\psi,\tau} \|Y_j(\psi, \tau, \varepsilon, \mu)\| \le \frac{\sigma_1}{\gamma} K n^2 \sup_{\psi,\tau} \|Y_{j-1}(\psi, \tau, \varepsilon, \mu)\|^2$$
$$+ \sigma_5 \mu^{\frac{1}{p-1}} + \left(\sigma_6 + \frac{2}{\gamma} K \sigma_4\right) \frac{\varepsilon}{\mu} + \sigma_4 \sigma_6 \frac{\varepsilon^2}{\mu^2}.$$

For $\varepsilon < \mu$, this yields

$$\sup_{\psi,\tau} \|Y_j\| \le \frac{\sigma_1}{\gamma} K n^2 \sup_{\psi,\tau} \|Y_{j-1}\|^2 + \sigma_5 \mu^{\frac{1}{p-1}} + \left(\sigma_6 + \frac{2}{\gamma} K \sigma_4 + \sigma_4 \sigma_6\right) \frac{\varepsilon}{\mu}. \tag{21.21}$$

We set $\mu^{\frac{1}{p-1}} = \dfrac{\varepsilon}{\mu}$, i.e., $\mu = \mu(\varepsilon) = \varepsilon^{\frac{p-1}{p}}$. Taking into account that $Y_0 \equiv 0$ and using (21.21), for

$$\varepsilon_0 \le \min\left\{\left(\frac{2}{\gamma} K n^2 \sigma_1 d_1\right)^{-p}; \left(\frac{\rho}{2d_1}\right)^p\right\},$$

$$d_1 = 2\left(\frac{2}{\gamma} K \sigma_4 + \sigma_5 + \sigma_4 \sigma_6 + \sigma_6\right)$$

we get

$$\|Y_j(\psi,\tau,\varepsilon,\mu(\varepsilon))\| \le d_1\varepsilon^{\frac{1}{p}} \quad \forall(\psi,\tau,\varepsilon)\in G_1, j\ge 0.$$

Note that, for the value of $\mu = \mu(\varepsilon)$ chosen above, the last estimate is the best order estimate with respect to ε.

By analogy, using Lemmas 12.1 and 12.2, we can establish the last two estimates in (21.19). Note that, in this case, we essentially use the restrictions imposed on the Fourier coefficients of the functions $\widetilde{a}(x,\varphi,\tau)$ and $b(x,\varphi,\tau,\varepsilon)$. Estimates (21.20) follow from conditions (21.2), (21.3), (21.8), and (21.19) and the identity

$$\frac{\partial Y_j}{\partial\tau} + \frac{\partial Y_j}{\partial\psi}\Big[\frac{\omega(\tau)}{\varepsilon} + \widetilde{b}(\overline{x}(\tau)+Y_{j-1},\psi,\tau,\varepsilon,\varepsilon^{\frac{p-1}{p}})\Big]$$

$$= H(\tau)Y_j + F(Y_{j-1},\tau) + \delta(\overline{x}(\tau)+Y_{j-1},\psi,\tau,\varepsilon^{\frac{p-1}{p}})$$

$$+ \varepsilon v(\overline{x}(\tau)+Y_{j-1},\psi,\tau,\varepsilon^{\frac{p-1}{p}}) + \varepsilon A_1(\overline{x}(\tau)+Y_{j-1},\psi,\tau,\varepsilon,\varepsilon^{\frac{p-1}{p}}),$$

where $Y_l = Y_l(\psi,\tau,\varepsilon,\varepsilon^{\frac{p-1}{p}})$ for $l=j-1,j$. Theorem 21.1 is proved.

The theorem presented below solves the problem of the existence and smoothness of the integral manifold of system (21.1). Note that, in its proof, we use Theorem 21.1 and the scheme of the proof of Theorem 14.1. The only difference lies in the fact that the proof of the convergence of the sequences $\{Y_j\}$ and $\left\{\frac{\partial}{\partial\psi}Y_j\right\}$ is based not on the smallness of the norm of the matrix P, but on properties of the functions $h_\mu((\overline{k},\omega(t)))$. According to Lemma 21.1, the measure of the set of points of a time interval of length 2Δ for which $h_\mu((\overline{k},\omega(t)))\ne 0$ tends to zero as $\mu=\varepsilon^{\frac{p-1}{p}}\to 0$.

Theorem 21.2. *Suppose that conditions (21.2)–(21.4) are satisfied. Then, for sufficiently small* $\varepsilon_0>0$, *the following assertions are true:*

(i) in the $\overline{\sigma}_1\varepsilon^{\frac{1}{p}}$*-neighborhood of the curve* $\overline{x}=\overline{x}(\tau)$, *there exists the integral manifold* $x = X(\psi,\tau,\varepsilon)$ *of system (21.1), where* $(\psi,\tau,\varepsilon)\in G_1$,

$$X(\psi,\tau,\varepsilon) = \overline{x}(\tau) + Y(\psi,\tau,\varepsilon) + \varepsilon u(\overline{x}(\tau)+Y(\psi,\tau,\varepsilon),\psi,\tau,\varepsilon^{\frac{p-1}{p}}),$$

$$Y(\psi,\tau,\varepsilon) = \lim_{j\to\infty} Y_j(\psi,\tau,\varepsilon,\varepsilon^{\frac{p-1}{p}});$$

(ii) the function $X(\psi,\tau,\varepsilon)$ *is* 2π*-periodic in* ψ_ν, $\nu = \overline{1,m}$, *continuously differentiable with respect to* ψ *and* τ, *and such that*

$$\left\|\frac{\partial X}{\partial \psi}\right\| + \varepsilon\left\|\frac{\partial X}{\partial \tau}\right\| \le \overline{\sigma}_2\varepsilon^{\frac{1}{p}} \quad \forall(\psi,\tau,\varepsilon)\in G_1,$$

and $\dfrac{\partial X}{\partial \psi}$ *and* $\dfrac{\partial X}{\partial \tau}$ *satisfy the Lipschitz condition:*

$$\left\|\frac{\partial X(\psi,\tau,\varepsilon)}{\partial \psi} - \frac{\partial X(\overline{\psi},\overline{\tau},\varepsilon)}{\partial \psi}\right\| \le \overline{\sigma}_3\varepsilon^{\frac{1}{p}}\|\psi-\overline{\psi}\| + \overline{\sigma}_3\varepsilon^{\frac{1}{p}-1}\|\tau-\overline{\tau}\|,$$

$$\left\|\frac{\partial X(\psi,\tau,\varepsilon)}{\partial \tau} - \frac{\partial X(\overline{\psi},\overline{\tau},\varepsilon)}{\partial \tau}\right\| \le \overline{\sigma}_3\varepsilon^{\frac{1}{p}-1}\|\psi-\overline{\psi}\| + \overline{\sigma}_3\varepsilon^{\frac{1}{p}-2}\|\tau-\overline{\tau}\|;$$

(iii) on the integral manifold, system (21.1) takes the form

$$\frac{d\varphi}{d\tau} = \frac{\omega(\tau)}{\varepsilon} + b(X(\varphi,\tau,\varepsilon),\varphi,\tau,\varepsilon).$$

Here, $\overline{\sigma}_1$, $\overline{\sigma}_2$, *and* $\overline{\sigma}_3$ *are constants independent of* ε.

4. INVESTIGATION OF A DYNAMICAL SYSTEM IN A NEIGHBORHOOD OF A QUASIPERIODIC TRAJECTORY

22. Statement and General Description of the Problem

Let $C^r(\mathcal{T}_m)$ be the space of 2π-periodic functions $f = (f_1, \dots, f_n)$ of a variable $\varphi = (\varphi_1, \dots, \varphi_m)$ of smoothness $r \geq 0$, and let $\lambda = (\lambda_1, \dots, \lambda_m)$ be the frequency vector, i.e., a collection of m positive numbers that satisfy the condition of linear independence over the field of integer numbers $\mathbf{Z}^m$, namely

$$(k, \lambda) = \sum_{\nu=1}^{m} k_\nu \lambda_\nu \neq 0 \quad \forall k \in \mathbf{Z}^m \backslash \{0\}.$$

A function

$$F(t) = f(\lambda t), \quad t \in R, \tag{22.1}$$

where $f(\varphi) \in C(\mathcal{T}_m)$ and $C(\mathcal{T}_m) = C^0(\mathcal{T}_m)$, is called a quasiperiodic function, λ is called its frequency basis, and m is the dimension of the frequency basis.

By $C^r(\lambda)$, we denote the collection of all quasiperiodic functions (22.1) with frequency basis λ for which $f \in C^r(\mathcal{T}_m)$.

The true dimension of a frequency basis of the quasiperiodic function (22.1) is defined as the number m such that $F(t) \in C(\lambda)$ for a certain basis $\lambda = (\lambda_1, \dots, \lambda_m)$ and $F(t) \notin C(\omega)$ for an arbitrary basis $\omega = (\omega_1, \dots, \omega_s)$ for which $s < m$. Denote by $x(t, x^0)$ a solution of the system of equations

$$\frac{dx}{dt} = X(x) \tag{22.2}$$

such that $x(0, x^0) = x^0$, where $x = (x_1, \dots, x_n)$ and $x^0 = (x_1^0, \dots, x_n^0)$ are points of the n-dimensional Euclidean space R^n. Assume that the function $X = X(x)$ is r times continuously differentiable in R^n.

For a set $M \subset R^n$, we denote by $x(t, M)$ solutions $x(t, x^0)$ for an arbitrary fixed $x^0 \in M$.

Assume that system (22.1) has a quasiperiodic solution $x = x(t, x^0)$. By definition, we have

$$x(t, x^0) = f(\lambda t + \psi^0) \tag{22.3}$$

for a certain function $f \in C(\mathcal{T}_m)$ and a certain basis $\lambda = (\lambda_1, \ldots, \lambda_m)$. Assume that m is the true dimension of the frequency basis. The closure of the trajectory that passes through the point x^0 consists of points $M \in R$ defined by the equation

$$x = f(\varphi), \quad \varphi \in \mathcal{T}_m, \tag{22.4}$$

where $\mathcal{T}_m$ denotes an m-dimensional torus. The set M is invariant because it is the closure of the trajectory of a dynamical system.

According to the equation of motion (22.2), we have

$$f(\lambda t + \psi^0) = f(\psi^0) + \int_0^t X(f(\lambda\tau + \psi^0))d\tau, \quad t \in R.$$

Hence, passing to the limit in the sequence of functions $f(\lambda t + \lambda t_n)$, $\lim_{n\to\infty} \lambda t_n = \varphi \bmod 2\pi$, we obtain the identity

$$f(\lambda t + \varphi) = f(\varphi) + \int_0^t X(f(\lambda\tau + \varphi))d\tau,$$

which proves that

$$x(t, f(\varphi)) = f(\omega t + \varphi) \quad \forall t \in R, \quad \varphi \in \mathcal{T}_m.$$

Thus, the set M is filled with quasiperiodic trajectories of the dynamical system (22.2) with the same frequency basis. The geometric structure of the set M in the space R^n is described by the following statement [Sam4]:

Theorem 22.1. *If $x(t, x^0) \in C^s(\lambda)$, where $s \leq r$, then the set M is C^s-homeomorphic to an m-dimensional torus.*

In what follows, a set $M \subset R^n$ C^s-homeomorphic to an m-dimensional torus is called an m-dimensional toroidal manifold of smoothness s, or, briefly, an m-dimensional torus of smoothness s.

It follows from the arguments presented above that the system of equations (22.2) on M reduces to a dynamical system on $\mathcal{T}_m$ of the form

$$\frac{d\varphi}{dt} = \lambda. \tag{22.5}$$

We pose the problem of the investigation of the behavior of solutions of system (22.2) that originate in a small neighborhood of manifold (22.4). For this purpose, it is natural to represent the neighborhood of the manifold M in the form of a product $\mathcal{T}_m \times K_\delta$, where K_δ is the $(n-m)$-dimensional cube with side δ, and introduce the local coordinates $\varphi = (\varphi_1, \ldots, \varphi_m)$ on $\mathcal{T}_m$ and $h = (h_1, \ldots, h_{n-m})$ in K_δ instead of the Euclidean coordinates $x = (x_1, \ldots, x_n)$. In the coordinates φ, h, the equation of the manifold M takes the form

$$h = 0, \quad \varphi \in \mathcal{T}_m, \tag{22.6}$$

and the system of equations (22.2) takes the form (22.5).

Assume that $x(t, x_0) \in C^s(\lambda)$ and $s \geq 1$. According to [Sam4], we have

$$\operatorname{rank} \frac{\partial f(\varphi)}{\partial \varphi} = m \quad \forall \varphi \in \mathcal{T}_m,$$

where $f(\varphi)$ is a function from (22.4).

The problem of the introduction of local coordinates (h, φ) reduces in this case to the following algebraic problem: Find a matrix $B(\varphi)$ whose columns belong to the space $C^s(\mathcal{T}_m)$ and for which the $n \times n$ matrix $\left(\frac{\partial f(\varphi)}{\partial \varphi}, B(\varphi)\right)$ is nondegenerate for all $\varphi \in \mathcal{T}_m$; here, $\left(\frac{\partial f(\varphi)}{\partial \varphi}, B(\varphi)\right)$ denotes the matrix m columns of which are the columns of the matrix $\frac{\partial f(\varphi)}{\partial \varphi}$, and $n-m$ columns are the columns of the matrix $B(\varphi)$. This problem is called the problem of the complementation of the m-frame $\frac{\partial f(\varphi)}{\partial \varphi}$ to a 2π-periodic basis in R^n, and it has a solution [Sam4] for $n = m+1$ or $n \geq 2m+1$. Assume that the matrix $\frac{\partial f(\varphi)}{\partial \varphi}$ can be complemented to a 2π-periodic basis in R^n, and $B(\varphi)$ is the complementing matrix.

Under the assumptions imposed above on the manifold M, the local coordinates (φ, h) of a point x from the neighborhood of M are determined by the equality

$$x = f(\varphi) + B(\varphi)h. \tag{22.7}$$

Consider the matrix

$$\Gamma_0(\varphi) = B^T(\varphi)B(\varphi), \quad \varphi \in \mathcal{T}_m, \tag{22.8}$$

where $B^T(\varphi)$ is the transpose of $B(\varphi)$. The matrix $\Gamma_0(\varphi)$ is the Gram matrix of the linearly independent columns of the matrix $B(\varphi)$; therefore, the eigenvalues of this matrix are positive for all $\varphi \in \mathcal{T}_m$. The periodicity of the matrix $\Gamma_0(\varphi)$ in φ enables one to estimate these eigenvalues from below and from above by positive constants γ_0 and γ^0 independent of φ. In this case, the quadratic form $(\Gamma_0(\varphi)h, h)$ satisfies the inequalities

$$\gamma_0\|h\|^2 \leq (\Gamma_0(\varphi)h, h) \leq \gamma^0\|h\|^2 \quad \forall \varphi \in \mathcal{T}_m, h \in R^{n-m}, \tag{22.9}$$

where $\|h\|^2 = (h, h)$. Inequalities (22.9) yield

$$\gamma_0\|h\|^2 \leq \|x - f(\varphi)\|^2 \leq \gamma^0\|h\|^2. \tag{22.10}$$

In view of these estimates, under the transformation $x \to (\varphi, h)$ defined by equality (22.7) a small neighborhood of the set $M \subset R^n$ turns into the small neighborhood

$$\|h\| \leq \delta, \quad \varphi \in \mathcal{T}_m \tag{22.11}$$

of the set $h = 0$, $\varphi \in \mathcal{T}_m$ in the space $R^{n-m} \times \mathcal{T}_m$. Using this fact, we rewrite the equations of motion of system (22.2) (originating in a neighborhood of the manifold M) in the local coordinates. To do this, we differentiate relation (22.7) as a formula of a change of variables in system (22.2). As a result, instead of (22.2), we obtain the system of equations

$$\begin{aligned} \frac{d\varphi}{dt} &= L_1(\varphi, h)X(f(\varphi) + B(\varphi)h), \\ \frac{dh}{dt} &= L_2(\varphi, h)X(f(\varphi) + B(\varphi)h), \end{aligned} \tag{22.12}$$

where $L_1(\varphi, h)$ and $L_2(\varphi, h)$ are blocks of the matrix inverse to the matrix $\left(\frac{\partial f(\varphi)}{\partial \varphi} + \frac{\partial B(\varphi)h}{\partial \varphi}, B(\varphi)\right)$, and (φ, h) are points of domain (22.11) with sufficiently small positive δ. Note that expressions for the matrices $L_1(\varphi, h)$ and $L_2(\varphi, h)$ in terms of the matrices $\frac{\partial f}{\partial \varphi}$ and B can be obtained using the Frobenius formula [Lan] for the construction of the inverse matrix for the matrix composed of the following blocks:

$$L_1(\varphi,h)=\Big[\Big(\frac{\partial f}{\partial\varphi}+\frac{\partial Bh}{\partial\varphi}\Big)^T(E-B\Gamma_0^{-1}B^T)\Big(\frac{\partial f}{\partial\varphi}+\frac{\partial Bh}{\partial\varphi}\Big)\Big]^{-1}$$
$$\times\Big(\frac{\partial f}{\partial\varphi}+\frac{\partial Bh}{\partial\varphi}\Big)^T(E-B\Gamma_0^{-1}B^T),$$
$$L_2(\varphi,h)=\Big\{B^T\Big[E-\Big(\frac{\partial f}{\partial\varphi}+\frac{\partial Bh}{\partial\varphi}\Big)\Gamma_1^{-1}\Big(\frac{\partial f}{\partial\varphi}+\frac{\partial Bh}{\partial\varphi}\Big)^T\Big]B\Big\}^{-1}$$
$$\times B^T\Big\{E-\Big(\frac{\partial f}{\partial\varphi}+\frac{\partial Bh}{\partial\varphi}\Big)^T\Gamma_1^{-1}\Big(\frac{\partial f}{\partial\varphi}+\frac{\partial Bh}{\partial\varphi}\Big)^T\Big\},\qquad(22.13)$$

where $\Gamma_0=\Gamma_0(\varphi)$ is matrix (22.8), E is the identity matrix, and

$$\Gamma_1=\Gamma_1(\varphi,h)=\Big(\frac{\partial f}{\partial\varphi}+\frac{\partial Bh}{\partial\varphi}\Big)^T\Big(\frac{\partial f}{\partial\varphi}+\frac{\partial Bh}{\partial\varphi}\Big).\qquad(22.14)$$

Taking into account that manifold (22.6) is invariant for the system of equations (22.12) with the flow of trajectories on it defined by system (22.5), we rewrite (22.12) in the form

$$\frac{d\varphi}{dt}=\lambda+L_1(\varphi,h)\Big[X(f(\varphi)+B(\varphi)h)-X(f(\varphi))-\frac{\partial B(\varphi)}{\partial\varphi}\lambda h\Big],$$
$$\frac{dh}{dt}=L_2(\varphi,h)\Big[X(f(\varphi)+B(\varphi)h)-X(f(\varphi))-\frac{\partial B(\varphi)}{\partial\varphi}\lambda h\Big],\qquad(22.15)$$

where

$$\frac{\partial B}{\partial\varphi}\lambda=\sum_{\nu=1}^{m}\frac{\partial B}{\partial\varphi_\nu}\lambda_\nu.$$

Parallel with system (22.15), we write the following auxiliary system of equations obtained from (22.15) by omitting the terms of order $\|h\|$ for φ and of order $\|h\|^2$ for h on the right-hand sides of these equations:

$$\frac{d\varphi}{dt}=\lambda,\quad\frac{dh}{dt}=P(\varphi)h,\qquad(22.16)$$

where

$$P(\varphi)=L_2(\varphi,0)\Big[\frac{\partial X(f(\varphi))}{\partial x}-\frac{\partial B(\varphi)}{\partial\varphi}\lambda\Big].\qquad(22.17)$$

Denote by $\Omega_0^t(\varphi)$, $\Omega_0^0(\varphi)=E$, the normal fundamental matrix of solutions of the second equation in (22.16). It is clear that $\Omega_0^t(\varphi)\in C^{s-1}(\mathcal{T}_m)$ for every

$t \in R$. Moreover, for arbitrary $t \in R$, $\theta \in R$, and $\varphi \in \mathcal{T}_m$, the following identity is true [Sam4]:

$$\Omega_0^t(\varphi_\theta(\varphi)) = \Omega_0^{t+\theta}(\varphi), \tag{22.18}$$

where $\varphi_\theta(\varphi) = \lambda\theta + \varphi$.

Assume that the condition

$$\|\Omega_0^t(\varphi)\| \leq \mathcal{L}e^{-\gamma t} \tag{22.19}$$

is satisfied for all $t \in R_+ = [0, \infty)$ and $\varphi \in \mathcal{T}_m$ and certain positive constants $\mathcal{L}$ and γ. We rewrite the system of equations (22.15) in the form

$$\frac{d\varphi}{dt} = \lambda + A(\varphi, h)h, \quad \frac{dh}{dt} = P(\varphi, h)h, \tag{22.20}$$

where

$$A(\varphi, h) = L_1(\varphi, h)\Big[\int_0^1 \frac{\partial X(f(\varphi) + \tau B(\varphi)h)}{\partial x} d\tau B(\varphi) + \frac{\partial B(\varphi)}{\partial \varphi}\lambda\Big],$$

$$P(\varphi, h) = L_2(\varphi, h)\Big[\int_0^1 \frac{\partial X(f(\varphi) + \tau B(\varphi)h)}{\partial x} d\tau B(\varphi) + \frac{\partial B(\varphi)}{\partial \varphi}\lambda\Big]. \tag{22.21}$$

It follows from relations (22.13), (22.14), and (22.21) that A and P are $s-1$ times continuously differentiable functions of their variables in their domain of definition (22.11).

Let $C^p_{Lip}(\mathcal{T}_m \times K_\mu)$ denote the space of functions of (φ, h) that are defined in the domain $\mathcal{T}_m \times K_\mu$, $K_\mu = \{h\colon \|h\| \leq \mu\}$, have (in this domain) continuous partial derivatives up to the order p inclusive, and are such that their pth derivatives satisfy the Lipschitz condition with respect to (φ, h). The meaning of the notation $C^p(\mathcal{T}_m \times K_\mu)$ is analogous for finite and infinite values of p.

23. Theorem on Reducibility

In what follows, we preserve the same notation as in Section 22. The statement below is the main result of the present section.

Theorem 23.1. *Suppose that the matrices $A(\varphi, h)$ and $P(\varphi, h)$ belong to the space $C^p(\mathcal{T}_m \times K_\delta)$ for $p \geq 1$, and the fundamental matrix of solutions*

$\Omega_0^t(\varphi)$ *of system (22.16) satisfies inequality (22.19). Then one can find* $\mu > 0$ *and a matrix* $\Phi(\psi, h)$ *belonging to the space* $C_{Lip}^{p-1}(\mathcal{T}_m \times K_\mu)$ *such that the change of variables*

$$\varphi = \psi + \Phi(\psi, h)h \tag{23.1}$$

reduces the system of equations (22.20) to the form

$$\frac{d\psi}{dt} = \lambda, \quad \frac{dh}{dt} = P(\psi + \Phi(\psi, h)h, h) \tag{23.2}$$

for $(\psi, h) \in \mathcal{T}_m \times K_\mu$.

Proof. Let us prove the theorem for $p = 1$. We write the following equation for the determination of the matrix $\Phi = \Phi(\psi, h)$:

$$\frac{\partial \Phi}{\partial \psi}\lambda + \frac{\partial \Phi}{\partial h}P(\psi + \Phi h, h)h + \Phi P(\psi + \Phi h, h) = A(\psi + \Phi h, h), \tag{23.3}$$

where

$$\frac{\partial \Phi}{\partial h}Ph = \sum_{\nu=1}^{n-m} \frac{\partial \Phi}{\partial h_\nu}(Ph)_\nu$$

and (Ph_ν) is the νth coordinate of the vector Ph. To solve Eq. (23.3), we use the method of passing from (23.3) to an operator equation, which is based on the ideas of the method of integral manifolds [Bog, BoM1, MiLy]. Let $C(M, K)$ denote the set of matrix functions $F = F(\psi, h)$ defined for all $\psi \in \mathcal{T}_m$ and $h \in R^{n-m}$ and satisfying the inequalities

$$\|F(\psi, h)\| \leq M, \quad \|F(\psi', h') - F(\psi, h)\| \leq K(\|\psi' - \psi\| + \|h' - h\|)$$

for any $(\psi, h) \in \mathcal{T}_m \times R^{n-m}$ and $(\psi', h') \in \mathcal{T}_m \times R^{n-m}$.

We define a scalar function $z(\tau)$ of a scalar variable τ as follows:

$$z(\tau) = \begin{cases} 0 & \text{for } |\tau| \geq 2\mu, \\ -(\tau + 2\mu) & \text{for } \tau \in (-2\mu, -\mu), \\ \tau & \text{for } |\tau| \leq \mu, \\ 2\mu - \tau & \text{for } \tau \in (\mu, 2\mu). \end{cases}$$

We set $g(h) = (z(h_1), \ldots, z(h_{n-m}))$. It is clear that, for all h' and h from R^{n-m}, we have

$$\|g(h)\| \leq \mu(n-m)^{\frac{1}{2}}, \quad \|g(h') - g(h)\| \leq \|h' - h\|.$$

Let

$$\psi_t = \lambda t + \psi, \quad \psi \in \mathcal{T}_m.$$

Denote by $X_t^F = \Omega_0^t(\psi, g, F)$ a solution of the equation

$$\frac{dX}{dt} = P(\psi_t + F(\psi_t, Xg)Xg, Xg)X \tag{23.4}$$

that takes the value of the identity matrix E for $t = 0$. Here, $g = g(h)$ and $h \in R^{n-m}$. Since

$$X_t^F = \Omega_0^t(\psi) + \int_0^t \Omega_s^t(\psi)\big[P(\psi_s + F(\psi_s, X_s^F g)X_s^F g, X_s^F g) - P(\psi_s, 0)\big]X_s^F ds,$$

we conclude that the following estimate holds for X_s^F:

$$\|X_t^F\| \le \mathcal{L}e^{-\gamma t}\Big[1 + \mu a \int_0^t e^{\gamma s}\|X_s^F\|^2 ds\Big], \quad t \in R_+, \tag{23.5}$$

where $a = \overline{K}_1(1+M)(n-m)^{\frac{1}{2}}$ and $\overline{K}_1$ is a constant independent of the first derivatives of the matrix $P(\psi, h)$ for $\psi \in \mathcal{T}_m$ and $h \in K_\delta$.

Assume that $\|X_t^F\| \le \mathcal{L}(1+\mu)$ for $t \in [0, T)$, where $[0, T)$ is the maximum half-interval on which X_t^F satisfies the above inequality. It follows from (23.5) that

$$\|X_t^F\| \le \mathcal{L}e^{-\gamma t}\Big[1 + \mu a\mathcal{L}(1+\mu) \int_0^t e^{\gamma s}\|X_s^F\| ds\Big]$$

for $t \in [0, T)$. Therefore, according to the Gronwall–Bellman inequality, we have

$$\|X_t^F\| \le \mathcal{L}e^{-\gamma_1 t} \quad \forall t \in [0, T), \tag{23.6}$$

where

$$\gamma_1 = \gamma - \mu a\mathcal{L}^2(1+\mu) \ge \frac{\gamma}{2} \tag{23.7}$$

for sufficiently small μ. It follows from (23.6) that $\|X_t^F\| \le \mathcal{L}(1+\mu)$ $\forall t \in R_+$, and, hence,

$$\|X_t^F\| \le \mathcal{L}e^{-\gamma_1 t}, \quad t \in R_+. \tag{23.8}$$

On the set of functions $C(M,K)$, we define an operator $S\colon F \to SF = W$ according to the formula

$$W(\psi,h) = SF$$
$$= -\int_0^\infty A\big[\psi_s + F(\psi_s, X_s^F g(h))X_s^F g(h), X_s^F g(h))\big] X_s^F \, ds. \quad (23.9)$$

Let us prove that, for properly chosen M, K, and μ, the operator S maps the set $C(M,K)$ into itself.

To do this, we estimate the difference

$$R_t = X_t^F(\psi', h') - X_t^F(\psi, h), \quad t \in R_+,$$

where (ψ', h') and (ψ, h) are arbitrary points from $\mathcal{T}_m \times K_\mu$. It follows from the equation for X_t^F that

$$R_t = \int_0^t \Omega_s^t(P')[P_s' - P_s]X_s^F \, ds, \quad (23.10)$$

where

$$\Omega_s^t(P') = X_t^F(\psi', h')\big[X_s^F(\psi', h')\big]^{-1}, \quad \psi_t' = \lambda t + \psi',$$
$$P_t' - P_t = P\big[\psi_t' + F(\psi_t', X_t^F(\psi', h')g(h'))$$
$$\times X_t^F(\psi', h')g(h'), X_t^F(\psi', h')g(h')\big]$$
$$- P\big[\psi_t + F(\psi_t, X_t^F(\psi, h)g(h))X_t^F(\psi, h)g(h), X_t^F(\psi, h)g(h)\big],$$
$$X_s^F = X_s^F(\psi, h).$$

According to (22.18), we have $\Omega_s^t(\psi) = \Omega_0^{t-s}(\psi_s)$. Therefore, for $\Omega_s^t(P')$, the following estimate of the form (23.8) is true:

$$\|\Omega_s^t(P')\| \le \mathcal{L}e^{-\gamma_1(t-s)}, \quad t \ge s \ge 0. \quad (23.11)$$

Let us estimate the quantity $\mathcal{I} = \|P_s' - P_s\|$. Using obvious notation, we get

$$\mathcal{I} \le \overline{K}_1\big[\delta + \|F'X'g' - FXg\| + \|X'g' - Xg\|\big]$$
$$\le \overline{K}_1\big[\delta + \|X'g' - Xg\|M + \|F - F'\|\|X_g\| + \|X'g' - Xg\|\big]$$

$$\leq \overline{K}_1\big[(1+\mu K\mathcal{L}(n-m)^{\frac{1}{2}})\delta + (1+M+\mu K\mathcal{L}(n-m)^{\frac{1}{2}})\|X'g'-Xg\|\big]$$

$$\leq \overline{K}_1\big[(1+\mu K\mathcal{L}(n-m)^{\frac{1}{2}})\delta + (1+M+\mu K\mathcal{L}(n-m)^{\frac{1}{2}})\mathcal{L}\delta_1$$

$$+ (1+M+\mu K\mathcal{L}(n-m)^{\frac{1}{2}})\mu(n-m)^{\frac{1}{2}}\|X'-X\|\big]$$

$$\leq \mathcal{L}\overline{K}_1(1+M+\mu K\mathcal{L}(n-m)^{\frac{1}{2}})\Big[\frac{\|\psi'-\psi\|}{\mathcal{L}} + \|h'-h\|$$

$$+ \mu\frac{(n-m)^{\frac{1}{2}}}{\mathcal{L}}\|X_t^F(\psi',h') - X_t^F(\psi,h)\|\Big], \quad t \geq 0. \tag{23.12}$$

We set

$$b = \mathcal{L}\overline{K}_1\big(1+M+\mu K\mathcal{L}(n-m)^{\frac{1}{2}}\big)$$

and rewrite (23.12) in the form

$$\mathcal{I} \leq b\Big(\frac{\|\psi'-\psi\|}{\mathcal{L}} + \|h'-h\| + \mu(n-m)^{\frac{1}{2}}\Big)\frac{R_t}{\mathcal{L}}, \quad t \geq 0. \tag{23.13}$$

Taking into account inequality (23.13), we derive from (23.10) the following inequality:

$$\|R_t\| \leq \mathcal{L}be^{-\gamma_1 t}\int_0^t \big(\|\psi-\psi'\|\mathcal{L}^{-1} + \|h'-h\|$$

$$+ \mu(n-m)^{\frac{1}{2}}\mathcal{L}^{-1}\|R_s\|\big)ds, \quad t \geq 0. \tag{23.14}$$

Estimate (23.14) has the form of the inequality

$$y_t \leq be^{-\gamma_1 t}\int_0^t \big(\alpha + \mu(n-m)^{\frac{1}{2}}y_s\big)ds, \quad t \geq 0, \tag{23.15}$$

where $y_t = \|R_t\|\mathcal{L}^{-1}$ and $\alpha = \|\psi-\psi'\|\mathcal{L}^{-1} + \|h'-h\|$. Since $y_t \leq \overline{y}_t$ for $t \in R_+$, where $\overline{y}_t$ is a solution of the equation obtained from (23.15) by replacing the sign $\leq$ by $=$, we get

$$y_t \leq be^{-\gamma_1 t}\big(\alpha + \mu(n-m)^{\frac{1}{2}}c\big)t, \quad t \in R_+, \quad c = \sup_{t\in R_+}\overline{y}_t. \tag{23.15'}$$

Let us determine c. It follows from the equation for $\overline{y}_t$ that the value of c is attained at the point $\tau \in R_+$, where $\dfrac{d}{dt}\overline{y}_\tau = 0$. Writing the last equality in more detail, we get

$$-\gamma_1 c + be^{-\gamma_1 \tau}(\alpha + \mu(n-m)^{\frac{1}{2}} c) = 0.$$

This implies that, for $\mu b(n-m)^{\frac{1}{2}} < \gamma_1$, the number c can be estimated as follows:

$$c = \frac{b\alpha e^{-\gamma_1 \tau}}{\gamma_1 - \mu(n-m)^{\frac{1}{2}} be^{-\gamma_1 \tau}} \leq \frac{b\alpha}{\gamma_1 - \mu b(n-m)^{\frac{1}{2}}}. \tag{23.16}$$

If we set

$$\mu \leq \frac{\gamma_1}{2b(n-m)^{1/2}}, \tag{23.17}$$

then it follows from (23.15′) and (23.16) for all $t \in R_+$ that

$$y_t \leq 2b\alpha t e^{-\gamma_1 t}. \tag{23.18}$$

In view of the notation used, inequality (23.18) yields

$$\|R_t\| \leq 2\mathcal{L} bte^{-\gamma_1 t}(\|\psi' - \psi\|\mathcal{L}^{-1} + \|h' - h\|), \quad t \in R_+. \tag{23.19}$$

Further, we consider the difference $W(\psi', h') - W(\psi, h)$. Using (23.9), we obtain the following estimate for this difference:

$$\|W(\psi', h') - W(\psi, h)\| \leq \int_0^\infty [\|A_s'\|\|R_s\| + \|A_s' - A_s\|\|X_s^F\|]\,ds,$$

where A_t' and A_t are the expressions obtained from the expressions for P_t' and P_t by replacing P by A. It is clear that the difference $A_t' - A_t$ can be estimated by analogy with the difference $P_t' - P_t$. As a result, we obtain the following inequality of the form (23.13):

$$\|A_t' - A_t\| \leq \mathcal{L}\underline{K}_1(1 + M + \mu K\mathcal{L}(n-m)^{\frac{1}{2}})$$
$$\times (\|\psi' - \psi\|\mathcal{L}^{-1} + \|h' - h\| + \mu(n-m)^{1/2}\mathcal{L}^{-1}\|R_t\|), \quad t \geq 0,$$

where $\underline{K}_1$ is a positive constant that depends on the first derivatives of the matrix $A(\psi, h)$ for $\psi \in \mathcal{T}_m$ and $h \in K_\delta$. This yields

$$\|W(\psi', h') - W(\psi, h)\|$$

$$\leq \int_0^\infty \left[(M_1 + \mu\bar{b}(n-m)^{\frac{1}{2}})\|R_s\| + \bar{b}(\|\psi' - \psi\| + \mathcal{L}\|h' - h\|)e^{-\gamma_1 s}\right] ds$$

$$\leq (M_1 + \mu\bar{b}(n-m)^{\frac{1}{2}}) \int_0^\infty \|R_s\| ds + \frac{\bar{b}}{\gamma_1}\mathcal{L}(\|\psi' - \psi\| + \|h' - h\|), \quad (23.20)$$

where

$$M_1 = \max_{\psi, h} \|A(\psi, h)\|, \quad \bar{b} = \mathcal{L}\underline{K}_1(1 + M + \mu K \mathcal{L}(n-m)^{\frac{1}{2}}).$$

Inequality (23.19) yields

$$\int_0^\infty \|R_s\| ds \leq 2\mathcal{L}b\gamma_1^{-2}(\|\psi' - \psi\| + \|h' - h\|),$$

which, together with (23.20), guarantees the validity of the inequality

$$\|W(\psi', h') - W(\psi, h)\|$$

$$\leq \mathcal{L}\gamma_1^{-1}\left[2b\gamma_1^{-1}(M_1 + \mu\bar{b}(n-m)^{\frac{1}{2}}) + \bar{b}\right](\|\psi' - \psi\| + \|h' - h\|).$$

It also follows from (23.9) that

$$\|W(\psi, h)\| \leq \mathcal{L}M_1\gamma_1^{-1}.$$

Thus, $W \in C(M, K)$, provided that M, K, and μ satisfy inequalities (23.7) and (23.17) and the inequalities

$$\mathcal{L}M_1\gamma_1^{-1} \leq M, \quad \mathcal{L}\gamma_1^{-1}\left[2b\gamma_1^{-1}(M_1 + \mu\bar{b}(n-m)^{\frac{1}{2}}) + \bar{b}\right] \leq K.$$

For $\mu \leq 1$, the inequalities indicated are satisfied if

$$\mu(n-m)^{\frac{1}{2}}\mathcal{L}^2\overline{K}_1(1+M)\leq\frac{\gamma}{4},\quad 2\mathcal{L}M_1\gamma^{-1}\leq M,$$

$$\mu(n-m)^{\frac{1}{2}}\mathcal{L}\overline{K}_1\big(1+M+\mu\mathcal{L}K(n-m)^{\frac{1}{2}}\big)\leq\frac{\gamma}{4},$$

$$2\mathcal{L}^2\gamma^{-1}\big\{4\overline{K}_1\gamma^{-1}\big[M_1+\mu\overline{K}_1(1+M+\mu\mathcal{L}K(n-m)^{\frac{1}{2}})(n-m)^{\frac{1}{2}}\big]\underline{K}_1\big\}$$
$$\times\big(1+M+\mu\mathcal{L}K(n-m)^{\frac{1}{2}}\big)\leq K. \tag{23.21}$$

Inequalities (23.21) are satisfied due to the proper choice of M, K, and μ. Namely, M and K are chosen from the conditions

$$2\mathcal{L}\gamma^{-1}M_1\leq M,$$
$$2\mathcal{L}^2\gamma^{-1}(2+M)\big\{\underline{K}_1+4\overline{K}_1\gamma^{-1}[M_1+K_1(2+M)]\big\}\leq K, \tag{23.22}$$

and μ satisfies the inequalities

$$\mu(n-m)^{\frac{1}{2}}\mathcal{L}^2\overline{K}_1(1+M)\leq\frac{\gamma}{4},\quad \mu(n-m)^{\frac{1}{2}}\mathcal{L}\overline{K}_1(2+M)\leq\frac{\gamma}{4},$$
$$\mu(n-m)^{\frac{1}{2}}\mathcal{L}K\leq 1. \tag{23.23}$$

We fix M and K so large that conditions (23.22) are satisfied. Then, for these values of M and K, one can choose $\mu_0>0$ so that inequalities (23.23) hold for any $\mu\in(0,\mu_0]$.

Assume that M, K, and μ_0 are chosen as indicated above. In this case, $W=SF\in C(M,K)$, i.e., the operator S maps the set $C(M,K)$ into itself. In $C(M,K)$, we introduce a metric according to the formula

$$\rho(F_1,F_2)=\sup\|F_1(\psi,h)-F_2(\psi,h)\|,$$

where the supremum is taken over $(\psi,h)\in\mathcal{T}_m\times R^{n-m}$. Thus, $C(M,K)$ becomes a complete metric space. Let us prove that the operator S, which acts from $C(M,K)$ into itself, is a contraction operator. For this purpose, we consider the difference

$$W^F-W^{F'}=SF-SF',$$

where F and F' are arbitrary functions from $C(M,K)$. Equations (23.4) yield

$$X_t^F-X_t^{F'}=\int_0^t\Omega_s^t(F)[P_s(F)-P_s(F')]X_s^{F'}ds,$$

where

$$\begin{aligned}\Omega_s^t(F) &= X_t^F (X_s^F)^{-1}, P_s(F) - P_s(F') \\ &= P\big[\psi_s + F(\psi_s, X_s^F g) X_s^F g, X_s^F g\big] \\ &\qquad - P(\psi_s + F'(\psi_s, X_s^{F'} g) X_s^{F'} g, X_s^{F'} g)\big], \quad g = g(h). \quad (23.24)\end{aligned}$$

In view of the notation used, we get

$$\begin{aligned}&\|P_s(F) - P_s(F')\| \\ &\quad \le \mu(n-m)^{\frac{1}{2}} \overline{K}_1 \big[\|F(\psi_s, X_s^F g) X_s^F \\ &\qquad\qquad - F'(\psi_s, X_s^{F'} g) X_s^{F'}\| + \|X_s^F - X_s^{F'}\|\big] \\ &\quad \le \mu(n-m)^{\frac{1}{2}} \overline{K}_1 \big[\mathcal{L}\rho(F, F') e^{-\gamma_1 s} \\ &\qquad\qquad + (1 + M + \mu(n-m)^{\frac{1}{2}} K\mathcal{L}) \|X_s^F - X_s^{F'}\|\big]\end{aligned}$$

for all $s \ge 0$. The last inequality, together with inequality (23.11) for $\psi' = \psi$ and $h' = h$, yields

$$\begin{aligned}\|X_t^F - X_t^{F'}\| &\le \mu(n-m)^{\frac{1}{2}} \overline{K}_1 \mathcal{L}^2 \int_0^t e^{-\gamma_1(t-s)} \big[\mathcal{L}\rho(F, F') e^{-\gamma_1 s} \\ &\qquad + (2+M)\|X_s^F - X_s^{F'}\|\big] e^{-\gamma_1 s} ds \\ &\le \mu \overline{K}_1 \mathcal{L}^2 (n-m)^{\frac{1}{2}} \Bigg[\mathcal{L}\rho(F, F')\gamma_1^{-1} \\ &\qquad + (2+M)\int_0^t \|X_s^F - X_s^{F'}\| ds\Bigg] e^{-\gamma_1 t}, \quad t \in R_+. \quad (23.25)\end{aligned}$$

Solving inequality (23.25), we obtain

$$\|X_t^F - X_t^{F'}\| \le \mu 2\mathcal{L}^3 \overline{K}_1 (n-m)^{\frac{1}{2}} \gamma_1^{-1} e^{-\gamma_1 t} \rho(F, F') \qquad (23.26)$$

for all $t \in R_+$ and all μ that satisfy the inequality

$$\mu(n-m)^{\frac{1}{2}} \overline{K}_1 \mathcal{L}^2 (2+M)\gamma_1^{-1} \le \ln 2. \qquad (23.27)$$

For the difference $W^F - W^{F'}$, the following estimate is true:

$$\|W^F - W^{F'}\| \leq \int_0^\infty \left[M_1 \|X_s^F - X_s^{F'}\| + \|A_s(F) - A_s(F')\| \|X_s^{F'}\|\right] ds,$$

where $A_s(F) - A_s(F')$ is the expression obtained from (23.24) by replacing the matrix P by the matrix A.

Since

$$\|A_s(F) - A_s(F')\| \leq \mu(n-m)^{\frac{1}{2}} \underline{K}_1 \left[\mathcal{L}\rho(F, F') e^{-\gamma_1 s} + (2+M)\|X_s^F - X_s^{F'}\|\right]$$

for all $s \geq 0$, we have

$$\begin{aligned}
\|W^F - W^{F'}\| &\leq \left[M_1 + \mu(2+M)\underline{K}_1 \mathcal{L}(n-m)^{\frac{1}{2}}\right] \\
&\quad \times \int_0^\infty \|X_s^F - X_s^{F'}\| ds + \mu(n-m)^{\frac{1}{2}} \underline{K}_1 \mathcal{L}(2\gamma_1)^{-1} \rho(F, F') \\
&\leq \mu(n-m)^{\frac{1}{2}} \Big\{ 2\mathcal{L}^3 \overline{K}_1 \left[M_1 + \mu(2+M)\underline{K}_1 \mathcal{L}(n-m)^{\frac{1}{2}}\right] \gamma_1^{-2} \\
&\quad + \underline{K}_1 \mathcal{L}(2\gamma_1)^{-1} \Big\} \rho(F, F') \\
&= \mu d_1 \rho(F, F'),
\end{aligned} \tag{23.28}$$

where d_1 denotes the corresponding constant. For sufficiently small $\mu > 0$, inequality (23.28) yields

$$\rho(SF, SF') \leq \frac{1}{2} \rho(F, F'), \tag{23.29}$$

which proves that the operator S is contracting.

According to the principle of contracting mappings, the operator S has a unique fixed point $F(\psi, h) = \Phi(\psi, h)$ in $C(M, K)$. This means that

$$\Phi(\psi, h) = -\int_0^\infty A[\psi_s + \Phi(\psi_s, X_s g) X_s g, X_s g] X_s ds, \tag{23.30}$$

where $X_t = X_t^\Phi$ is a solution of Eq. (23.4) for

$$F(\psi, h) = \Phi(\psi, h), \quad g = g(h).$$

Let us establish a relationship between the matrix $\Phi(\psi, h)$ and solutions of the system of equations (23.4). For this purpose, we substitute the following functions for ψ and h in (23.30):

$$\psi_t = \psi + \lambda t, \quad h_t = X_t(\psi, h)h,$$

where $t \in R_+$ and $h \in K_{\mu(\mathcal{L})^{-1}}$. As a result, we get

$$\Phi(\psi_t, h_t) = -\int_0^\infty A\big[\psi_{s+t} + \Phi(\psi_{s+t}, X_s(\psi_t, h_t)h_t)X_s(\psi_t, h_t)h_t,$$
$$X_s(\psi_t, h_t)h_t\big] X_s(\psi_t, h_t)ds.$$

It follows from the equation for $X_s(\psi, h)$ that, for $s \geq t \geq 0$, the function $X_s(\psi_t, h_t)h_t$ is a solution of the equation

$$\frac{dy}{ds} = P(\psi_{s+t} + \Phi(\psi_{s+t}, y)y, y)y \tag{23.31}$$

that takes the value

$$y_0 = h_t = X_t(\psi, h)h, \quad t \in R_+, \tag{23.32}$$

for $s = 0$. According to Eq. (23.4), the function

$$y = X_{s+t}(\psi, h)h, \quad s \geq t \geq 0,$$

is also a solution of the Cauchy problem (23.31), (23.32). It follows from the uniqueness of a solution of the Cauchy problem (23.31), (23.32) that

$$X_s(\psi_t, h_t)h_t = X_{s+t}(\psi, h)h$$

for all $s \geq t \geq 0$, $\psi \in \mathcal{T}_m$, and $h \in K_{\mu(\mathcal{L})^{-1}}$. Taking this identity into account, we conclude that

$$\Phi(\psi_t, h_t) = -\int_t^\infty A(\psi_\tau + \Phi(\psi_\tau, X_\tau h)X_\tau h, X_\tau h)X_\tau X_t^{-1} d\tau \tag{23.33}$$

for all $t \in R_+$.

We set

$$u(\psi, h) = \Phi(\psi, h)h$$

for $\psi \in \mathcal{T}_m$ and $h \in K_{\mu(\mathcal{L})^{-1}}$. Then it follows from (23.33) that

$$u(\psi_t, h_t) = \Phi(\psi_t, h_t)h_t$$
$$= -\int_t^{\infty} A(\psi_\tau + u(\psi_\tau, h_\tau), h_\tau)h_\tau d\tau, \quad t \in R_+. \tag{23.34}$$

Differentiating (23.34) with respect to t, we establish that $u(\psi_t, h_t)$ satisfies the equation

$$\frac{du}{dt} = A(\psi_t + u, h_t)h_t, \quad t \in R_+.$$

This implies that $u(\psi_t, h_t), h_t$ is a solution of the system of equations

$$\frac{du}{dt} = A(\psi_t + u, h)h, \quad \frac{dh}{dt} = P(\psi_t + u, h)h, \quad t \in R_+,$$

i.e., a solution φ_t, h_t of system (22.20) whose initial value φ, h is chosen from the conditions $h \in K_{\mu(\mathcal{L})^{-1}}$,

$$\varphi = \psi + u(\psi, h) \tag{23.35}$$

is determined by the change of variables (23.1) with $\varphi = \varphi_t$ and $h = h_t$ for all $t \in R_+$.

To complete the proof of the theorem for $p = 1$, it remains to establish that the mapping $(\psi, h) \to (\varphi, h)$ determined by (23.35) is a Lipschitz homeomorphism of the domain $\mathcal{T}_m \times K_\mu$ onto itself. Since the mapping indicated transforms h identically, it remains to prove that, using (23.35), one can find

$$\psi = \varphi + v(\varphi, h), \tag{23.36}$$

where the function $v = v(\varphi, h)$ satisfies the Lipschitz condition with respect to φ and h in the domain $\mathcal{T}_m \times K_\mu$. Substituting (23.36) into (23.35), we obtain the following equation for v:

$$v = -u(\varphi + v, h), \quad (\varphi, h) \in \mathcal{T}_m \times K_\mu.$$

The function $u(\varphi, h)$ satisfies the inequalities

$$\|u(\varphi + v, h)\| \leq \mu M,$$

$$\|u(\varphi' + v', h') - u(\varphi + v, h)\|$$
$$\leq \mu K(\|\varphi' - \varphi\| + \|v' - v\|) + (M + \mu K)\|h' - h\| \tag{23.37}$$

for arbitrary φ, h, v and φ', h', v' from the domain $\mathcal{T}_m \times K_\mu \times R^m$. In the space $C_{Lip}(\mathcal{T}_m \times K_\mu)$, we select the subspace of functions $v = v(\varphi, h)$ for which

$$\|v(\varphi, h)\| \leq \mu M,$$
$$\|v(\varphi', h') - v(\varphi, h)\| \leq (2M+1)(\|\varphi' - \varphi\| + \|h' - h\|). \tag{23.38}$$

Here, (φ, h) and (φ', h') are arbitrary points of the domain $\mathcal{T}_m \times K_\mu$. It follows from (23.37) and (23.38) that the operator

$$S_1 \colon v \to -u(\varphi + v, h)$$

maps the subspace indicated into itself for $\mu K \leq \dfrac{1}{2}$ and is a contraction operator:

$$\rho(S_1 v, S_1 v') \leq \mu K \rho(v, v') \leq \frac{1}{2}\rho(v, v'),$$

where

$$\rho(v, v') = \max_{\varphi, h} \|v(\varphi, h) - v'(\varphi, h)\|.$$

This is sufficient for the equation for v to have a unique solution $v = v(\varphi, h)$ defined for $(\varphi, h) \in \mathcal{T}_m \times K_\mu$ and satisfying inequalities (23.38). For $p = 1$, Theorem 23.1 is proved.

Remark 1. To prove this theorem for $p \geq 2$, it remains to investigate the smoothness of the change of variables (23.35) in the domain $\mathcal{T}_m \times K_\mu$. The case $p = 2$ is principal for this investigation because, beginning with this case, the function $\Phi(\psi, h)$ becomes continuously differentiable and turns into the classical solution of Eq. (23.1) for ψ, h from the domain $\mathcal{T}_m \times K_\mu$. It is clear that, for $p \geq 2$, it is necessary to take a p times continuously differentiable function $g(h)$. To study the smoothness of the function $\Phi(\psi, h)$, we use the fact that $\Phi(\psi, h)$ can be obtained as the limit (as $j \to \infty$) of the successive approximations $\Phi_j(\psi, h)$, $j \geq 1$, defined by the equality

$$\Phi_j(\psi, h) = -\int_0^\infty A\big[\psi_s + \Phi_{j-1}(\psi_s, X_s^{(j-1)} g) X_s^{(j-1)} g, X_s^{(j-1)} g\big] X_s^{(j-1)} ds,$$

where $\Phi_0(\psi, h) \equiv 0$ and $X_t^{(j-1)}$ is a solution of the equation

$$\frac{dX}{dt} = P[\psi_t + \Phi_{j-1}(\psi_t, Xg) Xg, Xg] X, \quad j = 1, 2, \ldots,$$

that takes the value E for $t = 0$. One can prove [Sam6] that the functions $\Phi_j(\psi, h)$ and their partial derivatives with respect to ψ and h up to the order p are continuous on the set $\mathcal{T}_m \times K_\mu$ and uniformly bounded for all $j \geq 0$, $\psi \in \mathcal{T}_m$, and $h \in K_\mu$ by a constant c_0. Then it follows from the Arzelà theorem [KoF] that the sequence $\Phi_j(\psi, h)$ is compact in $C^{p-1}(\mathcal{T}_m \times K_\mu)$ and the $(p-1)$th derivatives of the limit function satisfy the Lipschitz condition. These arguments complete the proof of Theorem 23.1.

24. Variational Equation and Theorem on Attraction to Quasiperiodic Solutions

Consider the following variational equation for a solution (22.3) of system (22.2):

$$\frac{dy}{dt} = \frac{\partial X(f(\lambda t + \psi^0))}{\partial x} y. \tag{24.1}$$

Assume that $f \in C^1(\mathcal{T}_m)$ and the matrix $\dfrac{\partial f(\varphi)}{\partial \varphi}$ can be complemented to a 2π-periodic basis in R^n. It follows from Eq. (24.1) that the system of functions $\dfrac{\partial f(\lambda t + \psi^0)}{\partial \varphi_\nu}$, $\nu = \overline{1, m}$, forms a system of linearly independent solutions of Eq. (24.1). We perform a change of variables in (24.1), namely, instead of y, we introduce new variables $(c, h) = (c_1, \ldots, c_m, h_1, \ldots, h_{n-m})$ according to the formulas

$$y = \frac{\partial f(\lambda t + \psi^0)}{\partial \varphi} c + B(\lambda t + \psi^0) h, \tag{24.2}$$

where $B(\varphi)$ is the matrix from formula (22.7). As a result, we obtain the system of equations

$$\frac{dc}{dt} = Q(\lambda t + \psi^0) h, \quad \frac{dh}{dt} = P(\lambda t + \psi^0) h, \tag{24.3}$$

where $P(\varphi)$ is matrix (22.17). The relationship between systems (24.3) and (22.16) is obvious, namely, the equations for h in system (24.3) are obtained from (22.16) for $\varphi = \lambda t + \psi^0$, and, vice versa, the closure of the second equation in system (24.3) with respect to t leads to system (22.16).

In what follows, the system of equations (22.16) is called the variational system of equations for the invariant torus (22.4) of the dynamical system (22.2) that is filled with the quasiperiodic trajectory of this system. The statement below describes the dependence of the variational equation (22.16) on the matrix $B(\varphi)$.

Theorem 24.1. *Under the assumptions made above, two arbitrary equations (22.16) are C^{p-1}-equivalent.*

Proof. Consider two changes of variables of the form (24.2) defined by matrices $B \in C^p(\mathcal{T}_m)$ and $B_1 \in C^p(\mathcal{T}_m)$, respectively. Denote by $P(\varphi)$ and $P_1(\varphi)$ the matrices of Eq. (24.3) obtained as a result of these changes of variables. Since both systems of equations are obtained with the use of changes of variables of the form (24.2) on the basis of the same equation (24.1), their fundamental matrices of solutions are related by the identity

$$\left[\frac{\partial f(\psi_t)}{\partial \varphi}, B(\psi_t)\right]\begin{pmatrix} E & Q \\ 0 & \Omega_0^t(P) \end{pmatrix} = \left[\frac{\partial f(\psi_t)}{\partial \varphi}, B_1(\psi_t)\right]\begin{pmatrix} E & Q_1 \\ 0 & \Omega_0^t(P_1) \end{pmatrix} C_1, \quad (24.4)$$

where C_1 is a nondegenerate constant matrix.

Let

$$\begin{pmatrix} L_1(\psi_t) \\ L_2(\psi_t) \end{pmatrix}$$

be the matrix inverse to $\left[\frac{\partial f(\psi_t)}{\partial \varphi}, B(\psi_t)\right]$, i.e.,

$$L_1(\psi_t)\frac{\partial f(\psi_t)}{\partial \varphi} = E, \qquad L_1(\psi_t)B(\psi_t) = O,$$

$$L_2(\psi_t)\frac{\partial f(\psi_t)}{\partial \varphi} = O, \qquad L_2(\psi_t)B(\psi_t) = E,$$

where E and O are, respectively, the identity matrix and the zero matrix of the corresponding dimensions. Using (24.4), one can easily obtain the following identity:

$$\begin{pmatrix} E & Q \\ 0 & \Omega_0^t(P) \end{pmatrix} = \begin{pmatrix} E & L_1(\psi_t)B_1(\psi_t) \\ 0 & L_2(\psi_t)B_1(\psi_t) \end{pmatrix}\begin{pmatrix} E & Q_1 \\ 0 & \Omega_0^t(P_1) \end{pmatrix} C_1. \quad (24.5)$$

Analyzing relation (24.5), we conclude that the matrix $L_2(\psi_t)B_1(\psi_t)$ is nondegenerate and

$$\Omega_0^t(P) = L_2(\psi_t)B_1(\psi_t)\Omega_0^t(P_1)C_2, \quad (24.6)$$

where C_2 is a nondegenerate constant matrix. Identity (24.6) proves that the change of variables

$$h = L_2(\varphi_t)B_1(\varphi_t)h_1 \tag{24.7}$$

transforms system (22.16) with the matrix $P(\varphi)$ to system (22.16) with the matrix $P_1(\varphi)$. Since $L_2B_1 \in C^{p-1}(\mathcal{T}_m)$, the change of variables (24.7) realizes a C^{p-1}-homeomorphism from (φ, h) onto (φ, h_1).

Theorem 24.1 is proved.

Let us clarify the behavior of solutions originating in a neighborhood of the torus $M\colon x = f(\varphi), \varphi \in \mathcal{T}_m$. For this purpose, we define the distance from the point y^0 to M by the formula

$$\rho(y^0, M) = \inf_{x \in M} \|y^0 - x\|.$$

Theorem 24.2. *Suppose that the smoothness conditions for the function $X = X(x)$ presented in Section 22 are satisfied and the system of equations (22.2) has a quasiperiodic solution $x = f(\lambda t) \in C^s(\lambda)$ for $r \geq s \geq 2$. Also assume that the matrix $\dfrac{\partial f(\varphi)}{\partial \varphi}$ can be complemented to a 2π-periodic basis in R^n, and the variational equation for the invariant torus M satisfies the condition of exponential stability (22.19).*

Then one can find a sufficiently small positive $\delta > 0$ such that, for every y^0 satisfying the inequality $\rho(y^0, M) \leq \delta$, there exist $\varphi^0 \in \mathcal{T}_m$ and $\psi^0 \in \mathcal{T}_m$ such that

$$\|x(t, y^0) - f(\varphi_t)\| \leq Ke^{-\gamma_1 t}\|y^0 - f(\varphi^0)\| \tag{24.8}$$

for all $t \in R_+$ and certain $K > 0$ and $\gamma_1 > 0$, where $\gamma_1 = \gamma_1(\delta) \to 0$ and $\|\psi^0 - \varphi^0\| \to 0$ as $\delta \to 0$.

Proof. We choose $\delta > 0$ so small that the inequality $\rho(y^0, M) \leq \delta$ yields the inequality $\|h^0\| \leq \mu$, where μ is the constant from Theorem 23.1 and (φ^0, h^0) are the local coordinates of the point y^0, i.e.,

$$y^0 = f(\varphi^0) + B(\varphi^0)h^0. \tag{24.9}$$

The possibility of such a choice of δ is guaranteed by inequality (22.10). Let

$$\psi^0 = \varphi^0 + v(\varphi^0, h^0), \tag{24.10}$$

where the function $v = v(\varphi, h)$ is defined by (23.36), and let $\psi_t = \lambda t + \psi^0$, where ψ^0 is defined by (24.10).

Let us estimate the difference $x(t, y^0) - f(\varphi_t)$. It follows from (24.9) that

$$x(t, y^0) = f(\varphi_t) + B(\varphi_t)h_t, \quad t \in R_+, \tag{24.11}$$

where φ_t, h_t is a solution of system (22.12) that takes the value φ^0, h^0 for $t = 0$.

It follows from (24.10) that

$$\varphi^0 = \psi^0 + u(\varphi^0, h^0),$$

where $u(\varphi, h)$ is the function of the change of variables (23.1), which reduces system (22.20) to the form (23.2).

According to Theorem 23.1, the solutions $\varphi_t, h(t)$ and ψ_t, h_t are related as follows:

$$\varphi_t = \psi_t + u(\psi_t, h_t), \quad h(t) = h_t, \quad t \in R_+. \tag{24.12}$$

Then relations (24.11) and (24.12) yield the following estimate for all $t \in [0, \infty)$:

$$\begin{aligned} \|x(t, y^0) - f(\psi_t)\| &\leq \|f(\psi_t + u(\psi_t, h_t)) - f(\psi_t)\| + \|B(\varphi_t)\|\|h_t\| \\ &\leq K\|u(\psi_t, h_t)\| + c_1\|h_t\| \leq c_2\|h_t\| \end{aligned}$$

where c_2 is a certain positive constant.

Since $h_t = X_t^{\Phi} h^0$, where X_t^{Φ} is a solution of Eq. (23.4) for $F = \Phi(\psi, g)$, $g = h$, we conclude that h_t satisfies the following estimate of the form (23.8):

$$\|h_t\| \leq \mathcal{L}e^{-\gamma_1 t}\|h^0\|.$$

This yields

$$\|x(t, y^0) - f(\psi_t)\| \leq \mathcal{L}c_2 e^{-\gamma_1 t}\|h^0\|, t \in R_+. \tag{24.13}$$

It follows from (24.9) that

$$\|h^0\| = \|\Gamma_0^{-1}(\varphi^0)B^T(\varphi^0)[y^0 - f(\varphi^0)]\| \leq c_3\|y^0 - f(\varphi^0)\|, \tag{24.14}$$

where c_3 is a certain positive constant. Inequalities (24.13) and (24.14) yield estimate (24.8). Since $\gamma_1 = \gamma_1(\mu) \to 0$ and $\|\psi^0 - \varphi_0\| \to 0$ as $\mu \to 0$, we have $\gamma_1 \to \gamma$ and $\|\psi^0 - \varphi^0\| \to 0$ as $\delta \to 0$, which completes the proof of Theorem 24.2.

Below, we present two corollaries of Theorem 24.2.

Corollary 1. *Under the conditions of Theorem 24.2, the quasiperiodic solutions* $x = f(\lambda t + \psi)$, $\psi \in \mathcal{T}_m$, *are Lyapunov stable.*

Indeed, let y^0 be an arbitrary point of the ball $\|y - x^0\| < \delta$, where δ is a sufficiently small positive number and $x^0 = f(\psi^0)$. The local coordinates (φ^0, h^0) of the point y^0 are determined from the equation

$$y^0 - f(\psi^0) = f(\varphi) - f(\psi^0) + B(\psi^0)h + \big[B(\varphi) - B(\psi^0)\big]h.$$

We rewrite this equation in the form

$$y^0 - f(\psi^0) = \frac{\partial f(\psi^0)}{\partial \varphi}(\varphi - \psi^0) + B(\psi^0)h + D(\varphi, h), \tag{24.15}$$

where $D(\varphi, h)$ denotes a value of higher order of smallness as compared with $\|\varphi - \psi^0\| + \|h\|$. It follows from the inequality $\|y^0 - x^0\| \leq \delta$, where δ is small, that the solution $\varphi = \varphi^0$, $h = h^0$ of Eq. (24.15) satisfies the conditions

$$\|\varphi^0 - \psi^0\| \leq \delta_1(\delta), \quad \|y^0 - f(\varphi^0)\| \leq \delta_1(\delta),$$

where $\delta_1(\delta) \to 0$ as $\delta \to 0$.

According to Theorem 24.2, the following inequality holds:

$$\|x(t, y^0) - f(\lambda t + \overline{\psi}^0)\| \leq K e^{-\gamma_1 t}\|y^0 - f(\varphi^0)\|, \quad t \in R_+,$$

where $\|\overline{\psi}^0 - \varphi^0\| \to 0$ as $\delta \to 0$.

This yields the following estimate for $t \in R_+$:

$$\begin{aligned}\|x(t, y^0) - x(t, x^0)\| &\leq \|x(t, y^0) - f(\lambda t + \overline{\psi}^0)\| + \|f(\lambda t + \overline{\psi}^0) \\ &\quad - f(\lambda t + \varphi^0)\| + \|f(\lambda t + \varphi^0) - f(\lambda t + \psi^0)\| \\ &\leq K\|y^0 - f(\varphi^0)\| + K_1(\|\overline{\psi}^0 - \varphi^0\| + \|\varphi^0 - \psi^0\|) \\ &\leq (K + K_1)\delta_1(\delta) + K_1\|\overline{\psi}^0 - \varphi^0\|, \end{aligned} \tag{24.16}$$

where K_1 is a constant that depends only on the first-order derivatives of the function $f(\varphi)$. Taking into account that $\|\overline{\psi}^0 - \varphi^0\| \to 0$ as $\delta \to 0$ and using inequality (24.16), for any $\varepsilon > 0$ one can choose $\delta = \delta(\varepsilon) > 0$ so small that the right-hand side of (24.16) is less than ε. Thus, the stability of the quasiperiodic solution $x(t, x^0) = f(\lambda t + \psi^0)$ is proved.

Corollary 2. *Under the conditions of Theorem 24.2, for an arbitrary function* $F = F(x)$ *satisfying the Hölder condition and an arbitrary solution* $x = x(t, y^0)$ *for which* $\rho(y^0, M) \le \delta$, *the following limit relation holds uniformly in* $t \in R_+$:

$$\lim_{T\to\infty} \frac{1}{T} \int_t^{t+T} F(x(t, y^0))dt = (2\pi)^{-m} \int_0^{2\pi} \dots \int_0^{2\pi} F(f(\varphi))d\varphi_1 \dots d\varphi_m. \tag{24.17}$$

Note that the property of the dynamical system (22.2) expressed by equality (24.17) characterizes the ergodicity of semitrajectories in a neighborhood of the manifold M.

To prove relation (24.17), it is necessary to use estimate (24.8), which implies that

$$\|F(x(t, y^0)) - F(f(\psi_t))\| \le c_4 e^{-\gamma_1 \beta t}, \quad t \in R_+, \tag{24.18}$$

where β is the Hölder index of the function $F(x)$. Inequality (24.18) yields

$$\frac{1}{T} \int_t^{t+T} F(x(t, x^0))dt = \frac{1}{T} \int_t^{t+T} F(f(\psi_t))dt + \alpha(t, T), \tag{24.19}$$

where

$$\|\alpha(t, T)\| \le \frac{c_4}{T} \int_t^{t+T} e^{-\gamma_1 \beta t} dt = \frac{c_4}{T} \frac{e^{-\gamma_1 \beta t}(1 - e^{-\gamma_1 \beta T})}{\gamma_1 \beta}. \tag{24.20}$$

Inequality (24.20) implies that the following relation holds uniformly in $t \in R_+$:

$$\lim_{T\to\infty} \alpha(t, T) = 0.$$

Equality (24.17) can easily be obtained from relations (24.19) and (24.20), and the mean-value theorem for a quasiperiodic function.

25. Behavior of Trajectories under Small Perturbations of a Dynamical System

Let $Y = Y(x)$ be a function of the same smoothness as $X(x)$ and let ε be a small positive parameter. Parallel with system (22.2), we consider the system of equations

$$\frac{dx}{dt} = X(x) + \varepsilon Y(x), \tag{25.1}$$

which is called perturbed with respect to (22.2).

We pose the problem of the investigation of the behavior of solutions of system (25.1) originating in a small neighborhood of the manifold $M\colon x = f(\varphi)$, $\varphi \in \mathcal{T}_m$. This problem includes the investigation of the problem of the existence of an invariant torus $M(\varepsilon)\colon x = f(\varphi, \varepsilon)$, $\varphi \in \mathcal{T}_m$, that tends to M as $\varepsilon \to 0$, and the investigation of the behavior of trajectories of system (25.1) originating in the manifold $M(\varepsilon)$ and in its neighborhood.

To solve the problem posed, we introduce local coordinates (φ, h) in the neighborhood of M according to formula (22.7). As a result, we obtain the equations

$$\frac{d\varphi}{dt} = \lambda + A(\varphi, h)h + \varepsilon L_1(\varphi, h)Y(f(\varphi) + B(\varphi)h),$$

$$\frac{dh}{dt} = P(\varphi, h)h + \varepsilon L_2(\varphi, h)Y(f(\varphi) + B(\varphi)h), \tag{25.2}$$

where A, P, L_1, and L_2 are the functions that define the system of equations (22.20), $\varphi \in \mathcal{T}_m$, and $h \in K_\delta$.

For the system of equations (25.2), we study the problem of the existence of an invariant manifold

$$h = u(\varphi, \varepsilon), \quad \varphi \in \mathcal{T}_m, \tag{25.3}$$

that tends to the trivial one $h = 0$, $\varphi \in \mathcal{T}_m$, as $\varepsilon \to 0$.

Theorem 25.1. *Suppose that the conditions of Theorem 23.1 are satisfied. Then there exists* $\varepsilon_0 = \varepsilon_0(r) > 0$ *such that, for any* $\varepsilon \in [0, \varepsilon_0]$, *the system of equations (25.2) has the invariant torus (25.3) with a function* $u(\varphi, \varepsilon) \in C^r_{Lip}(\mathcal{T}_m)$, *where* $r = p - 1$ *for finite* p *and* $r < p$ *for* $p = \infty$, *that satisfies the condition*

$$\lim_{\varepsilon \to 0} \|u(\varphi, \varepsilon)\|_{r,Lip} = 0. \tag{25.4}$$

Here, $\|u(\varphi, \varepsilon)\|_{r,Lip} = \|u(\varphi, \varepsilon)\|_r + K_r$, $K_r = K_r(\varepsilon)$ *is the Lipschitz constant of the* r*th derivatives of the function* $u(\varphi, \varepsilon)$ *with respect to* φ, *and* $\|u(\varphi, \varepsilon)\|_r$ *is the norm of the function* $u(\varphi, \varepsilon)$ *as an element of the space* $C^r(\mathcal{T}_m)$.

Theorem 25.1 follows from the perturbation theory of invariant manifolds, which was constructed in works of numerous authors [BMS, MiR, Sam4, Hal1, Kup, Sac1]; in particular, it is a direct corollary of Theorem 1 in [Sam4] (Chapter 4, Section 3).

Let $r \geq 1$. The change of variables

$$h = u(\varphi, \varepsilon) + z$$

brings system (25.2) to the form

$$\frac{d\varphi}{dt} = \lambda + F(\varphi, \varepsilon) + A(\varphi, z, \varepsilon)z,$$
$$\frac{dz}{dt} = P(\varphi, z, \varepsilon)z, \tag{25.5}$$

where $F(\varphi, \varepsilon)$ denotes the function

$$F(\varphi, \varepsilon) = A(\varphi, u(\varphi, \varepsilon))u(\varphi, \varepsilon) + \varepsilon L_1(\varphi, u(\varphi, \varepsilon))Y(f(\varphi) + B(\varphi)u(\varphi, \varepsilon)), \tag{25.6}$$

and $A(\varphi, z, \varepsilon)$ and $P(\varphi, z, \varepsilon)$ are the matrices defined as follows:

$$A(\varphi, z, \varepsilon) = A(\varphi, u(\varphi, \varepsilon) + z) + \int_0^1 \frac{\partial}{\partial h} A(\varphi, u(\varphi, \varepsilon) + tz)u(\varphi, \varepsilon)dt$$
$$+ \varepsilon \int_0^1 \frac{\partial}{\partial h}\Big\{L_1(\varphi, u(\varphi, \varepsilon) + tz)Y\Big[f(\varphi) + B(\varphi)(u(\varphi, \varepsilon) + tz)\Big]\Big\}dt,$$

$$P(\varphi, z, \varepsilon) = P(\varphi, u(\varphi, \varepsilon) + z) + \int_0^1 \frac{\partial}{\partial h} P(\varphi, u(\varphi, \varepsilon) + tz)u(\varphi, \varepsilon)dt$$
$$+ \varepsilon \int_0^1 \frac{\partial}{\partial h}\Big\{L_2(\varphi, u(\varphi, \varepsilon) + tz)Y\Big[f(\varphi) + B(\varphi)(u(\varphi, \varepsilon) + tz)\Big]\Big\}dt$$
$$- \frac{\partial u(\varphi, \varepsilon)}{\partial \varphi} A(\varphi, z, \varepsilon). \tag{25.7}$$

It follows from relations (25.6) and (25.7) that, for every fixed $\varepsilon \in [0, \varepsilon_0]$, the functions F, A, and P belong to the space $C^{r-1}_{Lip}(\mathcal{T}_m \times K_{\delta_0})$, where $\delta_0 = \delta_0(\varepsilon_0)$ tends to δ as $\varepsilon_0 \to 0$. Moreover, relations (25.4), (25.6), and (25.7) yield

$$\lim_{\varepsilon \to 0} \|F(\varphi, \varepsilon)\|_r = 0, \quad \lim_{\varepsilon \to 0} \|A(\varphi, z, \varepsilon) - A(\varphi, z)\|_{r-1} = 0,$$
$$\lim_{\varepsilon \to 0} \|P(\varphi, z, \varepsilon) - P(\varphi, z)\|_{r-1} = 0. \tag{25.8}$$

Equalities (25.8) enable one to properly modify the reasoning used in the proof of Theorem 23.1 and obtain an analog of Theorem 23.1 for the system of equations (25.5).

Theorem 25.2. *Suppose that, for every $\varepsilon \in [0, \varepsilon_0]$, the functions $F(\varphi, \varepsilon)$, $A(\varphi, z, \varepsilon)$, and $P(\varphi, z, \varepsilon)$ belong to the space $C^{r-1}(\mathcal{T}_m \times K_{\delta_0})$ and satisfy conditions (25.8) for $r \geq 2$, and the fundamental matrix $\Omega_0^t(\varphi)$ of solutions of system (22.16) satisfies inequality (22.19).*

Then one can find $\mu > 0$, $\bar{\varepsilon}_0 = \bar{\varepsilon}_0(r) > 0$, and a matrix $\Phi(\varphi, z, \varepsilon)$ that belongs to the space $C_{Lip}^{r-2}(\mathcal{T}_m \times K_\mu)$ for every $\varepsilon \in [0, \bar{\varepsilon}_0]$ and satisfies the condition

$$\lim_{\varepsilon \to 0} \|\Phi(\varphi, z, \varepsilon) - \Phi(\varphi, z, 0)\|_{r-2, Lip} = 0$$

such that the change of variables

$$\varphi = \psi + \Phi(\psi, z, \varepsilon) z$$

reduces the system of equations (25.5) to the form

$$\frac{d\psi}{dt} = \lambda + F(\psi, \varepsilon), \quad \frac{dz}{dt} = P(\psi + \Phi(\psi, z, \varepsilon) z, z, \varepsilon) z. \tag{25.9}$$

Theorem 25.2 can be proved by analogy with Theorem 23.1 with the difference that one should take into account the dependence (more complicated than in the case of Theorem 23.1) of the solution $\psi_t = \psi_t(\psi, \varepsilon)$ of the first equation in (25.9) on the values of $\psi \in \mathcal{T}_m$. This dependence is such that the derivatives

$$D_\psi^l \psi_t(\psi, \varepsilon) = \frac{\partial^l \psi_t(\psi, \varepsilon)}{\partial \psi_1^{l_1} \ldots \partial \psi_m^{l_m}}, \quad l = l_1 + \ldots + l_m,$$

of order $l \leq r$ satisfy the estimate

$$\|D_\psi^l \psi_t(\psi, \varepsilon)\| \leq c e^{\eta l t}, \quad t \in R_+ \tag{25.10}$$

for a certain constant $c = c(l, m)$ and a certain $\eta = \eta(\varepsilon)$ that depends only on $\left\| \frac{\partial}{\partial \psi} F(\psi, \varepsilon) \right\|$ and is such that $\eta(\varepsilon) \to 0$ as $\varepsilon \to 0$. The corresponding modification of the proof of Theorem 23.1 caused by estimate (25.10) does not lead to substantial difficulties.

Note that the essential difference of Theorem 25.2 from Theorem 23.1 is the fact that $\bar{\varepsilon}_0$ depends on r. This dependence is such that $\bar{\varepsilon}_0(r) \to 0$ as $r \to \infty$.

Therefore, the case $r = \infty$ is not solved in Theorem 25.2. The question of whether this is caused by a fundamental reason or by the method of the proof of Theorems 23.1 and 25.2 remains open.

It follows from Eqs. (25.9) that the behavior of solutions of system (25.1) in the neighborhood of the manifold M depends essentially only on the behavior of solutions of this system on the manifold $M(\varepsilon)$, i.e., on the behavior of solutions of the system of equations

$$\frac{d\varphi}{dt} = \lambda + F(\varphi, \varepsilon). \tag{25.11}$$

For $m = 2$, a qualitative description of the trajectories of system (25.11) is given by the Poincaré–Denjoy theory [Poi, Den] and the theorem presented in [Her].

For $m \geq 3$, we can apply to system (25.11) the result of [BMS] concerning the rectification of the smooth flow of a trajectory on a torus. For system (25.11), this result can be formulated as follows:

Theorem 25.3. *Suppose that* $\lim\limits_{\varepsilon\to 0} \|F(\varphi,\varepsilon)\|_r = 0$ *and* $\lambda = \omega + \Delta$, *where* $\omega = (\omega_1, \ldots, \omega_m)$ *satisfies the following inequality for all integer-valued vectors* $k \neq 0$:

$$|(k, \omega)| \geq c\|k\|^{-(m+1)},$$

where $c > 0$ *is a certain constant. Then, for integer* $s \geq 1$, *one can find sufficiently small* $\varepsilon_0 = \varepsilon_0(c, s) > 0$ *and* $\Delta = \Delta(\varepsilon) = (\Delta_1, \ldots, \Delta_m)$ $\big(\|\Delta(\varepsilon)\| \to 0$ *as* $\varepsilon \to 0\big)$, *an integer* $l = l(s, m)$, *and a function* $\Psi = (\Psi_1, \ldots, \Psi_m) = \Psi(\varphi, \varepsilon)$ *belonging to the space* $C^s(\mathcal{T}_m)$ *for every* $\varepsilon = [0, \varepsilon_0]$ *and satisfying the condition*

$$\lim_{\varepsilon\to 0} \|\Psi(\varphi,\varepsilon)\|_s = 0$$

such that, for $l \leq r$, *the change of variables*

$$\varphi = \psi + \Psi(\psi, \varepsilon)$$

reduces the system of equations (25.11) with chosen $\Delta(\varepsilon)$ *to the system*

$$\frac{d\psi}{dt} = \omega.$$

In this form, Theorem 25.3 follows from Theorem 12 in [BMS] (Chapter 14, Section 17) .

In [Mos4], the dependence $l = l(s, m)$ was reduced to the form

$$l(s, m) \geq s + m + 2.$$

The method of the proof of Theorem 25.3 is the method with accelerated convergence of iterations of Newton type, which was developed in the course of the construction of the Kolmogorov–Arnol'd–Moser theory [Arn3, Mos5].

Theorem 25.3 specifies conditions under which the perturbed system of equations (25.1) has quasiperiodic solutions in the neighborhood of M. These conditions show that, as a rule, small perturbations destroy quasiperiodic solutions.

Completing the investigation of the dynamical system (25.1) in the neighborhood of a quasiperiodic solution of the unperturbed system of equations (22.2), we present the result concerning the phenomenon of attraction to solutions on manifold (25.3) that follows from Theorem 25.2 and is an analog of Theorem 24.2 for the case of the perturbed system (25.1).

Theorem 25.4. *If the conditions of Theorem 24.2 are satisfied and $Y(x)$ is r times continuously differentiable in R^n, then there exist sufficiently small $\delta > 0$ and $\varepsilon_0 > 0$ such that, for every y^0, $\rho(y^0, M(\varepsilon)) \leq \delta$, one can find $\psi^0 \in \mathcal{T}_m$ and $\varphi^0 \in \mathcal{T}_m$ such that the following inequality holds for $t \in R_+$:*

$$\|x(t, y^0, \varepsilon) - [f(\psi_t) + u(\psi_t, \varepsilon)]\| \leq \overline{c}e^{-\gamma_1 t}\|y^0 - [f(\varphi^0) + u(\varphi^0, \varepsilon)]\|,$$

where $K > 0$ and $\gamma_1 > 0$ are certain constants such that $\gamma_1 = \gamma_1(\delta, \varepsilon) \to \gamma$ as $\delta \to 0$ and $\varepsilon \to 0$, and $\|\psi^0 - \varphi^0\| \to 0$ as $\delta \to 0$.

Theorem 25.4 yields the principle of reducibility for solutions of system (25.1) originating on $M(\varepsilon)$. According to this principle, the stability of the solutions indicated follows from their stability on $M(\varepsilon)$.

26. The Case of a Toroidal Manifold Filled with Trajectories of General Form

In Sections 22–25, we have studied a dynamical system in the neighborhood of an invariant manifold filled with a quasiperiodic trajectory of the system. In what follows, we extend the results obtained in Sections 22–25 to the case where the invariant toroidal manifold of the system is filled with trajectories of the general form. In particular, we present a theorem on the reducibility of the dynamical system in the neighborhood of the invariant toroidal manifold M and a statement on the exponential attraction of solutions from the neighborhood of the manifold M to solutions on M and establish conditions for the invariance of the behavior of trajectories of the dynamical system in the neighborhood of the manifold M under small perturbations.

Consider the system

$$\frac{dx}{dt} = X(x), \tag{26.1}$$

where $x = (x_1, \ldots, x_n)$ is a point of the n-dimensional Euclidean space R^n and $X(x) \in C^r(R^n)$, $r \geq 1$. Let $f = f(\varphi)$ be a function from the space $C^s(\mathcal{T}_m)$, $s \leq r$, of 2π-periodic functions of $\varphi = (\varphi_1, \ldots, \varphi_m)$ of smoothness $s \geq 2$ and with values in R^n. Let

$$M\colon x = f(\varphi), \quad \varphi \in \mathcal{T}_m, \tag{26.2}$$

be the invariant set of system (26.1) and let

$$\operatorname{rank} \frac{\partial f(\varphi)}{\partial \varphi} = m, \quad \varphi \in \mathcal{T}_m. \tag{26.3}$$

According to [Sam4], the first condition for the set M is satisfied if

$$\left[\frac{\partial f(\varphi)}{\partial \varphi}\Gamma^{-1}(\varphi)\left(\frac{\partial f(\varphi)}{\partial \varphi}\right)^T - E\right] X(f(\varphi)) = 0, \quad \varphi \in \mathcal{T}_m,$$

where $\Gamma(\varphi) = \left(\frac{\partial f(\varphi)}{\partial \varphi}\right)^T \frac{\partial f(\varphi)}{\partial \varphi}$, and the second condition means that M is a toroidal manifold.

The system of equations (26.1) on M can be reduced to a dynamical system on the torus $\mathcal{T}_m$ of the form

$$\frac{d\varphi}{dt} = a(\varphi), \tag{26.4}$$

where, according to [Sam4], the function $a(\varphi)$ has the form

$$a(\varphi) = \Gamma^{-1}(\varphi)\left(\frac{\partial f(\varphi)}{\partial \varphi}\right)^T X(f(\varphi)), \quad \varphi \in \mathcal{T}_m.$$

Assume that the m-frame $\dfrac{\partial f(\varphi)}{\partial \varphi}$ can be complemented to a 2π-periodic basis in R^n, and $B(\varphi)$ is a complementing matrix from $C^s(\mathcal{T}_m)$. If we introduce the local coordinates

$$x = f(\varphi) + B(\varphi)h \tag{26.5}$$

in the neighborhood of the manifold M, then, taking into account the invariance of M and Eq. (26.4) for the flow of trajectories on M, we can rewrite the system of equations (26.1) in the neighborhood of M in the local coordinates φ, h as follows:

$$\frac{d\varphi}{dt} = a(\varphi) + L_1(\varphi, h)\Big[X(f(\varphi) + B(\varphi)h) - X(f(\varphi)) - \frac{\partial B(\varphi)}{\partial \varphi}a(\varphi)h\Big],$$

$$\frac{dh}{dt} = L_2(\varphi, h)\Big[X(f(\varphi) + B(\varphi)h) - X(f(\varphi)) - \frac{\partial B(\varphi)}{\partial \varphi}a(\varphi)h\Big]. \tag{26.6}$$

Here, $L_1(\varphi, h)$ and $L_2(\varphi, h)$ are blocks of the matrix inverse to the matrix

$$\Big[\frac{\partial f(\varphi)}{\partial \varphi} + \frac{\partial B(\varphi)}{\partial \varphi}h, B(\varphi)\Big], \qquad \frac{\partial B(\varphi)}{\partial \varphi}a(\varphi) \equiv \sum_{\nu=1}^{m} \frac{\partial B(\varphi)}{\partial \varphi_\nu}a_\nu(\varphi),$$

and φ and h are points from the domain

$$\varphi \in \mathcal{T}_m, \qquad \|h\| \le \delta, \tag{26.7}$$

where $\delta > 0$ is sufficiently small.

Consider the variational equation for the manifold M

$$\frac{d\varphi}{dt} = a(\varphi), \qquad \frac{dh}{dt} = P(\varphi)h, \tag{26.8}$$

where, by definition [Sam4],

$$P(\varphi) = L_2(\varphi, 0)\Big[\frac{\partial X(f(\varphi))}{\partial x} - \frac{\partial B(\varphi)}{\partial \varphi}a(\varphi)\Big]. \tag{26.9}$$

Let

$$\varphi = \psi_t(\varphi), \qquad \psi_0(\varphi) = \varphi \in \mathcal{T}_m, \tag{26.10}$$

be a solution of the first equation of system (26.8) and let $\Omega_0^t(P)$ be the fundamental matrix of solutions of the second equation of system (26.8) for $\varphi = \psi_t(\varphi)$. Using the matrix $P(\varphi)$, we define a function $\beta(\varphi)$ as follows:

$$\inf_{S \in N} \max_{\|h\|=1} \frac{\left\langle \Big[S(\varphi)P(\varphi) + \frac{1}{2}\frac{\partial S(\varphi)}{\partial \varphi}a(\varphi)\Big]h, h \right\rangle}{\langle S(\varphi)h, h\rangle} \le -\beta(\varphi), \tag{26.11}$$

where N is the set of $(n-m) \times (n-m)$ positive-definite symmetric matrices $S = S(\varphi) \in C^1(\mathcal{T}_m)$ and $\langle \cdot, \cdot \rangle$ is the scalar product in R^n.

Assume that

$$\beta_0 = \inf_{\varphi \in \mathcal{T}_m} \beta(\varphi) > 0. \tag{26.12}$$

Condition (26.12) is sufficient [Sam4] for the following inequality to be satisfied for all $t \in R_+$ and $\varphi \in \mathcal{T}_m$:

$$\|\Omega_0^t(P)\| \leq \mathcal{L}e^{-\gamma t}, \tag{26.13}$$

where γ is an arbitrary positive number satisfying the inequality $\gamma < \beta_0$, and $\mathcal{L} = \mathcal{L}(\gamma)$ is a certain positive constant.

Further, we define a function $\alpha_1(\varphi)$ by the inequality

$$\inf_{S_1 \in N_1} \max_{\|\psi\|=1} \frac{\left\langle \left[S_1(\varphi) \dfrac{\partial a(\varphi)}{\partial \varphi} + \dfrac{1}{2} \dfrac{\partial S_1(\varphi)}{\partial \varphi} a(\varphi) \right] \psi, \psi \right\rangle}{\langle S_1(\varphi)\psi, \psi \rangle} \leq \alpha_1(\varphi), \tag{26.14}$$

where N_1 is the set of m-dimensional square positive-definite symmetric matrices $S_1 = S_1(\varphi) \in C^1(\mathcal{T}_m)$.

Using $\alpha_1(\varphi)$, we can obtain the following estimate for the derivatives of the function $\psi_t(\varphi)$ with respect to φ [Sam4]:

$$\left\| \frac{\partial^l \psi_t(\varphi)}{\partial \varphi_1^{l_1} \dots \partial \varphi_m^{l_m}} \right\| \leq \mathcal{L}_1 \exp \left\{ \int_0^t l a(\psi_\tau(\varphi)) d\tau + \mu t \right\}, \quad t \in R_+, \tag{26.15}$$

where $l = l_1 + \ldots + l_m$, μ is an arbitrarily small positive number, and $\mathcal{L}_1 = \mathcal{L}_1(l, \mu)$ is a certain positive constant.

Assume that

$$\inf_{\varphi \in \mathcal{T}_m} [\beta(\varphi) - l\alpha_1(\varphi)] > 0 \tag{26.16}$$

for a certain integer $l \in [1, s-1]$.

The following analog of Theorem 23.1 for the system of equations (26.6) is true:

Theorem 26.1. *Suppose that the right-hand side of system (26.6) satisfies the smoothness conditions given above, and inequalities (26.12) and (26.16) are true. Then one can find a constant* $\mu > 0$ *and a matrix* $\Psi(\psi, h) \in C_{Lip}^{l-1}(\mathcal{T}_m \times K_\mu)$ *such that the change of variables*

$$\varphi = \psi + \Psi(\psi, h)h \tag{26.17}$$

reduces the system of equations (26.6) to the form

$$\frac{d\psi}{dt} = a(\psi), \quad \frac{dh}{dt} = P(\psi, h)h, \tag{26.18}$$

where $P(\psi, h)$ *is a matrix that belongs to* $C^{l-1}_{Lip}(\mathcal{T}_m \times K_\mu)$ *and coincides with* $P(\psi)$ *for* $h = 0$.

Theorem 26.1 is proved by analogy with Theorem 23.1 with the difference that one should take into account estimate (26.15) for the derivatives of the function $\psi_t(\varphi)$ and inequality (26.16).

The verification of conditions (26.12) and (26.16) encounters certain difficulties. These difficulties can be avoided if, for the fundamental matrices of solutions $\Omega^t_0(P)$ and $\Omega^t_0\Big(\dfrac{\partial a}{\partial \varphi}\Big)$ of the systems

$$\frac{dh}{dt} = P(\psi_t(\varphi))h \qquad \text{and} \qquad \frac{dg}{dt} = \frac{\partial a(\psi_t(\varphi))}{\partial \varphi} g, \tag{26.19}$$

respectively, estimates of the following form are known:

$$\begin{aligned} \|\Omega^t_0(P)\| &\le \mathcal{L} e^{-\beta_0 t}, \quad t \in R_+, \\ \Big\|\Omega^t_0\Big(\frac{\partial a}{\partial \varphi}\Big)\Big\| &\le \mathcal{L}_1 e^{+\alpha_1 t}, \quad t \in R_+, \end{aligned} \tag{26.20}$$

where β_0 and α_1 are positive constants. In this case, for inequality (26.16) to be satisfied, it is sufficient that

$$\frac{\beta_0}{\alpha_1} > l, \tag{26.21}$$

where $l \in [1, s-1]$.

Remark 2. Condition (26.16) can be satisfied for $a(\varphi) \not\equiv \text{const}$ only for a finite value of l.

In this case, the change of variables (26.17) has finite smoothness. However, if

$$a(\varphi) \equiv \text{const}, \tag{26.22}$$

then the value of l is equal to $s - 1$, and, for $s = \infty$, the change of variables (26.17) is infinitely differentiable. If condition (26.22) is satisfied, then Theorem 26.1 coincides with Theorem 23.1.

The theorem below enables one to describe the behavior of solutions of system (26.1) originating in the neighborhood of M.

Theorem 26.2. *Suppose that $X(x) \in C^r(R^n)$ and the system of equations (26.1) has the invariant toroidal manifold (26.2), where $f(\varphi) \in C^s(\mathcal{T}_m)$ for $r \geq s \geq 2$.*

Also assume that the following conditions are satisfied:

(i) the matrix $\dfrac{\partial f(\varphi)}{\partial \varphi}$ can be complemented to a 2π-periodic basis in R^n;

(ii) the variational equation for the manifold M satisfies the condition of exponential stability (26.12);

(iii) inequality (26.16) is true.

Then one can indicate a sufficiently small $\delta > 0$ such that, for every y^0, $\rho(y^0, M) \leq \delta$, one can find $\psi^0 \in \mathcal{T}_m$ and $\varphi^0 \in \mathcal{T}_m$ such that

$$\|x(t, y^0) - f(\psi_t(\psi^0))\| \leq \mathcal{L}_2 e^{-\gamma_1 t} \|y^0 - f(\varphi^0)\| \tag{26.23}$$

for all $t \in R_+$ and certain $\mathcal{L}_2 > 0$ and $\gamma_1 > 0$, where $\gamma_1 = \gamma_1(\delta) \to \gamma$ and $\|\varphi^0 - \psi^0\| \to 0$ as $\delta \to 0$.

Inequality (26.23) proves that a solution of system (26.1) originating in a small neighborhood of the manifold M is exponentially attracted as $t \to +\infty$ to the corresponding solution of this system originating on M. The proof of Theorem 26.2 repeats the proof of Theorem 24.2 word for word.

Parallel with the system of equations (26.1), we consider the perturbed system of equations

$$\frac{dy}{dt} = X(y) + \varepsilon Y(y), \tag{26.24}$$

where $Y \in C^r(R^n)$ and ε is a small positive parameter. Let us clarify the behavior of solutions of this system originating in a small neighborhood of the manifold M. In the local system of coordinates (φ, h), system (26.24) takes the form

$$\begin{aligned} \frac{d\varphi}{dt} &= a(\varphi) + A(\varphi, h)h + \varepsilon L_1(\varphi, h) Y(f(\varphi) + B(\varphi)h), \\ \frac{dh}{dt} &= P(\varphi, h)h + \varepsilon L_2(\varphi, h) Y(f(\varphi) + B(\varphi)h), \end{aligned} \tag{26.25}$$

which coincides with (26.6) for $\varepsilon = 0$.

We define a function $\alpha(\varphi)$ by the inequality

$$\sup_{S_1 \in N_1} \min_{\|\psi\|=1} \frac{\left\langle \left[S_1(\varphi) \frac{\partial a(\varphi)}{\partial \varphi} + \frac{1}{2} \frac{\partial S_1(\varphi)}{\partial \varphi} a(\varphi) \right] \psi, \psi \right\rangle}{\langle S_1(\varphi)\psi, \psi \rangle} \geq \alpha(\varphi) \qquad (26.26)$$

and require that the following condition be satisfied for a certain integer $p \in [1, s-1]$:

$$\inf_{\varphi \in \mathcal{T}_m} [\beta(\varphi) + p\alpha(\varphi)] > 0. \qquad (26.27)$$

According to the perturbation theory of invariant toroidal manifolds, the validity of inequalities (26.12) and (26.27) is a sufficient condition [Sam4] for the system of equations (26.25) to have the invariant torus

$$h = u(\varphi, \varepsilon), \quad \varphi \in \mathcal{T}_m, \qquad (26.28)$$

for all $\varepsilon \in [0, \varepsilon_0]$, where $\varepsilon_0 > 0$ is sufficiently small, $u \in C^{p-1}_{Lip}(\mathcal{T}_m)$, and

$$\lim_{\varepsilon \to 0} \|u\|_{p-1, Lip} = 0. \qquad (26.29)$$

Let $p \geq 2$. Then the change of variables

$$h = u(\varphi, \varepsilon) + z,$$

where u is the function from (26.28), reduces the system of equations (26.25) to the form

$$\frac{d\varphi}{dt} = a(\varphi) + F(\varphi, \varepsilon) + A(\varphi, z, \varepsilon)z, \quad \frac{dz}{dt} = P(\varphi, z, \varepsilon)z, \qquad (26.30)$$

where $F \in C^{p-1}_{Lip}(\mathcal{T}_m)$, $(A, P) \in C^{p-2}_{Lip}(\mathcal{T}_m \times K_{\delta_0})$ for every $\varepsilon \in [0, \varepsilon_0]$, $\delta_0 = \delta_0(\varepsilon_0) \to 0$ as $\varepsilon_0 \to 0$, and

$$\lim_{\varepsilon \to 0} \|F\|_{p-1, Lip} = 0, \quad \lim_{\varepsilon \to 0} \|A(\varphi, z, \varepsilon) - A(\varphi, z)\|_{p-2, Lip} = 0,$$

$$\lim_{\varepsilon \to 0} \|P(\varphi, z, \varepsilon) - P(\varphi, z)\|_{p-2, Lip} = 0. \qquad (26.31)$$

For every $\varepsilon \in [0, \varepsilon_0]$, the system of equations (26.30) has the form (26.25). It follows from inequalities (26.11) and (26.14), which determine the functions $\beta(\varphi)$ and $\alpha_1(\varphi)$, that, for small changes in the values of a, P, and $\frac{\partial a}{\partial \varphi}$, the changes in the values of β and α_1 are also small. Therefore, the limit relations

(26.31) imply that the functions $\beta(\varphi,\varepsilon)$ and $\alpha_1(\varphi,\varepsilon)$ defined with the use of $a(\varphi)+F(\varphi,\varepsilon)$ and $P(\varphi,\varepsilon,0)$ according to formulas (26.11) and (26.14) satisfy the limit relations

$$\lim_{\varepsilon\to 0}[\beta(\varphi,\varepsilon)-\beta(\varphi)] = \lim_{\varepsilon\to 0}[\alpha_1(\varphi,\varepsilon)-\alpha_1(\varphi)] = 0 \tag{26.32}$$

uniformly in $\varphi \in \mathcal{T}_m$. Consequently, one can find $\varepsilon_0 = \varepsilon_0(l) > 0$ such that inequalities (26.12) and (26.16) yield the following estimates for $l \in [1, s-1]$:

$$\inf_{\varepsilon\in[0,\varepsilon_0]} \quad \inf_{\varphi\in\mathcal{T}_m} \beta(\varphi,\varepsilon) > 0, \tag{26.33}$$

$$\inf_{\varepsilon\in[0,\varepsilon_0]} \quad \inf_{\varphi\in\mathcal{T}_m} [\beta(\varphi,\varepsilon) - l\alpha_1(\varphi,\varepsilon)] > 0. \tag{26.34}$$

Since the right-hand side of the system of equations (26.30) belongs to the space $C^{p-2}(\mathcal{T}_m \times K_{\delta_0})$ for every $\varepsilon \in [0,\varepsilon_0]$, we deduce from (26.34) the following inequality for $p \geq 3$:

$$\inf_{\varepsilon\in[0,\varepsilon_0]} \quad \inf_{\varphi\in\mathcal{T}_m} [\beta(\varphi,\varepsilon) - l_1\alpha_1(\varphi,\varepsilon)] > 0, \tag{26.35}$$

where

$$l_1 = \min\{l, p-2\} \geq 1. \tag{26.36}$$

For $p \geq 3$, these arguments allow us to apply Theorem 26.1 to the system of equations (26.30). As a result, we obtain the following statement:

Theorem 26.3. *Suppose that the conditions of Theorem 26.1 are satisfied and inequality (26.27) holds for* $p \geq 3$. *Then there exist positive constants* μ *and* $\varepsilon_0 = \varepsilon_0(p)$ *and a matrix* $\Psi(\varphi, z, \varepsilon)$ *that belongs to the space* $C^{l_1-1}_{Lip}(\mathcal{T}_m \times K_\mu)$ *for every* $\varepsilon \in [0,\varepsilon_0]$ *and satisfies the relations*

$$\lim_{\varepsilon\to 0} \|\Psi(\varphi,z,\varepsilon) - \Psi(\varphi,z,0)\|_{l_1-1,Lip} = 0 \tag{26.37}$$

such that the change of variables

$$\varphi = \psi + \Psi(\psi, z, \varepsilon) z \tag{26.38}$$

reduces the system of equations (26.30) to the form

$$\frac{d\psi}{dt} = a(\psi) + F(\psi,\varepsilon), \qquad \frac{dz}{dt} = P(\psi + \Psi(\psi,z,\varepsilon)z, z, \varepsilon) z. \tag{26.39}$$

Condition (26.12) guarantees that

$$\|z(t,\varepsilon)\| \le \mathcal{L}_3 e^{-\beta t}\|z(0,\varepsilon)\|, \quad t \in R_+, \tag{26.40}$$

for solutions of system (26.39) such that $\|z(0,\varepsilon)\| \le \mu$ for $\varepsilon \in [0,\varepsilon_0]$. Here, $\mathcal{L}_3 = \text{const}$ and $\beta = \beta(\mu,\varepsilon)$ are certain positive quantities and $\beta(\mu,\varepsilon) \to \beta_0$ as $(\mu,\varepsilon) \to 0$.

As noted above, the verification of inequalities (26.11), (26.14), and (26.26) encounters certain difficulties. Therefore, it is natural to express the conditions of Theorem 26.3 in terms of inequalities of the form (26.21). For this purpose, we assume that the fundamental matrices of solutions $\Omega_0^t(P)$ and $\Omega_0^t\left(\frac{\partial a}{\partial \varphi}\right)$ of the corresponding equations of systems (26.19) satisfy the inequalities

$$\begin{aligned} &\|\Omega_0^t(P)\| \le \mathcal{L}e^{-\beta_0 t}, \quad t \in R_+, \\ &\left\|\Omega_0^t\left(\frac{\partial a}{\partial \varphi}\right)\right\| \le \mathcal{L}_1 e^{\alpha_0|t|}, \quad t \in (-\infty,\infty), \end{aligned} \tag{26.41}$$

where β_0 and α_0 are positive constants.

To express the conditions that guarantee the reducibility of the system of equations (26.30) to the form (26.39) in terms of the parameters β_0 and α_0, we use the following statement:

Lemma 26.1. *If $a \in C^1(\mathcal{T}_m)$, $P \in C^1(\mathcal{T}_m)$, and inequalities (26.41) are satisfied, then*

$$\inf_{S\in N}\max_{\|h\|=1} \frac{\left\langle \left[S(\varphi)P(\varphi) + \frac{1}{2}\frac{\partial S(\varphi)}{\partial \varphi}a(\varphi)\right]h, h\right\rangle}{\langle S(\varphi)h, h\rangle} \le -\left[\beta_0 - \frac{\alpha_0}{2}(1-\mathcal{L}^{-2})\right].$$

Proof. For $\mu > 0$, the fundamental matrix $\Omega_0^t(P + (\beta_0 - \mu)E)$ of solutions of the equation

$$\frac{dx}{dt} = [P(\psi_t(\varphi)) + (\beta_0 - \mu)E]x$$

satisfies the inequality

$$\|\Omega_0^t(P + (\beta_0 - \mu)E)\| \le \mathcal{L}e^{-\beta_0 t}e^{(\beta_0-\mu)t} = \mathcal{L}e^{-\mu t}, \quad t \in R_+.$$

Let

$$S(\varphi) = \int_0^\infty (\Omega_0^\tau[P + (\beta_0 - \mu)E])^T \Omega_0^\tau[P + (\beta_0 - \mu)E]d\tau. \tag{26.42}$$

It follows from the calculations carried out in [Sam4] (Chapter 3, Section 5) that

$$\dot{S}(\varphi) = -P^T(\varphi)S(\varphi) - S(\varphi)P(\varphi) - 2(\beta_0 - \mu)S(\varphi) - E,$$

where

$$\dot{S}(\varphi) = \frac{d}{dt}S(\psi_t(\varphi))|_{t=0}.$$

For the matrix $Y_\nu(t) = \dfrac{\partial}{\partial \varphi_\nu}\Omega_0^t(P)$, we have

$$\begin{aligned}\|Y_\nu(t)\| &= \Big\| \int_0^t \Omega_\tau^t(P)\frac{\partial P(\psi_\tau(\varphi))}{\partial \psi}\frac{\partial \varphi_\tau(\varphi)}{\partial \varphi_\nu}\Omega_0^\tau(P)d\tau \Big\| \\ &\leq \Big(\mathcal{L}^2\mathcal{L}_1\overline{K}_1\frac{m}{\alpha_0}\Big)e^{-(\beta_0-\alpha_0)t}, \quad t \in R_+,\end{aligned}$$

where

$$\max_{\varphi \in \mathcal{T}_m}\Big\|\frac{\partial P(\varphi)}{\partial \varphi_j}\Big\| \leq \overline{K}_1, \quad j = \overline{1, m}.$$

Taking this estimate into account, we get

$$\begin{aligned}&\Big\|\frac{\partial}{\partial \varphi_\nu}\{\Omega_0^t(P + (\beta_0 - \mu)E)^T\Omega_0^t(P + (\beta_0 - \mu)E)\}\Big\| \\ &\qquad\qquad \leq 2\mathcal{L}^3\mathcal{L}_1\overline{K}_1\frac{m}{\alpha_0}e^{(\alpha_0 - 2\mu)t}, \quad t \in R_+. \qquad (26.43)\end{aligned}$$

Choosing μ from the condition

$$\alpha_0 < 2\mu, \qquad (26.44)$$

one can verify that the integral

$$I_\nu = \int_0^\infty \frac{\partial}{\partial \varphi_\nu}\{(\Omega_0^\tau(P + (\beta_0 - \mu)E))^T\Omega_0(P + (\beta_0 - \mu)E)\}d\tau$$

is majorized by a convergent one and, furthermore,

$$\|I_\nu\| \leq 2\mathcal{L}^3\mathcal{L}_1\overline{K}_1\frac{m}{\alpha_0(2\mu - \alpha_0)}.$$

This is sufficient for the matrix $S(\varphi)$ to belong to the space $C^1(\mathcal{T}_m)$. Then

$$\dot{S}(\varphi) = \frac{dS(\varphi)}{d\varphi} a(\varphi)$$

and

$$\Delta(\varphi) \equiv \max_{\|h\|=1} \frac{\left\langle \left[S(\varphi)P(\varphi) + \frac{1}{2}\frac{\partial S(\varphi)}{\partial \varphi} a(\varphi) \right] h, h \right\rangle}{\langle S(\varphi)h, h \rangle}$$

$$= -\left[\beta_0 - \mu + \frac{1}{2 \max\limits_{\|h\|=1} < S(\varphi)h, h >} \right]. \tag{26.45}$$

Since

$$\max_{\|h\|=1} \langle S(\varphi)h, h \rangle = \int_0^\infty \max_{\|h\|=1} \langle \Omega_0^\tau (P + (\beta_0 - \mu)E)h, \Omega_0^\tau (P + (\beta_0 - \mu)E) \rangle \, d\tau$$

$$\leq \mathcal{L}^2 \int_0^\infty e^{-2\mu\tau} d\tau = \frac{\mathcal{L}^2}{2\mu},$$

it follows from (26.45) that

$$\Delta(\varphi) \leq -[\beta_0 - \mu(1 - \mathcal{L}^{-2})].$$

Passing to the limit in the last estimate as $\mu \to \frac{1}{2}\alpha_0$, we get

$$\inf_{S \in N} \max_{\|h\|=1} \frac{\left\langle \left[S(\varphi)P(\varphi) + \frac{1}{2}\frac{\partial S(\varphi)}{\partial \varphi} a(\varphi) \right] h, h \right\rangle}{\langle S(\varphi)h, h \rangle} \leq -\left[\beta_0 - \frac{\alpha_0}{2}(1 - \mathcal{L}^{-2}) \right].$$

The positive definiteness of the matrix $S(\varphi)$ was proved in [Sam4] (Chapter 3, Section 5).

According to Lemma 26.1, we can take the constant $\beta_0 - \frac{1}{2}\alpha_0(1 - \mathcal{L}^{-2})$ as the function $\beta(\varphi)$ in inequalities (26.27) and (26.34). It is also obvious that the values $\alpha_1(\varphi)$ and $\alpha(\varphi)$ in these inequalities can be replaced by the constant α_0.

Thus, in order that the statement of Theorem 26.3 be true, it is sufficient to require that the following inequalities be satisfied:

$$\beta_0 - \frac{1}{2}\alpha_0(1 - \mathcal{L}^{-2}) - l\alpha_0 > 0 \qquad \text{for} \quad l \in [1, s-1],$$

$$\beta_0 - \frac{1}{2}\alpha_0(1 - \mathcal{L}^{-2}) - p\alpha_0 > 0 \qquad \text{for} \quad p \in [3, s-1].$$

In order that both inequalities be satisfied, it suffices to set $l = p$ and require that

$$\frac{\beta_0}{\alpha_0} > p + \frac{1}{2}(1 - \mathcal{L}^{-2}) \tag{26.46}$$

for integer $p \in [3, s-1]$.

Corollary 3. *Suppose that the smoothness conditions presented in Theorem 26.3 are satisfied and inequalities (26.41) with constants satisfying condition (26.46) are true. Then, for any* $\varepsilon \in [0, \varepsilon_0]$, *the change of variables (26.38) with matrix* $\Psi(\psi, z, \varepsilon) \in C_{Lip}^{p-3}(\mathcal{T}_m \times K_\mu)$ *reduces the system of equations (26.30) to the form (26.39).*

As in the case of Theorem 26.1, Theorem 26.3 can be formulated in the form of a statement related to the system of equations (26.24) and its invariant torus $M(\varepsilon)$, namely

$$y = f(\varphi) + u(\varphi, \varepsilon), \quad \varphi \in \mathcal{T}_m, \quad \varepsilon \in [0, \varepsilon_0].$$

Hence, one can easily establish the principle of reducibility in problems of stability of solutions of system (26.24) originating on $M(\varepsilon)$. According to this principle, the stability or the asymptotic stability of these solutions is determined by their stability or asymptotic stability on $M(\varepsilon)$.

27. Discrete Dynamical System in the Neighborhood of a Quasiperiodic Trajectory

Let $x = x(n, x^0)$, $x^0 \in R^q$, $n = 0, \pm 1, \pm 2, \ldots$, denote a solution of the system of difference equations

$$x(n+1) - x(n) = X(x(n)), \quad x(0) = x^0, \tag{27.1}$$

where $n = 0, \pm 1, \ldots$ is discrete time, $x \in R^q$, and $X(x) \in C^r(R^q)$. Assume that system (27.1) has an invariant surface

$$M\colon x = f(\varphi), \quad \varphi \in \mathcal{T}_m, \tag{27.2}$$

where $f(\varphi) \in C^r(\mathcal{T}_m)$, filled with quasiperiodic trajectories

$$x(n, f(\varphi)) = f(\omega n + \varphi), \quad n = 0, \pm 1, \ldots, \quad \varphi \in \mathcal{T}_m. \tag{27.3}$$

Here, $\omega = (\omega_1, \ldots, \omega_m)$ is a frequency basis of the quasiperiodic function $f(\omega t)$, and φ is an arbitrary point in $\mathcal{T}_m$.

Assume that

$$\operatorname{rank} \frac{\partial f(\varphi)}{\partial \varphi} = m, \quad \varphi \in \mathcal{T}_m, \tag{27.4}$$

and the matrix $\dfrac{\partial f(\varphi)}{\partial \varphi}$ can be complemented to a 2π-periodic basis, i.e., there exists a matrix $B(\varphi) \in C^r(\mathcal{T}_m)$ such that

$$\det\left[\frac{\partial f(\varphi)}{\partial \varphi}, B(\varphi)\right] \neq 0, \quad \varphi \in \mathcal{T}_m. \tag{27.5}$$

Under the assumptions made above, we investigate the behavior of trajectories of system (27.1) originating in a small neighborhood of the manifold M. First, note that the invariance of the manifold M and the quasiperiodicity of trajectories on it require that the following identity be satisfied:

$$f(\varphi + \omega) = f(\varphi) + X(f(\varphi)), \quad \varphi \in \mathcal{T}_m. \tag{27.6}$$

Assumptions (27.4) and (27.5) guarantee the introduction of local coordinates $(\varphi, h) = (\varphi_1, \ldots, \varphi_m, h_1, \ldots, h_{q-m})$ in the neighborhood of M according to the formula

$$x = f(\varphi) + B(\varphi)h \tag{27.7}$$

and representation (27.1) in the neighborhood of M in the form

$$\begin{aligned} \varphi(n+1) - \varphi(n) &= \omega + A(\varphi(n), h(n))h(n), \\ h(n+1) - h(n) &= P(\varphi(n), h(n))h(n), \end{aligned} \tag{27.8}$$

where the matrices $A(\varphi, h)$ and $P(\varphi, h)$ of the corresponding dimensions are 2π-periodic in φ and sufficiently smooth with respect to φ and h in the domain

$$\|h\| \leq \delta, \quad \varphi \in \mathcal{T}_m. \tag{27.9}$$

Here, $n = 0, \pm 1, \pm 2, \ldots$ and δ is a sufficiently small positive number. To establish this fact, we perform the change of variables (27.7) in (27.1). As a result, we obtain the following system of equations for the determination of the matrices $A = A(\varphi, h)$ and $P = P(\varphi, h)$:

$$f(\varphi + \omega + A(\varphi, h)h) + B(\varphi + \omega + A(\varphi, h)h)[h + P(\varphi, h)h] - [f(\varphi) + B(\varphi)h] = X(f(\varphi) + B(\varphi)h),$$

or, with regard for identity (27.6),

$$f(\varphi + \omega + Ah) - f(\varphi + \omega) + B(\varphi + \omega + Ah)(h + Ph) - B(\varphi)h = X(f(\varphi) + B(\varphi)h) - X(f(\varphi)).$$

We represent the last equation in the form

$$\frac{\partial f(\varphi + \omega)}{\partial \varphi} Ah + B(\varphi + \omega + Ah)Ph = X(f(\varphi) + B(\varphi)h) - X(f(\varphi)) - \Big\{\Big[f(\varphi + \omega + Ah) - f(\varphi + \omega) - \frac{\partial f(\varphi + \omega)}{\partial \varphi} Ah\Big] + [B(\varphi + \omega + Ah) - B(\varphi + \omega)]h + [B(\varphi + \omega) - B(\varphi)]h\Big\}.$$

This yields

$$\frac{\partial f(\varphi + \omega)}{\partial \varphi} A + B(\varphi + \omega + Ah)P = \int_0^1 \frac{\partial X(f(\varphi) + tB(\varphi)h)}{\partial x} dt B(\varphi) - \Big\{\int_0^1 \Big[\frac{\partial f(\varphi + \omega + tAh)}{\partial \varphi} - \frac{\partial f(\varphi + \omega)}{\partial \varphi}\Big] dt A + [B(\varphi + \omega + Ah) - B(\varphi + \omega)] + [B(\varphi + \omega) - B(\varphi)]\Big\}. \tag{27.10}$$

According to assumption (27.5), for fixed $M > 0$ there exists $\delta = \delta(M) > 0$ such that

$$\det\Big[\frac{\partial f(\varphi+\omega)}{\partial\varphi}, B(\varphi+\omega+Ah)\Big] \neq 0 \tag{27.11}$$

for all φ and h from domain (27.9) and an arbitrary matrix A satisfying the condition

$$\|A\| \le M, \tag{27.12}$$

where the norm of the matrix A is consistent with the norm of the vector h.

Therefore, Eq. (27.10) has a solution of the form

$$A = L_1(\varphi, Ah)Q(\varphi, h, A), \quad P = L_2(\varphi, Ah)Q(\varphi, h, A), \tag{27.13}$$

for all φ and h from domain (27.9) and A from domain (27.12). Here, $Q = Q(\varphi, h, A)$ is the matrix function defined by the right-hand side of Eq. (27.10), and $L_1(\varphi, Ah)$ and $L_2(\varphi, Ah)$ are the blocks of the matrix inverse to the matrix $\Big[\frac{\partial f(\varphi+\omega)}{\partial\varphi}, B(\varphi+\omega+Ah)\Big]$.

The matrix Q admits the following representation:

$$Q = B(\varphi+\omega) - B(\varphi) + \frac{\partial X(f(\varphi))}{\partial x}B(\varphi) + Q_1(\varphi, h, A), \tag{27.14}$$

where $Q_1 = Q_1(\varphi, h, A)$ is a matrix defined in the domain $\|h\| \le \delta,\ \varphi \in \mathcal{T}_m$, $\|A\| \le M,\ r-1$ times continuously differentiable with respect to its variables and such that

$$Q_1(\varphi, 0, A) = 0. \tag{27.15}$$

The matrices $L_1(\varphi, Ah)$ and $L_2(\varphi, Ah)$ possess properties analogous to properties of the matrix Q; moreover, $L_1(\varphi, 0)$ and $L_2(\varphi, 0)$ are the blocks of the matrix inverse to the matrix $\Big[\frac{\partial f(\varphi+\omega)}{\partial\varphi}, B(\varphi+\omega)\Big]$.

In the space $C_{Lip}(\mathcal{T}_m \times K_\mu)$, we define the subset $C(M, K)$ of matrix functions $A = A(\varphi, \mu)$ that satisfy the conditions

$$\|A(\varphi, h)\| \le M, \quad \|A(\varphi', h') - A(\varphi, h)\| \le K(\|\varphi' - \varphi\| + \|h' - h\|)$$

for any (φ', h') and (φ, h) from $\mathcal{T}_m \times K_\mu$.

Let us prove that the first equation in (27.13) has a solution in $C(M, K)$ for the properly chosen constants M, K, and μ. To do this, we define an operator $S\colon A \to SA = L_1(\varphi, Ah) \times Q(\varphi, h, A)$ on the set $C(M, K)$. For $r \ge 2$, this operator maps $C(M,K)$ into a subset of the space $C_{Lip}(\mathcal{T}_m \times K_\mu)$. Furthermore, $SA = (SA)(\varphi, h)$ satisfies the following estimates for arbitrary (φ, h) and (φ', h') from $\mathcal{T}_m \times K_\mu$:

$$\|(SA)(\varphi,h)\| \leq c_1(1+\mu+\mu M^2),$$

$$\|(SA)(\varphi',h')-(SA)(\varphi,h)\|$$
$$\leq c_2(1+\mu MK+M^2)(\|\varphi'-\varphi\|+\|h'-h\|), \qquad (27.16)$$

where c_1 and c_2 are positive constants independent of M, K, and $\mu \leq \dfrac{\delta}{K}$. By a proper choice of sufficiently large M and K and sufficiently small μ, one can guarantee that inequality (27.16) implies that SA belongs to the set $C(M,K)$.

For a pair of matrix functions $A = A(\varphi, h)$ and $A_1 = A_1(\varphi, h)$ from the set $C(M,K)$, we have

$$\|SA - SA_1\| \leq c_3\mu(1+M)\|A - A_1\|,$$

where c_3 is a constant independent of M, K, and μ. For sufficiently small μ, this estimate implies that S is a contraction operator on $C(M,K)$. According to the principle of contracting mappings, the equation $A = SA$ has a unique solution on the set $C(M,K)$.

The last equation coincides with the first equation in (27.13), and, for sufficiently small $\mu > 0$, its solution determines in $C_{Lip}(\mathcal{T}_m \times K_\mu)$ the unique matrix $A = A(\varphi, h)$ of the right-hand side of system (27.8). The implicit-function theorem guarantees that $A(\varphi, h)$ belongs to the space $C^{r-1}_{Lip}(\mathcal{T}_m \times K_\mu)$.

With the use of the obtained matrix $A(\varphi, h)$, the second equation in (27.13) determines the matrix $P = P(\varphi, h)$ of the right-hand side of system (27.8), which also belongs to the space $C^{r-1}_{Lip}(\mathcal{T}_m \times K_\mu)$.

Thus, in the small neighborhood of the manifold M, the dynamical system (27.1) reduces to the form (27.8) with the matrices A and P from $C^{r-1}_{Lip}(\mathcal{T}_m \times K_\mu)$. The problem is to find conditions under which there exists a change of variables $\varphi \to \psi$ that transforms system (27.1) into the quasiperiodic system

$$\psi(n+1) - \psi(n) = \omega,$$
$$h(n+1) - h(n) = R(\psi(n), h(n))h(n). \qquad (27.17)$$

Theorem 27.1. *Suppose that the conditions presented above are satisfied and the matrix* $P(\varphi, 0)$ *satisfies the inequality*

$$\|E + P(\varphi, 0)\| \leq d < 1, \quad \varphi \in \mathcal{T}_m. \qquad (27.18)$$

Then one can find $\mu > 0$ *and a matrix* $U(\varphi, h) \in C^{r-2}_{Lip}(\mathcal{T}_m \times K_\mu)$, $2 \leq r < \infty$, *such that the change of variables*

$$\varphi = \psi + U(\psi, h)h \tag{27.19}$$

reduces system (27.8) to the form (27.17) with the matrix

$$R(\psi, h) = P(\psi + U(\psi, h)h, h). \tag{27.20}$$

Proof. We define the required transformation $\varphi \to \psi$ by the formula

$$\psi = \varphi + V(\varphi, h)h, \tag{27.21}$$

where $V = V(\varphi, h)$ is a matrix function from $C(\mathcal{T}_m \times K_\mu)$. According to Eqs. (27.8) and (28.17) and the change of variables (27.21), we obtain the following relation for the determination of the matrix V:

$$V(\varphi(n) + \omega + A(\varphi(n), h(n))h(n), P_1(\varphi(n), h(n))h(n))P_1(\varphi(n),$$
$$h(n))h(n) - V(\varphi(n), h(n))h(n) + A(\varphi(n), h(n))h(n) = 0.$$

Therefore, V satisfies the equation

$$V(\varphi, h) = V(\varphi + \omega + A(\varphi, h)h, P_1(\varphi, h)h)P_1(\varphi, h) + A(\varphi, h), \tag{27.22}$$

where $P_1(\varphi, h) = E + P(\varphi, h)$. We set

$$\varphi_1(\varphi, h) = \varphi + \omega + A(\varphi, h)h, \quad h_1(\varphi, h) = P_1(\varphi, h)h \tag{27.23}$$

and rewrite (27.21) in the form

$$V(\varphi, h) = V(\varphi_1(\varphi, h), h_1(\varphi, h))P_1(\varphi, h) + A(\varphi, h). \tag{27.24}$$

For sufficiently small $\mu > 0$, condition (27.18) yields the following inequality for all $(\varphi, h) \in \mathcal{T}_m \times K_\mu$:

$$\|P_1(\varphi, h)\| \leq d_1 = \text{const} < 1. \tag{27.25}$$

This reasoning leads to the successive approximations for a solution of Eq. (27.24)

$$V_1(\varphi, h) = A(\varphi, h),$$
$$V_{i+1}(\varphi, h) = V_i(\varphi_1(\varphi, h), h_1(\varphi, h))P_1(\varphi, h) + A(\varphi, h), \quad i \geq 1, \tag{27.26}$$

and the estimates

$$\|V_1(\varphi, h)\| \le \|A(\varphi, h)\| \le \max_{\mathcal{T}_m \times K_\mu} \|A(\varphi, h)\| = M_1,$$

$$\|V_{i+1}(\varphi, h)\| \le M_1 \sum_{\nu=0}^{i} d_1^\nu, \quad i \ge 1.$$

By virtue of the last estimates, sequence (27.26) converges uniformly in $(\varphi, h) \in \mathcal{T}_m \times K_\mu$, and its limit function

$$V(\varphi, h) = \lim_{i \to \infty} V_i(\varphi, h)$$

is a solution of Eq. (27.22) that belongs to the space $C(\mathcal{T}_m \times K_\mu)$.

Let us study the problem of the smoothness of the function $V(\varphi, h)$. For this purpose, we consider the function $W = W(\varphi, h, \mu) = V(\varphi, \mu h)$ for $(\varphi, h) \in \mathcal{T}_m \times K_\mu$ and sufficiently small $\mu > 0$. This function satisfies the equation

$$W(\varphi, h, \mu) = W(\varphi_1(\varphi, \mu h), h_1(\varphi, \mu h), \mu) P_1(\varphi, \mu h) + A(\varphi, \mu h) \tag{27.27}$$

and is the limit of the successive approximations

$$W_1(\varphi, h, \mu) = A(\varphi, \mu h), \tag{27.28}$$

$$W_{i+1}(\varphi, h, \mu) = W_i(\varphi_1(\varphi, \mu h), h_1(\varphi, \mu h), \mu) P_1(\varphi, \mu h) + A(\varphi, \mu h), \quad i \ge 1.$$

Differentiating expressions (27.28), we obtain the following equalities for derivatives:

$$\frac{\partial W_1(\varphi, h, \mu)}{\partial \varphi_\nu} = \frac{\partial A(\varphi, \mu h)}{\partial \varphi_\nu}, \quad \frac{\partial W_1}{\partial h_\nu} = \mu \frac{\partial A(\varphi, \mu h)}{\partial (\mu h)_\nu},$$

$$\begin{aligned}\frac{\partial W_{i+1}(\varphi, h, \mu)}{\partial \varphi_\nu} = {}& \Big[\sum_{j=1}^{m} \frac{\partial W_i(\varphi_1(\varphi, \mu h), h_1(\varphi, \mu h), \mu)}{\partial \varphi_{1,j}} \frac{\partial \varphi_{1,j}(\varphi, \mu h)}{\partial \varphi_\nu} \\ & + \sum_{j=1}^{q-m} \frac{\partial W_i(\varphi_1(\varphi, \mu h), h_1(\varphi, \mu h), \mu)}{\partial h_{1,j}} \frac{\partial h_{1,j}(\varphi, \mu h)}{\partial \varphi_\nu}\Big] P_1(\varphi, \mu h) \\ & + W_i(\varphi_1(\varphi, \mu h), h_1(\varphi, \mu h), \mu) \frac{\partial P_1(\varphi, \mu h)}{\partial \varphi_\nu} \\ & + \frac{\partial A(\varphi, \mu h)}{\partial \varphi_\nu},\end{aligned} \tag{27.29}$$

$$\frac{\partial W_{i+1}(\varphi,h,\mu)}{\partial h_\nu} = \mu\Big[\sum_{j=1}^{m} \frac{\partial W_i(\varphi_1(\varphi,\mu h),h_1(\varphi,\mu h),\mu)}{\partial \varphi_{1,j}} \frac{\partial \varphi_{1,j}(\varphi,\mu h)}{\partial(\mu h)_\nu}$$

$$+ \sum_{j=1}^{q-m} \frac{\partial W_i(\varphi_1(\varphi,\mu h),h_1(\varphi,\mu h),\mu)}{\partial h_{1,j}} \frac{\partial h_{1,j}}{\partial(\mu h)_\nu}\Big] P_1(\varphi,\mu h)$$

$$+ \mu W_i(\varphi_1(\varphi,\mu h),h_1(\varphi,\mu h),\mu)\frac{\partial P_1(\varphi,\mu h)}{\partial(\mu h)_\nu}$$

$$+ \mu \frac{\partial A(\varphi,\mu h)}{\partial(\mu h)_\nu}, \quad i \ge 1,$$

where $\varphi_{1,j}$ and $h_{1,j}$ are the jth coordinates of the vectors $\varphi_1(\varphi,\mu h)$ and $h_1(\varphi,\mu h)$. For $\mu = 0$, equalities (27.29) take the form

$$\frac{\partial W_1(\varphi,h,0)}{\partial \varphi_\nu} = \frac{\partial A(\varphi,0)}{\partial \varphi_\nu}, \qquad \frac{\partial W_1(\varphi,h,0)}{\partial h_\nu} = 0,$$

$$\frac{\partial W_{i+1}(\varphi,h,0)}{\partial \varphi_\nu}$$

$$= \frac{\partial W_i(\varphi+\omega,h,0)}{\partial \varphi_\nu} P_1(\varphi,0) + W_1(\varphi+\omega,h,0)\frac{\partial P_1(\varphi,0)}{\partial \varphi_\nu} + \frac{\partial A(\varphi,0)}{\partial \varphi_\nu},$$

$$\frac{\partial W_{i+1}(\varphi,h,0)}{\partial h_\nu} = 0, \quad i \ge 1, \tag{27.30}$$

and yield the following estimate:

$$\max_{\mathcal{T}_m \times K_\mu} \|W'_{i+1}\| \le d_1 \max_{\mathcal{T}_m \times K_\mu} \|W'_i\| + M_1, \quad i \ge 1, \tag{27.31}$$

where W'_i is the matrix of derivatives of the iteration W_i, and M_1 is a certain constant.

It follows from (27.31) that

$$\max_{\mathcal{T}_m \times K_\mu} \|W'_{i+1}\| \le \frac{M_1}{1-d_1}, \quad i \ge 1. \tag{27.32}$$

For $\mu \ne 0$, relations (27.29) have the form of the matrix equalities

$$W'_{i+1}(\varphi,h,\mu) = W'_i(\varphi_1(\varphi,\mu h),h_1(\varphi,\mu h),\mu)P_2(\varphi,\mu,h)$$

$$+ W_i(\varphi_1(\varphi,\mu h),h_1(\varphi,\mu h),\mu)P'_1(\varphi,h,\mu)$$

$$+ A'_1(\varphi,h,\mu), \quad i = 0,1,2,\ldots, \tag{27.33}$$

where W_0' and W_0 are zero matrices, W_i' is the matrix of derivatives of the iteration W_i, P_1', A_1', and P_2 are matrix functions of the variables φ, h, and μ that continuously depend on these variables for $\varphi \in \mathcal{T}_m$, $h \in K_\mu$, and $\mu \in [0, \mu_0]$, and μ_0 is a sufficiently small positive number.

Relations (27.30) are a particular case of relations (27.33) and coincide with them for $\mu = 0$. Therefore, in the case $\mu = 0$, we obtain the following expressions for the matrices P_2, P_1', and A_1':

$$P_2(\varphi, h, 0) = \operatorname{diag}\{P_1(\varphi, 0), \ldots, P_1(\varphi, 0), 0, \ldots, 0\},$$

$$P_1'(\varphi, h, 0) = \left\{\frac{\partial P_1(\varphi, 0)}{\partial \varphi_1}, \ldots, \frac{\partial P_1(\varphi, 0)}{\partial \varphi_m}, 0, \ldots, 0\right\},$$

$$A_1'(\varphi, h, 0) = \left\{\frac{\partial A(\varphi, 0)}{\partial \varphi_1}, \ldots, \frac{\partial A(\varphi, 0)}{\partial \varphi_m}, 0, \ldots, 0\right\}.$$

The first of these expressions yields

$$\|P_2(\varphi, h, 0)\| = \|P_1(\varphi, 0)\| \leq d_1, \tag{27.34}$$

which guarantees the following estimate for the norm of the matrix $P_2(\varphi, h, \mu)$:

$$\|P_2(\varphi, h, \mu)\| \leq d_1(\mu), \tag{27.35}$$

where $d_1(\mu) \to d_1$ as $\mu \to 0$.

Choosing $\mu_0 > 0$ sufficiently small, we get

$$d_1(\mu) \leq d_2 \leq \text{const} < 1 \tag{27.36}$$

for all $\mu \in [0, \mu_0]$. Then relations (27.33), (27.35), and (27.36) yield

$$\max_{\mathcal{T}_m \times K_\mu} \|W_{i+1}'(\varphi, h, \mu)\| \leq d_2 \max_{\mathcal{T}_m \times K_\mu} \|W_i'(\varphi, h, \mu)\| + M_2, \quad i = 0, 1, 2, \ldots .$$

Hence,

$$\max_{\mathcal{T}_m \times K_\mu} \|W_{i+1}'(\varphi, h, \mu)\| \leq \frac{M_2}{1 - d_2}, \quad i = 0, 1, 2, \ldots, \tag{27.37}$$

where M_2 is a certain positive constant.

Inequality (27.37) means that

$$\max_{\nu, j} \left\{ \left\|\frac{\partial W_i(\varphi, h, \mu)}{\partial \varphi_\nu}\right\|; \left\|\frac{\partial W_i(\varphi, h, \mu)}{\partial h_j}\right\| \right\} \leq \frac{M_2}{1 - d_2}$$

for all $i = 1, 2, \ldots$. Then

$$\max_{\nu,j} \left\{ \left\| \frac{\partial V_i(\varphi, h)}{\partial \varphi_\nu} \right\|; \left\| \frac{\partial V_i(\varphi, h)}{\partial h_j} \right\| \right\} \leq \frac{M_2}{\mu_0(1 - d_2)} \, .$$

Thus, the sequence of the first derivatives of approximations (27.26) is uniformly bounded. By analogy, one can establish the uniform boundedness of the sequence of arbitrary derivatives of approximations (27.26) up to the order $r-1$ inclusive. This is sufficient for $V = V(\varphi, h)$ to belong to the space $C^{r-2}_{Lip}(\mathcal{T}_m \times K_\mu)$.

To complete the proof of the theorem, it remains to solve relations (27.21) with respect to φ in the form

$$\varphi = \psi + U(\psi, h)h, \tag{27.38}$$

where $U = U(\psi, h)$ is a function from the space $C^{r-2}_{Lip}(\mathcal{T}_m \times K_\mu)$. Substituting (27.38) into (27.21), we obtain the following equation for the matrix U:

$$U = -V(\psi + Uh, h). \tag{27.39}$$

Equation (27.39) has the form of the first equation in (27.13). Therefore, the arguments used in the proof of the solvability of the first equation in (27.13) remain true in the case of Eq. (27.39). This yields the solvability of Eq. (27.39) in the space $C^{r-2}_{Lip}(\mathcal{T}_m \times K_\mu)$ and, hence, the solvability of Eq. (27.21) in the form (27.38).

In order to obtain expression (27.20) for R, it remains to replace φ by its value (27.38) in the matrix $P(\varphi, h)$, which determines the right-hand side of the second equation in system (27.8). Theorem 27.1 is proved.

The statement below characterizes the behavior of trajectories of the discrete dynamical system (27.1) originating in a small neighborhood of M.

Theorem 27.2. *Suppose that the conditions of Theorem 27.1 are satisfied. Then there exists a sufficiently small* $\delta > 0$ *such that, for every* y^0 *satisfying the inequality* $\rho(y^0, M) \leq \delta$, *one can find values* φ^0 *and* ψ^0 *from* $\mathcal{T}_m$ *such that*

$$\|x(n, y^0) - f(\omega n + \psi^0)\| \leq \overline{K}_1 d_3^n \|y^0 - f(\varphi^0)\| \tag{27.40}$$

for all $n = 0, 1, \ldots$ *and certain positive* $\overline{K}_1$ *and* d_3, *where* $d_3 = d_3(\delta) \to d_2$ *and* $\|\varphi^0 - \psi^0\| \to 0$ *as* $\delta \to 0$.

The proof of Theorem 27.2 is analogous to the proof of Theorem 24.2.

Corollary 4. *If the conditions of Theorem 27.1 are satisfied, then a quasiperiodic solution*

$$x = x(n, f(\varphi)) = f(\omega n + \varphi)$$

of system (27.1) is Lyapunov stable for any $\varphi \in \mathcal{T}_m$.

The proof of this corollary is analogous to the proof of Corollary 1 in Section 24.

Corollary 5. *Suppose that the conditions of Theorem 27.1 are satisfied and* $(k, \omega) \neq 0 \mod 2\pi$ *for every integer-valued vector* $k = (k_1, \ldots, k_m) \neq 0$. *Then, for an arbitrary function* $F = F(x)$ *continuous in the neighborhood of* M *and an arbitrary solution* $x = x(n, y^0)$ *of system (27.1) for which* $\rho(y^0, M) \leq \delta$, *the following relation is true:*

$$\lim_{n\to\infty} \frac{1}{n} \sum_{\nu=0}^{n-1} F(x(\nu, y^0)) = F_0$$

$$= (2\pi)^{-m} \int_0^{2\pi} \ldots \int_0^{2\pi} F(f(\varphi)) d\varphi_1 \ldots d\varphi_m. \tag{27.41}$$

Proof. We represent the function F in the form

$$F(x) = P(x, \varepsilon) + R(x, \varepsilon),$$

where $P(x, \varepsilon)$ is a polynomial that approximates F in the neighborhood of M to within an arbitrary fixed $\varepsilon > 0$, i.e., $|R(x, \varepsilon)| \leq \varepsilon \;\; \forall x \in U_\delta(M) \equiv \{x \colon \rho(x, M) \leq \varepsilon\}$. This yields

$$\frac{1}{n} \left| \sum_{\nu=0}^{n-1} [F(x(\nu, y^0)) - P(x(\nu, y^0), \varepsilon)] \right| \leq \varepsilon \tag{27.42}$$

for arbitrary $n = 1, 2, \ldots$. Using inequality (27.40), we get

$$\frac{1}{n}\Big|\sum_{\nu=0}^{n-1}[P(x(\nu,y^0),\varepsilon)-P(f(\omega\nu+\psi^0),\varepsilon)]\Big|$$

$$\leq \frac{K(\varepsilon)\overline{K}_1}{n}\sum_{\nu=0}^{n-1} d_3^{\nu}\|y^0-f(\varphi^0)\| \leq \frac{1}{n}\frac{K_2(\varepsilon)}{1-d_3}, \qquad (27.43)$$

where $K(\varepsilon)$ is the Lipschitz constant of the polynomial P for $x \in U_\delta$. We represent the function $P(f(\varphi),\varepsilon)$ in the form

$$P(f(\varphi),\varepsilon) = Q(\varphi,\varepsilon) + R_1(\varphi,\varepsilon),$$

where $Q(\varphi,\varepsilon)$ is a trigonometric polynomial that approximates $P(f(\varphi),\varepsilon)$ to within ε, i.e., $|R_1(\varphi,\varepsilon)| \leq \varepsilon \ \ \forall\varphi \in \mathcal{T}_m$. Therefore, the following estimate holds for arbitrary $n = 1, 2, \ldots$:

$$\frac{1}{n}\Big|\sum_{\nu=0}^{n-1}[P(f(\omega\nu+\psi^0),\varepsilon)-Q(\omega\nu+\psi^0,\varepsilon)]\Big| \leq \varepsilon. \qquad (27.44)$$

By definition,

$$Q(\varphi,\varepsilon) = \sum_{\|k\|\leq N} Q_k e^{i(k,\varphi)},$$

where $N = N(\varepsilon)$ is a sufficiently large integer and $Q_k = Q_k(\varepsilon)$ are the Fourier coefficients of the function $Q(\varphi,\varepsilon)$. This yields the equalities

$$\frac{1}{n}\sum_{\nu=0}^{n-1} Q(\omega\nu+\psi^0,\varepsilon) = Q_0 + \frac{1}{n}\sum_{\nu=0}^{n-1}\sum_{1\leq\|k\|\leq N} Q_k e^{i(k,\omega)\nu} e^{i(k,\psi^0)}$$

$$= Q_0 + \frac{1}{n}\sum_{1\leq\|k\|\leq N} Q_k\Big[\sum_{\nu=0}^{n-1} e^{i(k,\omega)\nu}\Big] e^{i(k,\psi^0)}$$

and estimates

$$\Big|\frac{1}{n}\sum_{\nu=0}^{n-1} Q(\omega\nu+\psi^0,\varepsilon) - Q_0\Big| \leq \frac{1}{n}\sum_{1\leq\|k\|\leq N} |Q_k|\Big|\sum_{\nu=0}^{n-1} e^{i(k,\omega)\nu}\Big|$$

$$\leq \max_{1\leq\|k\|\leq N}\frac{1}{n}\Big|\sum_{\nu=0}^{n-1} e^{i(k,\omega)\nu}\Big| \sum_{1\leq\|k\|\leq N}|Q_k|$$

$$= M(\varepsilon)\max_{1\leq\|k\|\leq N}\frac{1}{n}\Big|\sum_{\nu=0}^{n-1} e^{i(k,\omega)\nu}\Big|. \qquad (27.45)$$

For the last sum in inequality (27.45), the following estimate is true:

$$\Big|\sum_{\nu=0}^{n-1} e^{i(k,\omega)\nu}\Big| = \Big[\Big(\sum_{\nu=0}^{n-1} \cos\nu(k,\omega)\Big)^2 + \Big(\sum_{\nu=0}^{n-1} \sin\nu(k,\omega)\Big)^2\Big]^{\frac{1}{2}}$$

$$= \Big|\sin\frac{n(k,\omega)}{2}\Big|\Big|\operatorname{cosec}\frac{(k,\omega)}{2}\Big| \leq \Big|\operatorname{cosec}\frac{(k,\omega)}{2}\Big|,$$

$$(k,\omega) \neq 0 \mod 2\pi.$$

This estimate yields

$$\Big|\frac{1}{n}\sum_{\nu=0}^{n-1} Q(\omega\nu+\psi^0,\varepsilon) - Q_0\Big| \leq \frac{1}{n}M(\varepsilon)\max_{1\leq\|k\|\leq N}\Big|\operatorname{cosec}\frac{(k,\omega)}{2}\Big|. \tag{27.46}$$

We also have the inequality

$$|F_0 - Q_0| \leq |F_0 - P_0| + |P_0 - Q_0| \leq 2\varepsilon \tag{27.47}$$

for the averages F_0, Q_0, and P_0 of the functions $F(f(\varphi))$, $Q(\varphi,\varepsilon)$, and $P(f(\varphi),\varepsilon)$. Combining inequalities (27.42)–(27.47), we get

$$\begin{aligned}
\Big|\frac{1}{n}\sum_{\nu=0}^{n-1} F(x(\nu,y^0)) - F_0\Big| &\leq \frac{1}{n}\Big|\sum_{\nu=0}^{n-1}[F(x(\nu,y^0)) - P(x(\nu,y^0),\varepsilon)]\Big| \\
&\quad + \frac{1}{n}\Big|\sum_{\nu=0}^{n-1}[P(x(\nu,y^0),\varepsilon) - P(f(\omega\nu+\psi^0),\varepsilon)]\Big| \\
&\quad + \frac{1}{n}\Big|\sum_{\nu=0}^{n-1}[P(f(\omega\nu+\psi^0),\varepsilon) - Q(\omega\nu+\psi^0),\varepsilon)]\Big| \\
&\quad + \frac{1}{n}\Big|\sum_{\nu=0}^{n-1} Q(\omega\nu+\psi^0),\varepsilon) - Q_0\Big| + |Q_0 - F_0| \\
&\leq 4\varepsilon + \frac{1}{n}M_1(\varepsilon),
\end{aligned} \tag{27.48}$$

where

$$M_1(\varepsilon) = \frac{K_2(\varepsilon)}{1-d_3} + M(\varepsilon)\max_{1\leq\|k\|\leq N}\Big|\operatorname{cosec}\frac{(k,\omega)}{2}\Big|.$$

We choose $n_0 = n_0(\varepsilon)$ so large that the inequality $\frac{1}{n}M_1(\varepsilon) \leq \varepsilon$ holds for all $n \geq n_0$. Then relation (27.48) takes the form

$$\left|\frac{1}{n}\sum_{\nu=0}^{n-1} F(x(\nu, y^0)) - F_0\right| \leq 5\varepsilon \quad \forall n \geq n_0,$$

which yields the limit relation (27.41). Corollary 2 is proved.

At the end of the section, we consider the perturbed system of equations

$$x(n+1) - x(n) = X(x(n)) + \varepsilon Y(x(n)), \tag{27.49}$$

where $Y = Y(x) \in C^r(R^q)$ and ε is a small positive parameter. Upon the change of variables (27.7), this system of equations takes the following form in the neighborhood of the manifold M:

$$\begin{aligned} \varphi(n+1) - \varphi(n) &= \omega + \varepsilon a(\varphi(n)) + A(\varphi(n), h(n), \varepsilon)h(n), \\ h(n+1) - h(n) &= P(\varphi(n), h(n), \varepsilon)h(n) + \varepsilon b(\varphi(n)), \end{aligned} \tag{27.50}$$

where $a = a(\varphi)$, $A = A(\varphi, h, \varepsilon)$, $P = P(\varphi, h, \varepsilon)$, and $b = b(\varphi)$ are functions from the space $C^{r-1}_{Lip}(\mathcal{T}_m \times K_\mu)$ for all sufficiently small $\varepsilon > 0$.

Applying to system (27.50) the perturbation theory of invariant toroidal manifolds of discrete dynamical systems [MSM1, Nei] and the method of transformation of system (27.8) to the form (27.17) presented above, we establish the following statement:

Theorem 27.3. *Suppose that the conditions of Theorem 27.1 are satisfied. Then, for sufficiently small positive values of* μ *and* ε_0, *there exists a change of variables*

$$\varphi = \psi + U(\psi, z, \varepsilon)h, \quad h = u(\varphi, \varepsilon) + z$$

that reduces the system of equations (27.50) to the form

$$\psi(n+1) - \psi(n) = \omega + F(\psi(n), \varepsilon),$$

$$z(n+1) - z(n) = R(\psi(n), z(n), \varepsilon)z(n),$$

where the functions $u(\varphi, \varepsilon)$, $U(\varphi, h, \varepsilon)$, $F(\varphi, \varepsilon)$, *and* $R(\varphi, h, \varepsilon)$ *belong to the space* $C^{r-2}_{Lip}(\mathcal{T}_m \times K_\mu)$ *for every* $\varepsilon \in [0, \varepsilon_0]$ *and*

$$\lim_{\varepsilon\to 0}(\|u(\varphi, \varepsilon)\|_{r-2,Lip} + \|U(\varphi, h, \varepsilon) - U(\varphi, h)\|_{r-2,Lip}) = 0.$$

REFERENCES

Akulenko, L. D.

[Aku] *Asymptotic Methods of Optimal Control*, Nauka, Moscow (1987) [in Russian].

Andronov, A. A., Vitt, A. A., and Khaikin, S. É.

[AVK] *Oscillation Theory*, Moscow University, Moscow (1971) [in Russian].

Anosov, D. V.

[Ano] "Averaging in systems of ordinary differential equations with rapidly oscillating solutions," *Izv. Akad. Nauk SSSR, Ser. Mat.*, **24**, No. 5, 721–742 (1960).

Arnol'd, V. I.

[Arn1] "Small denominators and problems of stability of motion in classical and celestial mechanics," *Usp. Mat. Nauk*, **18**, No. 6, 91–192 (1963).

[Arn2] "Conditions of applicability and an estimate of the error of averaging method for systems that pass through resonances in the process of evolution," *Dokl. Akad. Nauk SSSR*, **161**, No. 1, 9–12 (1965).

[Arn3] *Mathematical Methods of Classical Mechanics*, Nauka, Moscow (1974) [in Russian].

[Arn4] *Additional Chapters of the Theory of Ordinary Differential Equations*, Nauka, Moscow (1979) [in Russian].

Astaf'eva, M. N.

[Ast] "On the problem of justification of averaging method for multifrequency oscillation systems with pulse influence," *Ukr. Mat. Zh.*, **41**, No. 8, 1124–1126 (1989).

Bainov, D. D., and Milusheva, S. D.

[BaM] "On some applications of averaging method to the solution of initial- and boundary-value problems for ordinary, differential, integro-differential, and functional-differential equations," in: *Proceedings of the VIII International Conference on Nonlinear Oscillations*, Prague (1978), pp. 771–879.

Bakai, A. S., and Stepanovskii, Yu. P.

[BaS] *Adiabatic Invariants*, Naukova Dumka, Kiev (1981) [in Russian].

Bakhtin, V. I.

[Bak1] "On averaging in multifrequency systems," *Funkts. Anal.*, **20**, No. 2, 1–7 (1986).

[Bak2] "Averaging in a one-frequency general-position system," *Differents. Uravn.*, **27**, No. 9, 1493–1505 (1991).

Banfi, C.

[Ban] "Sull'approssimazione di processi nonstazionari in meccanica nonlineare," *Bull. Unione Mat. Ital.*, **22**, No. 4, 442–450 (1967).

Bibikov, Yu. N.

[Bib] *Multifrequency Nonlinear Oscillations and Their Bifurcations*, Leningrad University, Leningrad (1991) [in Russian].

Bibikov, Yu. N., and Pliss, V. A.

[BiP] "On the existence of invariant tori in the neighborhood of the zero solution of a system of ordinary differential equations," *Differents. Uravn.*, No. 4, 785–797 (1968).

Bigun, Ya. I.

[Big] "Averaging method in a multifrequency system with infinite delay," in: *Constructive Methods for Investigation of Differential Equations*, Institute of Mathematics, Ukrainian Academy of Sciences, Kiev (1993), pp. 4–15 [in Ukrainian].

Blaquiére, A.

[Bla] *Nonlinear System Analysis*, Academic Press, New York–London (1966).

Bogolyubov, N. N.

[Bog] *On Statistical Methods in Mathematical Physics*, Ukrainian Academy of Sciences, Kiev (1945) [in Russian].

Bogolyubov, N. N., and Mitropol'skii, Yu. A.

[BoM1] "Methods of integral manifolds in nonlinear mechanics," in: *Proceedings of the International Symposium on Nonlinear Oscillations*, Institute of Mathematics, Ukrainian Academy of Sciences, Kiev (1963), pp. 93–154 [in Russian].

[BoM2] *Asymptotic Methods in the Theory of Nonlinear Oscillations*, Nauka, Moscow (1974) [in Russian].

Bogolyubov, N. N., Mitropol'skii, Yu. A., and Samoilenko, A. M.

[BMS] *Method of Accelerated Convergence in Nonlinear Mechanics*, Naukova Dumka, Kiev (1969) [in Russian].

Bogolyubov, N. N., and Zubarev, D. N.

[BoZ] "Method of asymptotic approximation for systems with rotating phase and its application to the motion of charged particles in magnetic field," *Ukr. Mat. Zh.*, **7**, No. 1, 5–17 (1955).

Boichuk, A. A.

[Boi] *Constructive Methods for the Analysis of Boundary-Value Problems*, Naukova Dumka, Kiev (1990) [in Russian].

Bryuno, A. D.

[Bry1] "Normal forms and averaging methods," *Dokl. Akad. Nauk SSSR*, **240**, No. 2, 257–260 (1976).

[Bry2] "On nondegeneracy conditions in the Kolmogorov theorem," *Dokl. Ros. Akad. Nauk*, **322**, No. 6, 1028–1032 (1992).

Butenin, N. V., Neimark, Yu. I., and Fufaev, N. A.

[BNF] *Introduction to the Theory of Nonlinear Oscillations*, Nauka, Moscow (1976) [in Russian].

Cesari, L.

[Ces] *Asymptotic Behavior and Stability Problems in Ordinary Differential Equations*, Springer, Berlin (1959).

Chang, K. W., and Howes, F. A.

[ChH] *Nonlinear Singular Perturbation Phenomena: Theory and Application*, Springer, New York (1984).

Chernous'ko, F. L.

[Che] "Asymptotic methods in some problems of optimal control," in: *Asymptotic Methods in the Theory of Nonlinear Oscillations*, Naukova Dumka, Kiev (1975), pp. 239–246 [in Russian].

Denjoy, A.

[Den] "Sur les courbes définies par les équations différentielles a là surface du tore," *J. Math. Pures Appl.* **11**, No. 3, 333–375 (1932).

Diliberto, S.

[Dil1] "Perturbation theorems for periodic surfaces. I. Definitions and main theorems," *Rend. Circolo Math. Palermo*, **9**, No. 3, 265–299 (1960).

[Dil2] "Perturbation theorems for periodic surfaces. II," *Rend. Circolo Math. Palermo*, **10**, No. 2, 111–161 (1961).

Dmitriev, M. G., and Klishevich, A. M.

[DmK] "Iteration methods for the solution of singularly perturbed boundary-value problems of conditionally stable type," *Zh. Vych. Mat. Mat. Fiz*, **27**, No. 12, 1812–1823 (1987).

Fatou, P.

[Fat] "Sur le mouvement d'un systéme soumis a des forces a courte période," *Bull. Soc. Math. France*, **56**, No. 1, 98–139 (1928).

Fedoryuk, M. V.

[Fed] *Asymptotic Methods for Linear Ordinary Differential Equations*, Nauka, Moscow (1983) [in Russian].

Filatov, A. N.

[Fil] *Asymptotic Methods in the Theory of Differential and Integro-Differential Equations*, FAN, Tashkent (1974) [in Russian].

Filatov, A. N., and Sharova, L. V.

[FiS] *Integral Inequalities and Theory of Nonlinear Oscillations*, Nauka, Moscow (1976) [in Russian].

Filatov, O. P.

[Fila] "Stability of multifrequency systems under resonance conditions without synchronization of phase variables," *Differents. Uravn.* **19**, No. 3, 543–546 (1983).

Filippov, A. K.

[Fili] *Differential Equations with Discontinuous Right-Hand Side*, Nauka, Moscow (1985) [in Russian].

Fodchuk, V. I., Bigun, Ya. I., Klevchuk, I. I., Cherevko, I. M., and Yakimov, I. V.

[FBKCY] *Regularly and Singularly Perturbed Functional-Differential Equations*, Institute of Mathematics, Ukrainian Academy of Sciences, Kiev (1996) [in Ukrainian].

Gabrel', O. M.

[Gab] *Solution of a Multipoint Problem for Ordinary Differential Equations with Small Nonlinearity by Projection-Iterative Method*, Preprint No. 82.49, Institute of Mathematics, Ukrainian Academy of Sciences, Kiev (1982) [in Russian].

Gantmakher, F. R.

[Gan] *The Theory of Matrices*, Nauka, Moscow (1967) [in Russian].

Golets, B. I., Golets, V. L., and Petryshyn, R. I.

[GGP1] "On averaging in oscillation systems passing through resonances," *Ukr. Mat. Zh.*, **32**, No. 4, 448–455 (1980).

[GGP2] "Averaging over fast variables in three-frequency systems of the second approximation," *Ukr. Mat. Zh.*, **37**, No. 4, 437–444 (1985).

Golets, V. L., and Petryshyn, R. I.

[GoP] "On nonlinear oscillation systems passing through resonances," *Prikl. Mekh.*, **23**, No. 4, 87–93 (1987).

Grebenikov, E. A.

[Gre] *Averaging Method in Applied Problems*, Nauka, Moscow (1986) [in Russian].

Grebenikov, E. A., and Ryabov, Yu. A.

[GrR1] *New Qualitative Methods in Celestial Mechanics*, Nauka, Moscow (1971) [in Russian].

[GrR2] *Resonances and Small Denominators in Celestial Mechanics*, Nauka, Moscow (1978) [in Russian].

[GrR3] *Constructive Methods for the Analysis of Nonlinear Systems*, Nauka, Moscow (1979) [in Russian].

Gurtovnykh, A. S., and Neimark, Yu. A.

[GuN] "On synchronization of dynamical systems," *Prikl. Mat. Mekh.*, **38**, No. 5, 800–809 (1974).

Hale, J. K.

[Hal1] "Integral manifolds of perturbed differential systems," *Ann. Math.*, **73**, No. 3, 496–531 (1961).

[Hal2] *Oscillations in Nonlinear Systems*, McGraw-Hill, New York (1963).

Hartman, P.

[Har] *Ordinary Differential Equations*, Wiley, New York (1964).

Herman, M. R.

[Her] "Conjugaison C^{∞} des diffeomorphismes du cercle dont le nombre de rotatione satisfait a une condition arithmétique," *C. R. Acad. Sci., Ser. A*, **282**, No. 10, 503–506 (1976).

Khapaev, M. M.

[Kha1] "On averaging in multifrequency systems," *Dokl. Akad. Nauk SSSR*, **217**, No. 5, 1021–1024 (1974).

[Kha2] *Averaging in Stability Theory*, Nauka, Moscow (1986) [in Russian].

Khapaev, M. M., and Filatov, O. P.

[KhF] "On averaging principle for systems with fast and slow variables," *Differents. Uravn.*, **19**, No. 9, 1640–1643 (1983).

Kiguradze, I. T.

[Kig] "Boundary-value problems for systems of ordinary differential equations," in: *Contemporary Problems in Mathematics. Recent Advances*, Vol. 30, Nauka, Moscow (1987), pp. 3–103 [in Russian].

Kolmogorov, A. N.

[Kol] "On preservation of conditionally periodic motions under small variation of the Hamilton function," *Dokl. Akad. Nauk SSSR*, **98**, No. 4, 527–530 (1954).

Kolmogorov, A. N., and Fomin, S. V.

[KoF] *Elements of the Theory of Functions and Functional Analysis*, Nauka, Moscow (1967) [in Russian].

Kostin, A. V., and Shchegolev, S. A.

[KoS] "On one class of solutions of a differential system with slowly varying parameters," *Ukr. Mat. Zh.*, **41**, No. 1, 101–103 (1989).

Krasnosel'skii, M. A., Burd, V. Sh., and Kolesov, Yu. S.

[KBK] *Nonlinear Almost Periodic Oscillations*, Nauka, Moscow (1970) [in Russian].

Krasnosel'skii, M. A., and Krein, S. G.

[KrK] "On averaging principle in nonlinear mechanics," *Usp. Mat. Nauk*, **10**, No. 3, 47–52 (1955).

Krylov, N. M., and Bogolyubov, N. N.

[KrB] *Introduction to Nonlinear Mechanics*, Ukrainian Academy of Sciences, Kiev (1937) [in Russian].

Kulik, V. L.

[Kul] "Alternating Lyapunov functions and preservation of invariant tori under perturbations," *Ukr. Mat. Zh.*, **39**, No. 1, 45–52 (1987).

Kupka, I.

[Kup] "Stabilité des variétés invariantes d'un champ de vecteurs pour les petites perturbations," *C. R. Acad. Sci.*, **258**, No. 17, 4197–4200 (1964).

Kurzweil, J.

[Kur] "Invariant manifolds of differential systems," *Differents. Uravn.*, **4**, No. 5, 785–797 (1968).

Lankaster, P.

[Lan] *Theory of Matrices*, Academic Press, New York (1969).

Lepin, A. Ya., and Lepin, L. A.

[LeL] *Boundary-Value Problems for Second-Order Ordinary Differential Equations*, Zinatne, Riga (1988) [in Russian].

Lika, D. K., and Ryabov, Yu. A.

[LiR] *Iteration Methods and Majorizing Lyapunov Equations in the Theory of Nonlinear Oscillations*, Shtiintsa, Kishinev (1974) [in Russian].

Lomov, S. A.

[Lom] *Introduction to the General Theory of Singular Perturbations*, Nauka, Moscow (1981) [in Russian].

Luchka, A. Yu.

[Luc] *Projection-Iterative Methods*, Naukova Dumka, Kiev (1993) [in Russian].

Lykova, O. B.

[Lyk] "Development of the Bogolyubov–Mitropol'skii method of integral manifolds," *Ukr. Mat. Zh.*, **44**, No. 1, 33–46 (1992).

Lykova, O. B., and Baris, Ya. S.

[LyB] *Approximate Integral Manifolds*, Naukova Dumka, Kiev (1993) [in Russian].

Martynyuk, D. I., and Samoilenko, A. M.

[MaS] "Existence of invariant manifolds of systems with delay," *Ukr. Mat. Zh.*, **26**, No. 5, 611–620 (1974).

Martynyuk, D. I., and Tsyganovskii, N. S.

[MaT] "Invariant manifolds of impulsive systems with delay," *Differents. Uravn.*, **15**, No. 10, 1783–1795 (1979).

Mikhlin, S. G.

[Mik] *Linear Partial Differential Equations*, Vysshaya Shkola, Moscow (1977) [in Russian].

Mishchenko, E. F., and Rozov, N. Kh.

[MiR] *Differential Equations with Small Parameter and Relaxation Oscillations*, Nauka, Moscow (1975) [in Russian].

Mitropol'skii, Yu. A.

[Mit1] *Problems of Asymptotic Theory of Nonstationary Oscillations*, Nauka, Moscow (1964) [in Russian].

[Mit2] *Averaging Method in Nonlinear Mechanics*, Naukova Dumka, Kiev (1971) [in Russian].

[Mit3] "Adiabatic invariant for a double pendulum with slowly varying lengths," *Mat. Fiz. Nelin. Mekh.*, **14**, 30–39 (1990).

[Mit4] *Nonlinear Mechanics. Asymptotic Methods*, Institute of Mathematics, Ukrainian Academy of Sciences, Kiev (1995) [in Russian].

Mitropol'skii, Yu. A., and Khoma, G. P.

[MiK] *Mathematical Justification of Asymptotic Methods of Nonlinear Mechanics*, Naukova Dumka, Kiev (1983) [in Russian].

Mitropol'skii, Yu. A., and Lopatin, A. K.

[MiLo] *Group-Theoretic Approach in Asymptotic Methods of Nonlinear Mechanics*, Naukova Dumka, Kiev (1988) [in Russian].

Mitropol'skii, Yu. A., and Lykova, O. B.

[MiLy] *Integral Manifolds in Nonlinear Mechanics* [in Russian], Nauka, Moscow (1973).

Mitropol'skii, Yu. A., and Samoilenko, A. M.

[MiS1] "On quasiperiodic oscillations in nonlinear mechanics," *Ukr. Mat. Zh.*, **24**, No. 2, 179–194 (1972).

[MiS2] *Asymptotic Investigation of Weakly Nonlinear Oscillation Systems*, Preprint No. 76.5, Institute of Mathematics, Ukrainian Academy of Sciences, Kiev (1976) [in Russian].

[MiS3] *Investigation of Second-Order Oscillation Systems*, Preprint No. 76.6, Institute of Mathematics, Ukrainian Academy of Sciences, Kiev (1976) [in Russian].

[MiS4] "On the problem of asymptotic expansions in nonlinear mechanics," *Ukr. Mat. Zh.*, **31**, No. 1, 42–53 (1979).

[MiS5] *General Problems in the Theory of Asymptotic Integration of Systems in Nonlinear Mechanics*, Preprint No. 87.41, Institute of Mathematics, Ukrainian Academy of Sciences, Kiev (1987) [in Russian].

Mitropol'skii, Yu. A., Samoilenko, A. M., and Kulik, V. L.

[MSK] *Investigation of Dichotomy of Linear Systems of Differential Equations Using Lyapunov Functions*, Naukova Dumka, Kiev (1990) [in Russian].

Mitropol'skii, Yu. A., Samoilenko, A. M., and Martynyuk, D. I.

[MSM1] *Systems of Evolution Equations with Periodic and Quasiperiodic Coefficients*, Naukova Dumka, Kiev (1985) [in Russian].

[MSM2] *Systems of Evolution Equations with Periodic and Quasiperiodic Coefficients*, Kluwer, Dordrecht (1993).

Moser, J.

[Mos1] "On invariant surfaces and almost periodic solutions for ordinary differential equations," *Notices Amer. Math. Soc.*, **12**, No. 1, 124 (1965).

[Mos2] "A rapidly convergent iteration method and nonlinear differential equations," *Ann. Scuola Norm. Sup. Pisa*, **20**, No. 3, 499–535 (1966).

[Mos3] "Convergent series expansions for quasiperiodic motions," *Math. Ann.*, 136–176 (1967).

[Mos4] "A rapidly convergent iteration method and nonlinear differential equations," *Usp. Mat. Nauk*, **23**, No. 4, 179–238 (1968).

[Mos5] *Lectures on Hamiltonian Systems*, Amer. Math. Soc. (1968).

Nayfeh, A. H.

[Nay] *Introduction to Perturbation Techniques*, Wiley, New York (1981).

Neimark, Yu. I.

[Nei] *Method of Point Mappings in the Theory of Nonlinear Oscillations*, Nauka, Moscow (1972) [in Russian].

Neishtadt, A. I.

[Neis1] "On passage through resonances in a two-frequency problem," *Dokl. Akad. Nauk SSSR*, **221**, No. 2, 301–304 (1975).

[Neis2] "On averaging in multifrequency systems," *Dokl. Akad. Nauk SSSR*, **226**, No. 6, 1295–1298 (1976).

Nekhoroshev, N. N.

[Nek1] "On the behavior of Hamiltonian systems close to integrable ones," *Funkts. Anal.*, **5**, No. 4, 82–83 (1971).

[Nek2] "Exponential estimate of the time of stability of Hamiltonian systems close to integrable ones," *Usp. Mat. Nauk*, **32**, No. 6, 5–66 (1977).

Palmer, K.

[Pal] "On the reducibility of almost periodic systems of linear differential equations," *J. Different. Equat.*, **36**, No. 2, 374–390 (1980).

Parasyuk, I. O.

[Par] *Coisotropic Invariant Tori of Hamiltonian Systems*, Doctoral-Degree Thesis (Physics and Mathematics), Kiev (1995) [in Ukrainian].

Perestyuk, N. A.

[Pere] "Invariant sets of one class of discontinuous dynamical systems," *Ukr. Mat. Zh.*, **36**, No. 1, 63–69 (1984).

Perestyuk, N. A., and Akhmetov, M. U.

[PeA] "On almost periodic solutions of impulsive systems," *Ukr. Mat. Zh.*, **39**, No. 1, 74–80 (1987).

Perov, A. I.

[Pero] *Variational Methods in the Theory of Nonlinear Oscillations*, Institute of Mathematics, Voronezh University, Voronezh (1981) [in Russian].

Petryshyn, R. I.

[Pet1] "Averaging with regard for resonance relations for frequencies in oscillation systems," *Ukr. Mat. Zh.*, **33**, No. 2, 262–267 (1981).

[Pet2] "On the problem on invariant manifolds of some nonlinear systems," *Dopov. Akad. Nauk Ukr. RSR, Ser. A*, No. 2, 22–24 (1981).

[Pet3] "Averaging in resonance boundary-value problems," *Dopov. Akad. Nauk Ukr. RSR, Ser. A*, No. 9, 11–14 (1988).

[Pet4] "Estimate of an error of the averaging method in multifrequency systems of higher approximation," in: *Asymptotic Methods in Equations of Mathematical Physics*, Institute of Mathematics, Ukrainian Academy of Sciences, Kiev (1989), pp. 100–104 [in Russian].

[Pet5] "Asymptotic expansion of an integral manifold of a multifrequency system," in: *Nonlinear Differential Equations and Their Applications*, Institute of Mathematics, Ukrainian Academy of Sciences, Kiev (1993), pp. 58–65 [in Ukrainian].

[Pet6] "Investigation of solutions of multifrequency systems by the averaging method," in: *Integral Transformations and Their Applications to Boundary-Value Problems*, Vol. 2, Institute of Mathematics, Ukrainian Academy of Sciences, Kiev (1993), pp. 189–202 [in Ukrainian].

[Pet7] “Reduction of a multifrequency system to the canonical form in the neighborhood of an integral manifold,” in: *Systems of Evolution Equations with Delay*, Institute of Mathematics, Ukrainian Academy of Sciences, Kiev (1994), pp. 74–88 [in Ukrainian].

[Pet8] “Investigation of second-order oscillation systems with slowly varying frequencies,” in: *Integral Transformations and Their Applications to Boundary-Value Problems*, Vol. 8, Institute of Mathematics, Ukrainian Academy of Sciences, Kiev (1995), pp. 158–167 [in Ukrainian].

[Pet9] “Averaging method in some problems of the theory of nonlinear oscillations,” *Ukr. Mat. Zh.*, **45**, No. 6, 801–810 (1995).

[Pet10] “On some problems for resonance oscillation systems with slowly varying frequencies,” in: *Integral Transformations and Their Applications to Boundary-Value Problems*, Vol. 9, Institute of Mathematics, Ukrainian Academy of Sciences, Kiev (1995), pp. 215–229 [in Ukrainian].

Petryshyn, R. I., and Lakusta, L. M.

[PeL] “On integral manifolds of oscillation systems with slowly varying frequencies,” *Ukr. Mat. Zh.*, **50**, No. 1, 87–93 (1998).

Petryshyn, R. I., and Pokhyla, O. M.

[PeP] “Estimate of an error of the averaging method on a semiaxis for a multifrequency resonance system,” *Ukr. Mat. Zh.*, **49**, No. 5, 685–690 (1997).

Pliss, V. A.

[Pli1] *Nonlocal Problems in the Theory of Oscillations*, Nauka, Moscow (1964) [in Russian].

[Pli2] *Integral Sets of Periodic Systems*, Nauka, Moscow (1977) [in Russian].

Plotnikov, V. A.

[Plo] *Averaging Method in Control Problems*, Lybid', Kiev–Odessa (1992) [in Russian].

Plotnikov, V. A., and Bardai, V. V.

[PlB] "Averaging of boundary-value problems for differential equations with slow and fast variables," *Ukr. Mat. Zh.*, **42**, No. 9, 1275–1278 (1990).

Plotnikov, V. A., and Larbani, M.

[PlL] "Justification of one scheme of partial averaging for systems with slow and fast variables," *Differents. Uravn.*, **28**, No. 3, 428–432 (1992).

Plotnikov, V. A., and Zverkova, T. S.

[PlZ] "Method of partial averaging in systems of standard form with discontinuous right-hand sides," *Ukr. Mat. Zh.*, **45**, No. 1, 140–142 (1993).

Poincaré, H.

[Poi] *Mémoire sur les Courbes Définies par Une Équation Différentielle*, Gauthier-Villars (1916).

Popov, D. A.

[Pop] "Estimates with constants for certain classes of integrals," *Usp. Mat. Nauk*, **52**, No. 1, 77–148 (1997).

Pronchatov, V. E.

[Pron] "On an estimate for the error of the averaging method in a two-particle problem," *Mat. Sb.*, **134**, No. 1, 28–41 (1987).

Proskuryakov, A. P.

[Pros] *Poincaré Method in the Theory of Nonlinear Oscillations*, Nauka, Moscow (1977) [in Russian].

Ptashnik, B. I.

[Pta] *Ill-Posed Boundary-Value Problems for Partial Differential Equations*, Naukova Dumka, Kiev (1984) [in Russian].

Sacker, R.

[Sac1] "A new approach to the perturbation theory of invariant surfaces," *Commun. Pure Appl. Math.*, **18**, No. 4, 717–732 (1965).

[Sac2] "A perturbation theorem for invariant manifolds and Hölder continuity," *J. Math. Mech.*, **18**, No. 8, 705–761 (1969).

Samoilenko, A. M.

[Sam1] "Averaging method in pulse systems," *Mat. Fiz.*, **9**, 101–117 (1971).

[Sam2] "On the smoothness of an invariant torus of a quasilinear system of differential equations with respect to parameter," *Ukr. Mat. Zh.*, **38**, No. 5, 605–618 (1986).

[Sam3] "Asymptotic expansions and differentiability of an invariant torus of a quasilinear system of differential equations with respect to a parameter," *Ukr. Mat. Zh.*, **39**, No. 1, 97–105 (1987).

[Sam4] *Elements of the Mathematical Theory of Multifrequency Oscillations. Invariant Tori*, Nauka, Moscow (1987) [in Russian].

[Sam5] "On the problem of justification of the averaging method for multifrequency oscillation systems," *Differents. Uravn.*, **23**, No. 2, 267–278 (1987).

[Sam6] *Investigation of a Dynamical System in a Neighborhood of a Quasiperiodic Trajectory*, Preprint No. 90.35, Institute of Mathematics, Ukrainian Academy of Sciences, Kiev (1990) [in Russian].

[Sam7] "Investigation of a dynamical system in a neighborhood of an invariant toroidal manifold," *Ukr. Mat. Zh.*, **42**, No. 4, 530–537 (1991).

[Sam8] *Elements of the Mathematical Theory of Multifrequency Oscillations*, Kluwer, Dordrecht (1991).

[Sam9] "Investigation of a discrete dynamical system in a neighborhood of a quasiperiodic trajectory," *Ukr. Mat. Zh.*, **44**, No. 2, 1702–1711 (1992).

[Sam10] "N. N. Bogolyubov and nonlinear mechanics," *Usp. Mat. Nauk*, **49**, No. 5, 103–146 (1994).

Samoilenko, A. M., and Perestyuk, N. A.

[SaP1] *Impulsive Differential Equations*, Vyshcha Shkola, Kiev (1987) [in Russian].

[SaP2] *Impulsive Differential Equations*, World Scientific, Singapore (1995).

Samoilenko, A. M., and Petryshyn, R. I.

[SPe1] "Investigation of resonance systems," *Dokl. Akad. Nauk Ukr. SSR, Ser. A*, No. 2, 11–14 (1985).

[SPe2] "Investigation of stability of two-frequency systems," *Ukr. Mat. Zh.*, **38**, No. 4, 483–487 (1986).

[SPe3] "Uniform estimates for one-dimensional oscillation integrals," *Dokl. Akad. Nauk Ukr. SSR, Ser. A*, No. 11, 12–15 (1987).

[SPe4] "Averaging method in multifrequency systems with slowly varying parameters," *Ukr. Mat. Zh.*, **40**, No. 4, 493–501 (1988).

[SPe5] "Averaging method in some boundary-value problems," *Differents. Uravn.*, **25**, No. 6, 956–964 (1989).

[SPe6] "On integral manifolds of multifrequency oscillation systems," *Izv. Akad. Nauk SSSR, Ser. Mat.*, **54**, No. 2, 378–395 (1990).

[SPe7] "Multipoint boundary-value problem for nonlinear oscillations," in: *Constructive Methods for the Investigation of Differential Equations*, Institute of Mathematics, Ukrainian Academy of Sciences, Kiev (1993), pp. 62–73 [in Ukrainian].

Samoilenko, A. M., and Petryshyn, Ya. R.

[SPet1] "Averaging method in multipoint problems of the theory of nonlinear oscillations," *Ukr. Mat. Zh.*, **48**, No. 8, 1036–1043 (1996).

[SPet2] "Boundary-value problems with parameters for a multifrequency oscillation system," *Ukr. Mat. Zh.*, **49**, No. 4, 581–589 (1997).

Samoilenko, A. M., and Ronto, N. I.

[SaR] *Numerical-Analytic Methods in the Theory of Boundary-Value Problems for Ordinary Differential Equations*, Naukova Dumka, Kiev (1992) [in Russian].

Samoilenko, A. M., and Svishchuk, M. Ya.

[SaS] "On the decomposition of a system of differential equations with slowly varying phase in the neighborhood of an asymptotically stable invariant torus," *Ukr. Mat. Zh.*, **37**, No. 6, 751–756 (1985).

Samoilenko, A. M., and Teplinskii, Yu. V.

[SaT] *Countable Systems of Differential Equations*, Institute of Mathematics, Ukrainian Academy of Sciences, Kiev (1993) [in Russian].

Sarybaev, M., and Shpakovich, V. P.

[SSh] "On an estimate of an error of the averaging method for multifrequency systems," in: *Differential Equations with Small Parameter*, Institute of Mathematics, Ukrainian Academy of Sciences, Kiev (1989), pp. 98–103 [in Russian].

Schauder, J.

[Sch] "Der Fixpunktsatz in Funktionalräumen," *Studia Math.*, **2**, 171–180 (1930).

Sell, G.

[Sel1] "The Floquet problem for almost periodic linear differential equations," *Lect. Notes Math.*, **415**, 239–251 (1974).

[Sel2] "Bifurcation of higher-dimensional tore," *Arch. Ration. Mech. Anal.*, **69**, No. 3, 199–230 (1979).

Shkil', N. I., Starun, I. I., and Yakovets, V. P.

[SSY] *Asymptotic Integration of Linear Systems of Ordinary Differential Equations*, Vyshcha Shkola, Kiev (1985) [in Russian].

Shkil', N. I., Voronoi, A. N., and Leifura, V. N.

[SVL] *Asymptotic Methods in Differential and Integro-Differential Equations*, Vyshcha Shkola, Kiev (1985) [in Russian].

Shpakovich, V. P.

[Shp] "Averaging method for multifrequency systems with delay," *Ukr. Mat. Zh.*, **37**, No. 4, 535–539 (1985).

Sobolev, V. A.

[Sob] "Integral manifolds, singular perturbations, and optimal control," *Ukr. Mat. Zh.*, **39**, No. 1, 111–116 (1987).

Strygin, V. V., and Sobolev, V. A.

[StS] *Separation of Motions by the Method of Integral Manifolds*, Nauka, Moscow (1988) [in Russian].

Teplinskii, Yu. V., and Avdeyuk, P. I.

[TeA] "On the exponential stability of an invariant torus of a nonlinear countable system of differential equations," *Ukr. Mat. Zh.*, **42**, No. 3, 401–405 (1990).

Trofimchuk, S. I.

[Tro] "Necessary condition for the existence of an invariant manifold of a linear extension of a dynamical system on a compact manifold," *Ukr. Mat. Zh.*, **36**, No. 3, 390–393 (1984).

Urabe, M.

[Ura] "Existence theorems of quasiperiodic solutions to nonlinear differential systems," *Funkcial. Ekvac.*, **15**, 75–100 (1972).

Vasil'ev, N. I., and Klokov, Yu. A.

[VaK] *Foundations of the Theory of Boundary-Value Problems for Ordinary Differential Equations*, Zinatne, Riga (1978) [in Russian].

Vasil'eva, A. B., and Butuzov, V. F.

[VaB] *Asymptotic Expansions of Solutions of Singularly Perturbed Equations*, Nauka, Moscow (1973) [in Russian].

Vasil'eva, A. B., and Dmitriev, M. G.

[VaD] "Singular perturbations in problems of optimal control," in: *VINITI Series in Mathematical Analysis*, Vol. 20, VINITI, Moscow (1982), pp. 3–78 [in Russian].

Volosov, V. M.

[Vol] "Averaging in systems of ordinary differential equations," *Usp. Mat. Nauk*, **17**, No. 6, 3–126 (1962).

Volosov, V. M., and Morgunov, B. I.

[VoM] *Averaging Method in the Theory of Nonlinear Oscillation Systems* [in Russian], Moscow University, Moscow (1971).

Wasow, W.

[Was] *Asymptotic Expansions for Ordinary Differential Equations*, Interscience (1965).

Yakubovich, V. A., and Starzhinskii, V. M.

[YaS] *Parametric Resonance in Linear Systems*, Nauka, Moscow (1987) [in Russian].

Zabareiko, P. P., and Ledovskaya, I. B.

[ZaL] "On the justification of the Bogolyubov–Krylov method for ordinary differential equations," *Differents. Uravn.*, **5**, No. 2, 240–253 (1969).

Zadiraka, K. V.

[Zad] "On a nonlocal integral manifold of a nonregularly perturbed differential system," *Ukr. Mat. Zh.*, **17**, No. 1, 47–63 (1965).

Zheleznyak, A. L., and Kogan, V. P.

[ZhK] "A method for the solution of one class of singularly perturbed boundary-value problems," *Zh. Vych. Mat. Mat. Fiz.*, **27**, No. 7, 1095–1100 (1987).

Zhukova, G. S.

[Zhu] *Asymptotic Integration of Linear Ordinary Differential Equations*, Voronezh University, Voronezh (1988) [in Russian].

MIX
Papier aus verantwortungsvollen Quellen
Paper from responsible sources
FSC® C105338

If you have any concerns about our products,
you can contact us on
ProductSafety@springernature.com

In case Publisher is established outside the EU,
the EU authorized representative is:
Springer Nature Customer Service Center GmbH
Europaplatz 3, 69115 Heidelberg, Germany

Printed by Libri Plureos GmbH
in Hamburg, Germany